22nd International Scientific Conference on Reshetnev Readings 2018

IOP Conference Series: Materials Science and Engineering
Volume 822

Krasnoyarsk, Russia
11 - 15 November 2018

ISBN: 978-1-7138-3162-4
ISSN: 1757-8981

Printed from e-media with permission by:

Curran Associates, Inc.
57 Morehouse Lane
Red Hook, NY 12571

proceedings.com

Some format issues inherent in the e-media version may also appear in this print version.

Institute of Physics
Dirac House, Temple Back
Bristol BS1 6BE UK

Phone: 44 1 17 929 7481
Fax: 44 1 17 920 0979

techtracking@iop.org

Additional copies of this publication are available from:

Curran Associates, Inc.
57 Morehouse Lane
Red Hook, NY 12571 USA
Phone: 845-758-0400
Fax: 845-758-2633
Email: curran@proceedings.com
Web: www.proceedings.com

TABLE OF CONTENTS

IOP Conf. Series: Materials Science and Engineering **822** (2020) 011001 doi:10.1088/1757-899X/822/1/011001

22nd International Scientific Conference

Reshetnev Readings-2018 (11–15 November 2018)

Preface to 22nd edition of Reshetnev Readings Conference Proceedings

The Russian Academician Mikhail Reshetnev (1924–1996) is one of the pioneers of the Soviet space program. His contributions were especially important to Soviet (later, Russian) satellite communication and navigation systems. As a sign of gratitude for Reshetnev's achievements, two institutions – our University and a satellite manufacturing company were named in his honor. 2018 is the 22nd time that the University has organized a scientific conference in memory of Academician Reshetnev.

Prior to 2017, the University was named Reshetnev Siberian State Aerospace University. After it merged with the Siberian University of Technology in 2016, its name was changed to Reshetnev Siberian State University of Science and Technology, or Reshetnev University. 2017 is, therefore, the first time that the University is holding the Conference in a new status. While the traditions sections of the Conference with the fields of aerospace technology and engineering, computer science and telecommunication systems remain as the nucleus of the Conference, new sections, including a significantly developed section or workshop on Materials Science and Engineering have been added. The scope of the section has been broadened, adding new areas of research such as forestry products and biomaterials in engineering, solid-state physics, composite materials, polymers, and other areas related to the science of developing and producing new materials. The Conference is traditionally held in the Siberian city of Krasnoyarsk. Over 200 participants annually take part in the Conference, making it one of the largest scientific events in the city.

In 2018, the topic of the Conference was expanded to house a growing demand for research reports in the field of Earth and environmental studies. The majority of these topics are related to remote sensing, geographical information systems, environmental issues, and ecology.

The Conference is multidisciplinary. However, the topics of the reports should be relevant to engineering and industry.

Pr. Yurii Loginov

Vice Rector for Science and Research Reshetnev University

IOP Conf. Series: Materials Science and Engineering **822** (2020) 011002 doi:10.1088/1757-899X/822/1/011002

Peer review statement

All papers published in this volume of *IOP Conference Series: Materials Science and Engineering* have been peer reviewed through processes administered by the proceedings Editors. Reviews were conducted by expert referees to the professional and scientific standards expected of a proceedings journal published by IOP Publishing.

Competition and management of wood-cement compositions among light concretes in the market of construction materials

S N Dolmatov, A V Nikonchuk and S V Gorbunova

Reshetnev Siberian State University of Science and Technology, 31 Krasnoyarsky Rabochy Av., Krasnoyarsk, 660037, Russia

E-mail: pipinaskus@mail.ru

Abstract. The paper describes the stages and motivation of a reasonable choice when purchasing construction materials for low-rise housing in B2C and B2B segments. Due to the optimal combination of indicators of strength, thermal conductivity, low cost in the market of construction materials for low-rise housing, light concretes took a stable position as a thermal insulation construction material. Light concretes are divided into aerated concretes and foam concretes, as well as concretes based on wood-cement compositions. The volumes of aerated and foam concrete production and consumption are hundreds times as high as the production of wood-cement compositions. The negative attitude of the consumers to the processed products and the high price of such materials are deterrence to the wide distribution of wood-cement compositions in low-rise housing. The policy of integrated use of renewable wood resources can be the basis for increasing the production and consumption of wood-cement compositions.

1. Introduction

The purpose of the study is the search of the methods of the effective management of construction materials based on wood-cement compositions.

Achieving the purpose implies the following tasks:

- analyzing the potential market segment, possible consumption of wood-cement compositions and decision-making algorithm;
- finding the reasons restricting the successful sales of wood-cement composition materials;
- determining the set of activities for market promotion of the wood-cement compositions.

At present the market of individual low-risc housing, light foam concretes are competing in this segment, develops in 2 directions. The 1st direction can be described as an individual contractor (B2C segment), aiming as a rule the construction of an individual housing project, private or summer house, garage.

2. Consumers segment- an individual contractor

It is quite difficult to forecast the moment of paying demand emergence and individual contractor motivation while choosing and purchasing of the precise type of construction materials, as every potential individual contractor is characterized by personal qualities formed in the process of education, bringing-up, personal experience, and belonging to the social level. The mentality development, individual contractor preferences are influenced by the information field available to comprehend and consume. Television, internet, telephones are the contemporary sources of information transport. The contractor activities algorithm consists of several conscious or non-conscious stages [1]:

- The stage of primary information. The person gets the primary information while looking through the magazine, watching TV programme or listening to the radio programme. The process of informing can be built in 2 mutually opposite directions. The plot of TV programme can describe the celebrity person life the rating of whom is the basis of format, timing and the story length. Precise construction materials, designs will be promoted in a hidden way (product placement technology) by means of a soft recommendation of an actor or a politician. While implementing of the 2nd direction the aggressive advertisement of the construction material or technology will be made. The person`s mentality and comprehension make it possible to watch the 1st format programmes for 30-40 minutes and even more. But a person will change the channel or close the browser tab after 40-60 seconds of aggressive advertisement watch.
- The stage of search of information. While searching the information the individual contractor uses the accessible means according to the personal characteristics of comprehension, motivation factors, age, and life style. At this stage the manufacturer has to supply the consumer with the full and understandable information of competitive benefits of the material, product or technology offered. The age, financial status and the life style of the consumer should be taken into consideration. For example, the quick loading/unloading, delivery of the materials can be determinant for the contractors of the active labour segment occupied at full-time job. And the main and motivating factor for the people of retirement age will be the accessibility of information by means of traditional mass media – newspapers, announcements, etc.
- The stage of options comparison and decision-making. While making a decision on construction materials purchase an individual contractor uses the information from advertisement websites, brochures, TV. The way the information is given is very important for great amount of technical information or subjective evaluation can cause the negative effect in decision-making. The bonus effect as well as an attractive price, for instance, of 1 m^2 of the wall without decoration, etc. are also effective. This is the way the information is given in the favourable light even if the material characteristics are not the main operational ones.
- The stage of purchase and consumption. The implementation of this stage is resulted from the previous one that is why the marketing technologies will be similar. The consumption in the individual contractor segment often goes beyond the recommended technology and targeted use of the material specified by the manufacturer. For instance, the use of the construction material in subnormal moisture conditions, wind load, insolation can cause its quick destruction and consumer`s discontent. And a discontent consumer will inevitably give negative comments, posts in social nets, forums and will influence the future sales in a negative way.
- The stage of service. Knowing the life cycle of the traded material and the contacts of the individual contractor the manufacturer can and should offer the optional service functions aiming to maintain the operating characteristics and to attach the new properties to the traded material. For instance, knowing the expiry date of the coating materials the manufacturer has to address the consumer and offer him the bonus programme of the materials suiting the earlier purchased goods.

3. Consumers segment- a professional contractor

The main characters who make decisions in the field of the professional building are executives and top managers of the construction companies. The direct decisions are made by the sales managers and heads of the logistics support departments. The important decisions on the innovative materials and technologies are made after the comprehensive analysis of the market for the B2B market [1] significantly differs from B2C, where the precise individual contractor-the house owner is of total responsibility of the wrong decision. In case of the wrong decisions in B2B market the whole division or even company is risky of insolvency. The task of the executive is to conduct the comprehensive analysis of precise situations aiming the support on maximum possible factors. The fewer factors are considered the less accuracy is in investments efficiency forecasting [2]. Personal and social responsibilities for the result of the decisions made complicate the task of the executive.

Nowadays marketing in the construction business becomes an important part of the increasing of the construction enterprise competitiveness all over the world. [3]

Due to the fact that companies – professional contractors consume a greater part of the materials, the manufacturers are of maximum interest in personal contacts with executives who make management decisions. The manufacturing companies are ready to support the individual interact with major contractors, the issue of the products nomenclature for the precise major customer.

The analysis of the technical and creative solutions of the construction site to ensure the comparability of alternative variants of management decision includes:

- the conditions and work schedules of the competitive goods manufacturer;
- the type, scales and maturity degree of the analyzed materials and goods production (small-scale, serial, full-scale);
- technological, construction peculiarities of produced goods (mass, thermal conductivity, moisture resistance, easy transportation, etc.);
- organizational value and technical merit of the manufacturer production (automation degree, quality certificates, quality management system, conditions of employment, etc.)
- image, reputation, culture of manufacturing company production;
- geographical situation, transport remoteness, existence of sustainable logistics linkages between raw sources and material production.

When choosing the strategy of management decisions the following aspects are analyzed: production conditions, competitiveness, financial risks, demand constraints and existence of production capacities, range of products, pricing policy, assets utilization, tendencies in construction market, raw materials accessibility and cost, economic and political aspects. Every precise case of a purchase decision making or investment needs a comprehensive analysis based on the marketing management principles in the field of construction, i.e. the process of company activity adjustment to the contemporary market conditions. The process consists of the key stages implementation: market research; market sampling; developing of marketing mix; implementing of the scheduled measures [4, 5].

Thus, successful promotion in the construction materials market in B2B and B2C segments demands from the manufactures of the wood-cement compositions the solution of a set of tasks, beginning with production and technology tasks and ending with the corporate positioning and forming of a new mentality attitude.

The questions of corporate positioning in the construction materials and technologies market is a key element of successful marketing and management. Material or product positioning is the parameters of the material or product perception by the individual contractor or construction company. At present there are 2 fundamentally different material positioning ways. The 1st allows positioning of the material as a competitor to the products existing in the market. The 2nd implies the developing of a brand new product or technology. It is more complicated, but in case of successful execution it enables a significant benefiting and sustainable occupying of the newly made niche in the construction materials market. The 1st way is of high risk either for it is difficult to conduct a competitive struggle with products or technologies already occupied their places in the market. Producing a common product (a building block, e.g.) on the initial stage the manufacturer decreases the expenses on the research and elaboration in the field of a brand new product, that becomes significantly valuable for the manufacturer and exposes an essential flaw for the consumers and the economics as a whole. Most companies usually take the 1st way. So, most manufacturers of the wood-cement compositions followed the rules and started the production of the traditional blocks and panels.

At first the technology of the wood-cement compositions production attracts greatly. Producing the construction materials for regions developers (construction industry, e.g.) is especially promising [6]. It will be a final product constituting milled wood particles joint by different binding agents. The feedstock can be produced from the wood wastes which are of large amounts at logging, sawmilling and wood processing industry functioning. These wastes reach substantial amounts. The volume of the logged wood in the Russian Federation in 2016 was 214 million m^3 and the volume of wastes constituted about 32 million m^3 [7]. Thus a wood-cement composition is a product of processing of sawmilling production wastes. The concept of full and integrated use of wood feedstock is declared in numerous development programmes and strategies. But does it meet a mental support of the final consumer?

We conducted an experimental research in the form of a questionnaire to assess the sustainability of the construction materials based on the wood-cement compositions. 329 respondents aging from 23 to 52 participated in the questionnaire. 63 % are university students, 21% - officials, 16% - workers. The form of a questionnaire implies the multiple-choice questions. 76 % of the respondents answered they would potentially build or purchase a private house for permanent living. Ownership of a property has a deep positive impact on health, forms a social behavior, satisfaction and general well-being [8]. The questionnaire participants chose the following key criteria for desired construction materials: construction materials durability (62%), high esthetics factors (14 %), factors of thermal efficiency and preservation (9 %), possibility of step-by-step building and living in an uncompleted house (7 %), low cost of the construction materials (6 %), other demands (2%). 96 % of the respondents announced positive attitude to the concept of full utilization of resources with the subsequent regenerating and processing. But only 7 % of the respondents agreed to apply such materials in the construction of their house. Thus, the vital requirement for the materials was their "novelty". In this regard the positioning of the wood-cement composition as a product based on the sawmilling wastes is fundamentally wrong and causes a negative perception of the product by the consumer.

Generally, the consumer comprehends that the application of wooden filler allows obtaining the wall constructions with high thermal insulation characteristics. Thermal and technical characteristics of the materials are inversely proportional to their volume weight and that is why light and extra light concretes are becoming increasingly important. They can be used both as thermal insulation and constructional load-bearing material for walls [9]. Wooden filler compositions should be related to aerated light concretes. Analyzing the competitor products of the similar application we can distinguish aerated and foam concretes, gas silicates. These are also the construction materials for the low-rise individual housing based on cement, sand, aluminum powder, gypsum, bleach, as well as on the industrial wastes such as lime and slags of metallurgy [10]. However, the manufacturers of aerated and foam concretes don't mention about the industrial wastes component in them while promoting and advertising. Russian contractors prefer to use such construction materials as brick, concrete, steel. Natural materials and recycled materials especially seem less applicable. Such a tendency of intensive environment management, ignoring of the integrated utilization and regeneration is typical for "developing" countries [11].

Within the dynamic development of aerated and foam concretes production there is still insufficient attention paid to the wood-cement compositions production that looks quite strange. The number of Russian enterprises producing aerated concrete exceeded 70 units, with an estimated total production of more than 16 million m^3 per year [12]. As forecasted [12, 13] the production of autoclaved cellular concrete in 2020 will reach 15 million m^3 and 8.1 million m^3 of non-autoclaved cellular concrete. In terms of square meters of total area it will constitute 36.16 million m^2 or 25% of the total housing commissioning. All the enterprises producing the cellular concretes are 10-15 years old. The largest enterprises of the Russian Federation produce more than 100 thousand m^3 products per year. The less productivity plants also operate. The "Agency of construction information" reports there are more than 150 large and small enterprises on aerated and foam concretes production in Russia [14]. The data regarding the number and production volumes of the wood-cement compositions are fragmentary and imprecise. Announced production volumes of the enterprises (Arbolit 33, Timfort, Sibar) don't exceed 100 thousand m^3 per year. In comparison with 16 million m^3 of aerated concrete it is a drop in a bucket. In general, we can say that the manufacturers of the wood-cement composition material operate in a format of individual contractors, enthusiastic persons, experimenters of B2B segment and cannot claim on developing of vast B2B market of construction materials for low-rise housing. At the current pricing of 60…70$ per 1 m^3 of aerated concrete and 80…90 $ per 1 m^3 of wood-cement composition materials the effective price competition of these materials is neither possible nowadays.

Since the wastes utilization issues are of special concern of both a precise manufacturer and the state on the whole the world has legislative basis examples of integrated wastes utilizing and processing. The system of grants, favorable tax treatment and direct legislative restriction is being applied. There is an established practice in European Union to favour the thermal energy industry on the ground of wooden biomass use over the ordinary industries of wooden wastes use [15]. The similar practice exists in the USA [16]. Unfortunately, such initiative in the field of effective wastes utilizing and processing didn't find practical use in Russia.

4. Conclusion

1. At present the materials based on the wood-cement compositions haven't become an effective product of B2B segment of construction materials market of low-rise housing. The production volumes of these materials are negligible as compared to aerated and foam concretes which are sustainably established in the corporate B2B segment of consumption.

2. It has been found that negative attitude to the material as to a utilization product (i.e. produced from wastes) is an important issue decreasing the competitive advantages of wood-cement compositions as compared to another light concretes. This constraint subjective perception of the participants of B2B market segment is due to the immature comprehension of the importance and significance of wastes utilizing and processing.

3. The practical impossibility of pricing competition of wood-cement compositions as compared to aerated and foam concretes. Effective promotion of wood-cement compositions can be feasible with legislative support of the productions based on the use of biologically renewable raw materials and wastes processing.

References

[1] Minett S 2002 *B2B Marketing: A Radically Different Aproach for Buisiness-to-Buisiness Marketers* (Harlow: Pearson Education) p 226

[2] Mescon M H, Albert M and Khedouri F 1988 *Mangement* (Harper&Row) p 777

[3] Veliev E F 2018 Marketing specificity in construction [Osobennosti marketinga v stroitelstve] *Young scientist* No 10 pp 92-94 available at: https://moluch.ru/archive/196/48733

[4] Laguta I V and Korotych I O 2017 Marketing specificity in construction. Introduction of marketing efforts [Spetsifika marketinga v stroitelstve. Vnedrenie marketingovykh meropriyatiy] *Regional development* No 6 available at: https://cyberleninka.ru/article/n/spetsifika-marketinga-v-stroitelstve-vnedrenie-marketingovyh-meropriyatiy

[5] Batrutdinov A S and Fedoseev I V 2008 *Basic models of the innovative process and qualifying indications of innovation* [Osnovnye modeli innovatsionnogo protsessa i kvalifikatsionnye priznaki innovatsii] *Problems of modern economics* Eurasian international scientific-analytical edition No 2 26 pp 27-31

[6] Zlobin D A 2005 Investigation of the economic growth factors of the entities of the Russian Federation, using the example of Krasnoyarsk region [Issledovaniya faktorov ekonomicheskogo razvitiya sub`ektov RF, na primere Krasnoyarskogo kraya] *Federal relations and regional social-economic policy* No 2 p 64-72

[7] Development strategy of the forest complex of the Russian Federation upto 2030/ the order of the Government No 1989 September, 9th 2018 available at: http://consultant.ru/document/cons_doc_LAW_307428/

[8] Onibokun A G 1977 Environmental Issues in Housing Habitability, Environment and Planning vol. 5 pp 461 - 476

[9] Rumyantsev B M, Zhukov A D and Smirnova T Yu 2012 *Thermal conductivity of highly porous materials* [Teploprovodnost vysokoporistykh materialov] Vestnik MGSU (Scientific and engineering journal for construction and architecture) No 3 pp 108-114

[10] Ignatova M V 2011 *Diversification of construction portfolio in Russia and the role of a brick as a building material in conditions of innovative development* [Diversifikatsiya stroitelnogo portfelya v Rossii i rol kirpicha kak stroitelnogo materiala v usloviyakh innovatsionnogo razvitiya] Materials of the conference IV International scientific practical forum "Innovative development of the Russian economics" part 3 (M:MESI) pp 93-99

[11] Woolley T 2004 The Role of Low Impact Building Materials in Sustainable Construction: The Potential of Hemp (Queens University Belfast School of Architecture, UK)

[12] Grinfeld G I, Vishnevskiy A A and Smirnova A S 2018 *Production of autoclaved aerated concrete in Russia in 2017* [Proizvodstvo avtoklavnogo gazobetona v Rossii v 2017 godu] Construction materials No 3 pp 62–64

[13] Pinsker V A 2004 Status and problems of cellular concretes production and application [Sostoyanie i problem proizvodstva i primeneniya yacheistykh betonov] *Cellular concretes in the modern construction: collection of the reports* International scientific practical conference Saint Petersburg

[14] Ukhova T A 2011 Present and future of cellular concretes in Russia [Nastoyaschee i buduschee yacheistykh betonov v Rossii] *LLC Stroy-Beton* available at: http://www.ibeton.ru/a198.php

[15] *Competition in wood waste*: *inventory of policies and markets* April 2013 (Colophon) available at:https://english.rvo.nl/sites/default/files/2013/12/Competition%20in%20wood%20waste%20June%202013.pdf

[16] Turner J, Buongiorno J, Zhu Sh and Prestemon J 2005 The U.S. forest sector in 2030: Markets and competitors *Forest Products Journal* No 55

Fruiting indicators of the apple tree growing in the creeping form in the Botanical garden named after Vs. M. Krutovsky

N V Moksina, O A Gerasimova and S V Gorbunova

Reshetnev Siberian State University of Science and Technology, 31 Krasnoyarsky Rabochy Av., Krasnoyarsk, 660037, Russia

E-mail: n.moksina2010@yandex.ru

Abstract. Study of the fruiting indicators of the apple tree growing in the creeping form in the Botanical garden named after Vs. M. Krutovsky resulted in the distinguishing of the cultivars characterized by the high annual yield, apples mass and sizes. The distinguished cultivars can be recommended for growing in the given and similar climate conditions.

1. Introduction

Apple tree is a key fruit crop in Siberian gardens that occupies 65-75 % of the fruit plantings, and the 1st in harvesting.

The quantity of the world grown cultivars varies from 25,000 to 35,000 in different ratings. There are nearly 350 apple tree cultivars appropriate to the geographical location on the territory of our country. Common Siberian assortment of apple trees can be divided into 3 different in genetic aspect groups of cultivars – rennets, semi-fruited (polukulturki), and large-fruited [2].

Apple tree fruits are consumed fresh and processed; they contain sugars, organic acids, mineral salts and microelements, they are endowed with biologically active substances as well. Apples contain vitamins and antioxidants [5]. They are also useful for humans because of fiber content [3].

Economic and biological properties of the fruit trees are determined not only by the genetic peculiarities of the apple tree species and cultivars, but also by the environment the phenotypes are formed in. Geographical factors (latitude, longitude, terrain of the region, etc.) significantly determine the zone climatic conditions influencing the fruiting, chemical content, keeping quality and other fruits properties [14]. Thus, from north-west to south-east within a country some cultivars increase their fruiting periodicity with the yield decreasing as a result [4].

The Botanical garden named after Vs. M. Krutovsky where the study was made in is located in Krasnoyarsk on the right bank of the Yenisey River in the estuary of the Laletina River. The garden territory is situated on the junction of the Kansk – Rybinskaya basin and forest – steppe zone of Western Siberian flatland with foothills of the Eastern Sayan. It causes the strongly continental climate conditions [1]. There was studied the memorial part of the garden the apple trees grow in a creeping form in. The form invented by Vs. M. Krutovsky in the beginning of 20th century is called Arctic or Krasnoyarsk stanian. The Botanical garden apple trees collection has 39 large – fruited cultivars of different geographical and selection origin [7, 8, 9, 10]. The trees are maximum 115 years old.

2. Methods and materials

The fruiting periodicity was determined according to the common method [13] by annual calculation of the fertile trees.

The yield estimation was done with method of model branches [11, 12] including the calculation of the fruits only on model branches average for the given tree and simultaneous clockwise determination of such branches on the tree. Fruits mass was determined by weighing method on the laboratory scale.

The tree yield (in kilogrammes) was determined by multiplying of an average fruits mass on a tree by their quantity.

Processing of the experimental material was done by the mathematical statistics methods with the help of Microsoft Excel programme.

3. Results and discussion

Vs. M. Krutovsky initiated the creation of the collection of domestic apple tree cultivars (Malus domestica Borkh) in 1904. Apple tree cultivars are differentiated in fruits size, form and mass, their gustatory qualities, ripening period and storage time, fruiting periodicity.

According to the ripening period and storage time the apple tree cultivars of the memorial part of the Botanical garden named after Vs. M. Krutovsky can be classified into winter cultivars (64 %) and summer ones (36 %). 6 cultivars (White Transparent, Bismarck, Grushovka moskovskaya, Golden schip, Nobilis, Papirovka) constitute more than a half of the total amount of the apple trees (55 %).

Apple tree is prone to fruiting periodicity that develops with aging. Among the apple tree cultivars presented in the collection of the Botanical garden named after Vs. M. Krutovsky Grushovka moskovskaya shows the abrupt periodicity according to literary data [6].

Analysis of the fruiting in 2013-2018 revealed the maximum of the fruiting trees (254 trees, 99 %) in 2018 (table 1). The year of 2017 is characterized by the lowest fruiting, adverse weather conditions in the beginning of vegetation (low temperatures and snowfalls).

Table 1. Analysis of the 6-year-period fruiting

	Quantity of the fruiting trees on annual basis, trees/%					
Cultivar	2013	2014	2015	2016	2017	2018
White Transparent	20 / 100	19 / 95	19 / 95	20 / 100	17 /85	20 / 100
Bismarck	28 / 97	29 / 100	29 / 100	28 / 97	23 /82	27 / 96
Grushovka moskovskaya	16 / 84	19 / 100	18 / 95	18 / 95	5 / 26	19 / 100
Golden schip	13 / 72	16 / 89	17 / 95	18 /100	8 / 40	20 / 100
Nobilis	19 / 86	18 / 82	22 / 100	22 / 100	3 / 13	22 / 100
Papirovka	31 / 91	21 / 62	34/ 100	33 / 97	21 / 62	34 / 100

Analyzing the percentage of the White Transparent fruiting trees we can state that 100 % of fruiting was observed in 2013, 2016 and 2018, in 2014 and 2018 95 % of trees gave fruits. Nobilis cultivar gave 100 % of fruiting 3 times within the studied period (in 2015, 2016 and 2018). In 2013 86 % of this cultivar trees gave fruits, in 2014 – 82 %, in 2018 – only 13%.

All the apple trees (100 %) of the Bismarck, Grushovka moskovskaya, Golden schip and Papirovka cultivars gave fruits twice: Bismarck in 2014 and 2015, Grushovka moskovskaya in 2014 and 2018, Golden schip in 2016 and 2018, Papirovka in 2015 and 2018.

The figure 1 shows that White Transparent and Bismarck have the most stable fruiting (more than 80 % of the trees have annual fruiting). Papirovka fails on this indicator (62 % of trees gave fruits for 2 years, the percentage of fruiting trees constituted 91 % and 97 % for 2 years and for 2 years – 100 %).

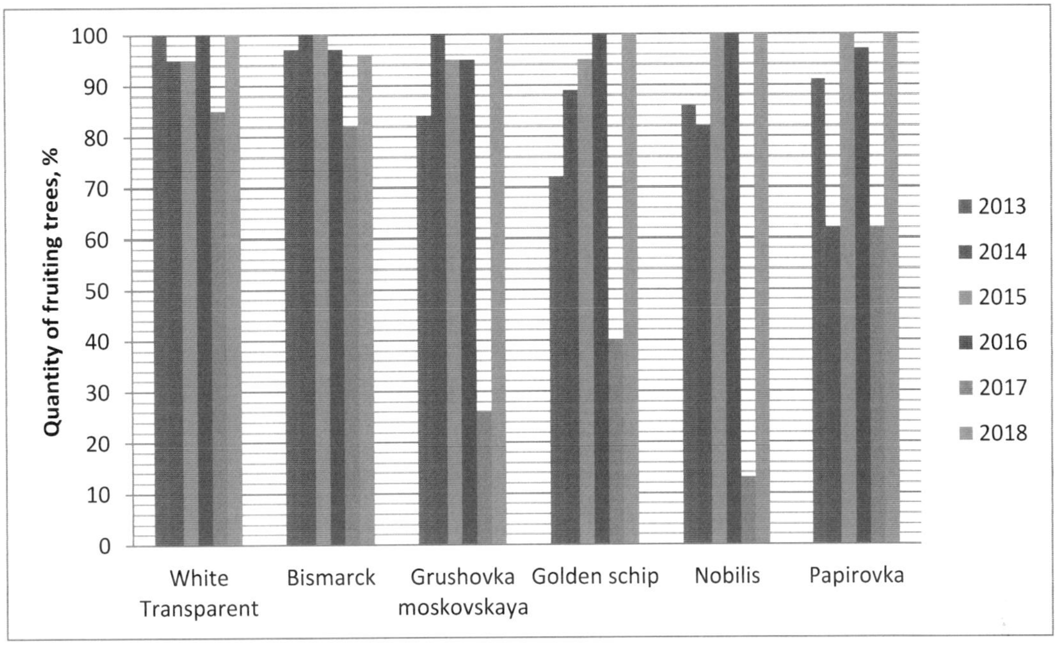

Figure 1. Percentage of fruiting trees in 2013 – 2018

The following scale was used for intensity determination: 0-200 fruits – low fruiting intensity, 201-400 fruits – average intensity, 400 and more fruits – abundant intensity. There was established that within the studied period of 2013-2018 the abundant fruiting was observed in 2016 and 2018 when the percentage of the trees with the quantity of more than 400 fruits constituted 40 % and 47 % respectively. The year of 2017 had the poorest harvest – only 128 trees gave fruits that constituted 50 % of the total trees amount. That year the low fruiting trees prevailed (58 %). The abundant fruiting trees constituted 8 %. The low fruiting prevailed in 2014 and 2015 (table 2).

Table 2. Analysis of apple tree fruiting intensity.

Year	Fruiting trees		Fruiting intensity					
			abundant intensity		average intensity		low intensity	
	Number of trees	%	Number of trees	%	Number of trees	%	Number of trees	%
2013	234	91	96	41	60	26	78	33
2014	220	86	39	18	66	30	115	52
2015	249	97	24	10	99	39	126	51
2016	250	98	102	40	61	25	87	35
2017	128	50	11	8	43	34	74	58
2018	254	99	119	47	92	36	43	17

While analyzing the yield of the large-fruited apple tree cultivars during the studied period there was revealed that the following cultivars - White Transparent, Bismarck and Papirovka- have higher fruiting intensity. The average yield of White Transparent varies from 1,29 to 3,02 kg/m^2 of crown projection, Bismarck – 1,63-4,27 kg/m^2 of crown projection, Papirovka – 0,93-2,97 kg/m^2 of crown projection (table 3).

Table 3. Apple tree yield in 2013 – 2018.

	Yield, kg/m^2 of crown projection					
Cultivar	2013	2014	2015	2016	2017	2018
White Transparent	1,30	1,37	1,64	1,57	1,29	3,02
Bismarck	2,34	2,32	1,63	2,57	2,18	4,27
Grushovka moskovskaya	1,43	1,04	0,50	1,28	0,45	1,22
Golden schip	1,26	0,72	0,67	1,14	0,79	1,23
Nobilis	2,29	0,40	1,40	1,94	0,99	2,95
Papirovka	1,89	1,02	1,37	1,47	0,93	2,97

Despite the fact that productive age of apple trees is 40-50 years old many trees showed a good yield being 114 years old (at time of the study).

Conclusion

As a result of the study there was established that all the apple trees of the Botanical garden named after Vs. M. Krutovsky have a high yield over the age of 100. There were selected the cultivars for growing in the given and similar natural and climatic conditions (White Transparent, Bismarck, Papirovka).

References

[1] Botanical garden named after Vs. M. Krutovsky, 2006 available at: http://sibgtu.sibsau.ru

[2] Vasilieva V N 1997 *Apple tree in Siberian garden* [Yablonya v sibirskom sadu] (Novosibirsk: SB RAS) p 101

[3] Vigorov L I 1976 *Garden of medicinal crops* [Sad lechebnykh kultur] (Sverdlovsk: Middle Ural Press) p 171

[4] Golikova N A 1979 Study of periodicity of fruiting and factors of its change of domestic apple tree in the Central regions of the Russian SFSR [Issledovanie periodichnosti plodonosheniya i faktorov eyo izmeneniya u yabloni domashnei v tsentralnykh oblastyakh RSFSR] dissertation ... of Doctor of Agricultural sciences 06.01.2007 (Kursk) p 346

[5] Golikova N A, Novikova O A and Ovchinnikova R I 2001 The heavy metals content in the apple tree fruits grown within the city of Kursk [Soderzhanie tyazhelykh metallov v plodakh yabloni vyrashchennykh v cherte goroda Kurska] *Agrarian bulletin of Urals* № 2 (81) (Ekaterinburg) p 43-45

[6] Guseva I N 1992 *Apple tree in your garden* [Yablonya v vashem sadu] (Moscow: Moscow University Press) p 189

[7] Matveeva R N, Butorova O F and Moksina N V 1998 *Selection studies in the Botanical garden named after Vs. M. Krutovsky* [Selektsionnye issledovaniya v Botanicheskom sadu im. Vs. M. Krutovskogo] (Krasnoyarsk: SibSTU) p 162

[8] Matveeva R N, Butorova O F, Moksina N V and Repyakh M V 2006 *Apple tree selection in the Botanical garden named after Vs. M. Krutovsky* [Selektsia yabloni v Botanicheskom sadu im. Vs. M. Krutovskogo] p 357

[9] Matveeva R N, Butorova O F and Moksina N V 2014 110th anniversary of the unique collection of apple tree in the Botanical garden named after Vs. M. Krutovsky [Unikalnoi kollektsii yabloni v Botanicheskom sadu im. Vs.M. Krutovskogo -110 let] *Perspectives of the introduction of ornamental plants in botanical gardens and arboreta* materials of the international scientific conference 23-25.09.2014 (Simferopol) p 136-139

[10] Moksina N V 2018 Fruiting periodicity of apple tree cultivars of I.V. Mitchurin selection in the Botanical garden named after Vs. M. Krutovsky [Periodichnost plodonosheniya sortov yabloni selektsii I.V. Michurina v Botanicheskom sadu im. Vs.M. Krutovskogo] *Horticulture, seed production, introduction of arboreal plants* (Krasnoyarsk: Reshetnev University) p 148-152

[11] Potapov V A 2000 *Horticulture* [Plodovodstvo] (Moscow: Kolos) p 432
[12] Potapov V A, Uliyanishchev A S and Krysanov Yu V 1991 *Low growth intensive garden* [Slaboroslyi intensivnyi sad] (Moscow: Posagropromizhdat) p 219
[13] 1973 *Programme and methodology of cultivar study of fruit, berry and nut crops* [Programma i metodika sortoizucheniya plodovykh, yagodnykh i orekhovoplodnykh kultur] ed. G.A. Lobanov (Michurinsk: VNIISKh Press) p 495
[14] Saburov N V and Antonov N V 1962 *Fruits and vegetables storage and processing* [Khranenie i pererabotka plodov i ovoshchei] (Moscow: Selkhozizdat) p 448

Modeling of biotic cycle formation in closed ecological systems

A V Brilkov[2], Y Y Loginov[1] and A V Babkin[2]

[1] Reshetnev Siberian State University of Science and Technology, 31 Krasnoyarsky Rabochy Av., Krasnoyarsk, 660037, Russia
[2] Siberian Federal University, Krasnoyarsk, 660041, Russia

E-mail: loginov@sibsau.ru

Abstract. The biotic turnover formation in closed ecological system (CES) is analyzed by means of mathematical and computer modeling. The model CES hierarchical structure formation is ruled by the energy flow through the ecosystem which increases after the introduction of a predator in the steady state of "production - reduction" systems. The model analysis shows that the presence of the primary predator under certain conditions leads to the acceleration of the biotic turnover because of its higher mineralization ability (in comparison with the same of decomposer). The most acceptable predator which is able to bring the system towards the higher biotic turnover rate and energy consumption can be found for each system.

1. Introduction

For mathematical analysis of closed ecological system (CES) the equations describing the population dynamics and resource concentration are usually used (known Lotka-Volterra equations) [1-3]. Unfortunately, the applicability of this standard model approach is limited. It does not allow to formulate the general criteria, which would help to predict the result of competition of several populations on resources, limits of turnover sustainability, the direction of succession. This indicates the narrowness of this model approach. Sequential application of the functional approach is the solution to this situation. Firstly, the analysis of level dynamics of one or more key factors limiting the population development is required. Generally, it is necessary to study the function of the system acting as a single whole [4-7].

This work considers biotic turnover in closed ecological system (CES) with an external flow of energy. We use ordinary differential equation systems to represent the main types of natural or laboratory ecosystems (biotic cycles) with a few interacting species [1, 8, 9]. These ecosystems are nutrient-limited and display mass balance.

2. Results and Discussion

The simplest biotic cycle we use to deal with a "producer-consumer-decomposer" food chain (Fig. 1). A nutrient limited plant or algae (X) consume a nutrient resource (N), two predators (Y_1, Y_2) consume a producer. Then it dies and is reduced by decomposers (D), therefore matter recycles and meanwhile initially captured energy dissipates (E_D).

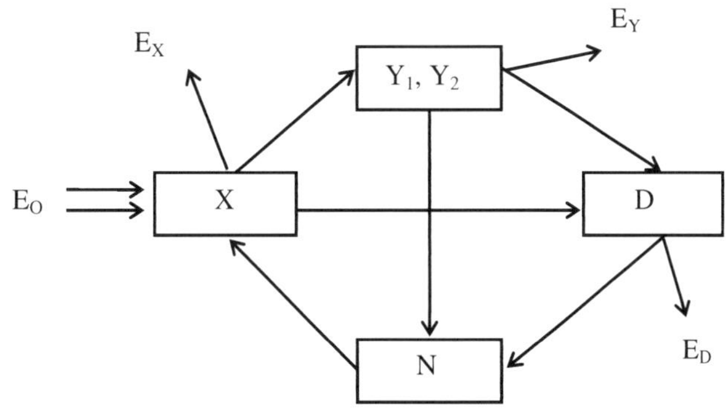

Figure 1. Scheme of ecosystem with "production- consumption- reduction" food chain.

The dynamics of this system is characterized by the four differential equations:

$$\begin{cases} \frac{dX}{dt} = \mu X - \varepsilon_X X - \nu Y \\ \frac{dY}{dt} = \nu Y - \varepsilon_Y Y - mY \\ \frac{dD}{dt} = \varepsilon_X X + \varepsilon_Y Y - dD \\ \frac{dN}{dt} = dD + mY - \mu X \end{cases} \quad (1)$$

where X, Y, D, N- concentrations of producers, predator, detritus and limiting substance, respectively; μ is specific growth rate of producers, n is specific rate of producer's consumption; ε_X, ε_Y - specific rates of mortality of producers and predators, respectively; d is a specific rate of dead biomass reduction; m - specific rate of predator's biomass expenditure for the processes of life support, and $m >> d$, because the growth rate of predators is supposed to be limited not with a substance, but the energy resources.

To gain more simplicity, instead of biomass values we use nutrient content in producers and detritus. Characteristic growth rate of the producers depends on the limiting nutrient concentration (the Monod function). To simplify the analytical study we use the following approximation:

$$\mu = \frac{\mu_m N}{K_N + N} \approx c_1 N \quad (2)$$

This equation is corresponds the case of deep limitation.

System (1) characterizes the matter flows and provides the closure of the ecosystem. As can be seen $d(X + D + N) = 0$, or $X + D + N = \text{const} = N_t$, where N_t is the total amount of limiting nutrient in ecosystem.

In this case only producers capture energy from outside the system. So, we can assume that the energy flow into ecosystem is proportional to the primary production of the ecosystem P, which is product of specific growth rate and biomass of producers:

$$\frac{dE_O}{dt} = eP = e\mu X \quad (3)$$

where e is characteristic energetic content of producers biomass. Producers bound much more energy into biomass, than dissipate. Therefore energetic activity of producers equals

$$a_p \approx -\frac{dE_O}{dt} = -eP \tag{4}$$

Energetic activity of decomposer link is

$$a_D = \frac{dE_D}{dt} \tag{5}$$

Equations (4) and (5) describe the energy flows, meanwhile the system (1) shows matter flows. Been brought together, assuming that characteristic energetic content of producers, consumers and decomposers is the same, we'll get the steady-state energy balance for the system (6):

$$a_D = \frac{dE_D}{dt} \tag{6}$$

It can be seen, that the system (1) contains both energy and matter conservation laws.

The growth of predators is supposed to be limited not with a substance, but the energy resources, bounded in producer's biomass. Predators consume much more prey biomass, than it is needed to build its own. The extra biomass is spent for the energy requirements. While the energy used for the processes of life support is dissipated, the correlated substance is immediately released into the medium, and therefore recycled. The nonliving predator's biomass is reduced by the decomposers.

Predator consumes the prey with specific rate:

$$v = \frac{v_m X}{K_X + X} \approx c_2 X \tag{7}$$

where v is the maximal specific rate of prey's consumption, K- prey's concentration, when the rate of its consumption is half of maximal. To simplify the analytical study we'll assume, that the rate of prey's consumption is proportional to the predator and prey concentrations (Lotka-Volterra model). Producer dies with specific rate ε_X and predator with ε_Y. Moreover, the part of predators biomass (properly its energetic content) is expended for the processes of life support and so on with the specific rate m, which is proportional to the energetic activity of predator according to (4):

$$mY = \frac{dN_Y}{dt} = k_Y a_Y \tag{8}$$

where k_Y is characteristic energy content of predator's biomass. Predator is releasing energy with the rate a_Y, meanwhile excreting extra substance with the rate μ_Y. We also can treat this system for the energy conservation law, and it will show that $E_O = E_D + E_Y$.

Solving the system (1) for the steady state we'll find the value of primary production:

$$\tilde{P} = \frac{\varepsilon_Y + m}{C_2} \cdot \frac{C_2 N_t - \varepsilon_Y - m + \varepsilon_X - m\varepsilon_X / d}{1 + C_2/C_1 + \varepsilon_Y / d}, \tag{9}$$

where N is total amount of limiting substance in ecosystem. Systems (1) were simulated on the computer with the values of coefficients common for the aqua systems (program Stella II, High Performance System, USA). Succession was modeled so that predators were introduced into the steady state of the "production - reduction" system and a new steady state were achieved. Dynamics of links and changes are shown in Fig. 2.

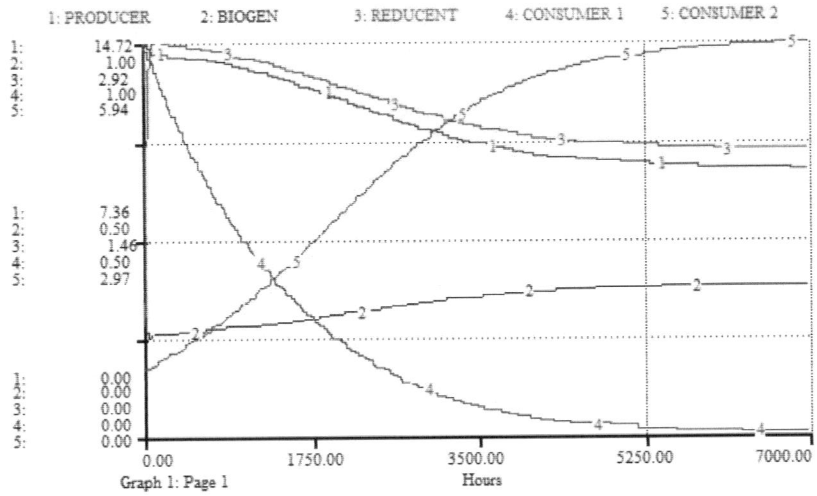

Figure 2. Computer simulation of ecosystem (1). The moment of predator's appearance is arrowed.

After appearance of predators in the ecosystem the primary production increases. Also number of decomposers decreases due to the lowering of their role in mineralization (recycling). Fig. 3 shows the dependence of ecosystem's primary production (formula (9)) on the characteristic energetic activity of predator $a_Y/Y = m/k_Y$, with the different values of N_t in the system. It can be seen, that productivity of the ecosystem increases and has a maximum (some optimal value). This graph can be called predator optimization curve.

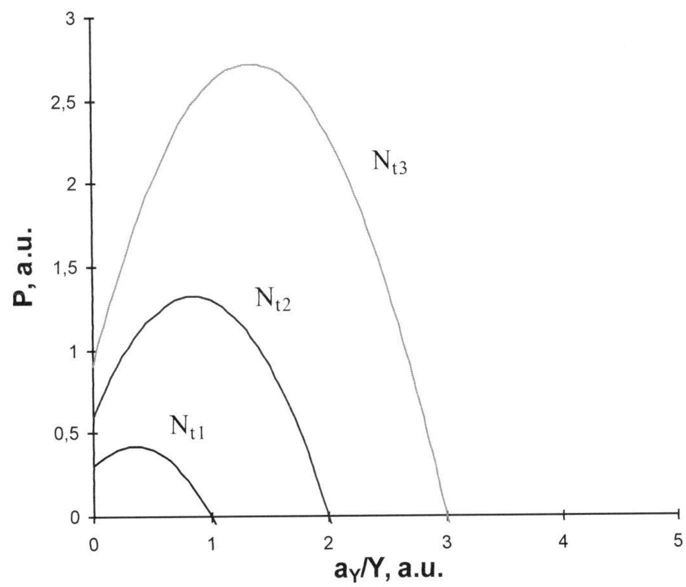

Figure 3. The dependence of the steady state primary production of production-consumption- reduction system on the characteristic activity of predator, shown for the different Nt in the system.

Having higher energetic activity, the predators are more active mineralizes, than decomposers. Therefore its appearance in the system leads to the increasing of the recycling rate, which is proportional to the primary production. We can assume that the process of predators' competition leads to the appearance of more active organisms.

3. Conclusion

The main difference between our model ecosystem and previously advanced models of closed ecosystems is that in our model the growth of consumers is limited by the available energy in preys but not the amount of nitrogen stored in preys [1, 8, 9]. Most of these previous estimates support the viewpoint that the growth of predators is constrained by energy but not by nutrient availability. Limitation of consumers by available energy enhances their importance in biotic cycles functioning because they allow a very rapid return of free nutrient to the environment whereas such return through mineralization of non-living matter is much slower process (see, for example, [3, 10, 11]). Due to this reason we cannot get rid of detritus and reducer chains by assuming the rate of decomposition to be high enough.

There are simulated elementary biotic cycles in the main types presented in the nature CES. It is shown, that in model CES hierarchical structure formation is ruled by the energy consumption: the primary production in the ecosystem increases after introduction of predator in the steady state of "production-reduction" system and achievement of the new steady state. The model analyses shows, that the presence of the primary predator under certain conditions leads to the acceleration of the biotic turnover, because of its higher mineralization ability (by a comparison with decomposer). This is the most general conclusion, which is interesting for solving the problems of predator introduction into the certain ecosystems.

References

[1] De Angelis D 1992 *Dynamics of nutrient cycling and food webs* (Chapman & Hall, London)

[2] Odum H 2007 *Environment, power, and society for the twenty-first century: the hierarchy of energy* (Columbia University Press, N.Y.)

[3] Ulanowicz R 1986 *Growth and development: Ecosystem Phenomenology* (Springer-Verlag, New York)

[4] Fred J, Broder B, Hauke R and Donald L 2011 *Perspectives in Ecological Modelling / In: Modelling Complex Ecological Dynamics. An Introduction in Ecological Modelling for Students, Teachers & Scientists* (Springer-Verlag Berlin Heidelberg)

[5] Hannon B 2010 *Ann. N.Y. Acad. Sci.* **1185** 30

[6] Morowitz H. and Smith E 2007 *Complexity*, **13** 51

[7] Gitelzon I, Pechurkin N and Brilkov A 1989 *Population problems in the biology of unicellular organism* (Harwood Academic Publ., London)

[8] Alexeev V, Krishev I, Polyakova M and Sazikina T 1978*Dynamics and statistical mechanics of biogeocenosis with fixed mass of limiting biogenic element / In: Man and Biosphere, V. 2* (MSU Press, Moscow)

[9] Gubanov V, Kovrov B and Fishtein G 1984 *Closed ecological systems- new test object for biophysical and ecological studies.- Biophysical methods of ecosystems study* (Nauka, Novosibirsk)

[10] Odum E 1975*Fundamentals of ecology* (Mir, Moscow)

[11] Volk T and Rummel J 1987 *Adv. Space Res.* **7** 141

Improving the performance characteristics of glued materials based on thermally modified veneer

A I Krivorotova and A A Orlov

Reshetnev Siberian State University of Science and Technology, 31 Krasnoyarsky Rabochy Av., Krasnoyarsk, 660037, Russia

E-mail: tkmkai@mail.ru

Abstract. The performance characteristics of glued materials based on thermo-modified veneer depend on the heat treatment mode. The choice of a mode that does not lead to a decrease in the strength of plywood allows reducing water absorption and swelling by a value from 20,4 to 33,5 %. Plywood based on thermo-modified veneer acquires higher strength and water resistance, which significantly expands the range of application of laminated wood based on urea-formaldehyde resins.

1. Introduction

The first studies in the field of high-temperature wood processing were conducted in the 30s of the last century in Germany Stamm A J. and other scientists. Further development of technologies for thermal modification of wood received in the United States and several European countries. A large amount of research in this area was conducted by the VTT Center in Finland [1]. A review of studies on the thermal modification of wood by Russian scientists is given in [2].

A common element of all technologies of thermal modification is the heating of wood in the temperature range of 150 - 230 °C in environments: water vapor, inert gas, flue gas or liquid.

Researchers [3-6] note a positive change in the physico-mechanical and decorative properties of the material. Treated wood acquires resistance to the effects of wood-destroying fungi and insect larvae, has low moisture absorption, dimensional stability of products, high thermal insulation characteristics and aesthetics. After heat treatment, the color changes, the texture is more clearly and effectively detected. Wood of low-value species (birch, aspen, etc.) acquires a resemblance to the wood of valuable hardwood.

But in the process of thermal modification, exposure to high temperatures leads to a decrease in a number of mechanical and operational properties of wood, which limits its scope. So Kollmann F. observed violations in the microstructure of spruce wood heated to 150 ° C. It has been found that a reduction in the volume of wood occurs due to the loss of a substance [7]. In [8], it was found that heating wood in the presence of moisture at temperatures of 180–200 ° C leads to a significant decrease in resistance to shock loads, modulus of elasticity, tensile strength at stretching and compressive strength. Researchers [9] found that the tensile strength at break decreases as the temperature of heat treatment decreases, whereas the temperature affects the modulus of elasticity only slightly. In [10], the relationship between loss of strength and darkening of wood when exposed to high temperatures during processing is shown.

A review of the work of researchers shows the different directions of changes in the properties of thermally modified wood. Strength properties are reduced, but the characteristics of the material are increased, allowing it to be operated in conditions of high humidity of the environment.

2. Experimental part and discussion

At the Department of Woodworking Technology SibGU them. Mf Reshetnev for the experiment on thermal modification of the veneer was mounted and tested the experimental setup. Taking into account that a birch veneer with a thickness of 2,0 mm was chosen for the study, the temperature range for heat treatment was adopted from 160 to 200 °C. The duration of the heat treatment process was 2 and 4 hours. Thermal modification consisted of the following stages: heating-high-temperature drying, heat treatment, cooling-conditioning.

For research was used plan full factor experiment. The required accuracy was ensured by a tenfold duplication of measurements.

After the termination of the thermal modification process, the veneer was packed in batches in batches and kept for 24 hours under a small load to level out possible stresses.

The plywood samples were pressed on the basis of thermally modified birch veneer using the standard method. A preliminary experiment showed that the pressing mode of plywood based on thermally modified veneer using urea-formaldehyde resin does not require additional adjustments and corresponds to the pressing mode of general-purpose plywood: specific pressure P = 1,6 MPa; temperature T = 120 °C; specific pressing time τ = 0,7 min/mm.

The size of the pressing for all samples of plywood corresponded to that adopted for birch plywood and ranged from 9,8 to 10,6 %.

Testing of the physicomechanical characteristics of plywood samples based on thermally modified veneer was made after exposure for 48 hours. The state-of-the-art testing machine UTS-30 was used for the test.

A summary report of the results of static bending and shearing strength on the adhesive layer is presented in tables 1-2.

For a comparative assessment of the strength characteristics of plywood based on modified birch veneer, samples of plywood were made from the original birch veneer. Birch veneer was selected from the same batch of veneer that was used for processing. The results of testing the physico-mechanical properties of plywood based on the original veneer are presented in table 3.

Table 1. The results of tests on static bending.

Processing temperature / processing time	Strength at static bending, MPa				
	Sample 1	Sample 2	Sample 3	Sample 4	Sample 5
160/2	74,22	72,75	69,8	73,9	72,96
160/4	75,77	78,058	76,42	77,03	76,12
200/2	75,86	78,6	81,36	77,62	79,33
200/4	57,12	53,609	65,83	56,7	59,36

Table 2. The results of the test when spalking on the adhesive layer.

Processing temperature / processing time	Strength at spallation, MPa				
	Sample 1	Sample 2	Sample 3	Sample 4	Sample 5
160/2	1,761	1,972	1,821	1,798	1,902
160/4	2,309	2,401	2,362	2,296	2,106
200/2	1,994	2,257	2,012	1,990	2,106
200/4	1,584	1,163	1,265	1,369	1,401

Table 3. The results of the testimony of ordinary plywood.

Name of the indicator	Value of the indicator
Strength at static bending, MPa	58,2
Strength of splitting on the adhesive layer, MPa	1,75
Water absorption, %	65,6
Swelling, %	18,2

From the presented data it is clear that all the samples of plywood, made on the basis of thermally modified veneer, have high strength and meet the requirements of the standard. Samples of plywood made of modified veneer with the following parameters have the highest durability at static bending: T = 160 °C, τ = 4 h and T = 200 °C, τ = 2 h.

A comparative analysis of the strength of plywood samples shows that the greatest increase in strength during static bending is observed for samples of plywood made of modified veneer at T = 200 ° C, τ = 2 h. The average tensile strength is 78,6 MPa. The smallest strength value in static bending for samples of plywood made of modified veneer at T = 200 ° C, τ = 4 h. The average tensile strength is 58,9 MPa. In this case, the strength of the samples is close to the strength of the samples based on the original birch veneer. Such changes in strength properties are probably associated with changes in the properties of wood in the process of thermal modification. As a result of thermal modification, the veneer surface becomes more dense and solid, the influence of irregularities and roughness of the surfaces being joined on the bonding strength is reduced, and the cellular structure of the wood changes.

For the indicator of the strength when splitting on the adhesive layer, a dependence similar to the ultimate strength in static bending is observed. Samples of plywood made of modified veneer at T = 160 °C, τ = 4 h and T = 200 °C, τ = 2 h have the highest rates of cleavage on the adhesive layer for the samples of plywood made of veneer modified at T = 200 °C, τ = 2 h. The average shear strength at spallation is 2,1 MPa, with the required standard 1,5 MPa. The smallest splitting strength of the adhesive layer in samples of plywood made of modified veneer at T = 200 °C, τ = 4 h. The average tensile strength is 1,4 MPa, which is lower than the standard requirements (1,5 MPa). This feature can be explained by the formation of a more dense layer on the veneer surface, which prevents the veneer surface from being wetted by glue, an increase in the so-called inactivation of the surface layer of wood, a decrease in roughness and, therefore, a decrease in adhesive contact between glue and wood. This assumption confirms the type of destruction of the sample of plywood under test. If in the first three types of samples the mixed type of failure was most often observed, then in the case of testing plywood based on veneer sheet modified at maximum heat treatment parameters, the type of separation was observed to be cohesive in the adhesive layer.

The obtained plywood samples were also tested for water absorption and swelling. In many works on thermal modification of wood, it is noted that thermal modification allows reducing the water absorption values of wood several times. The test results are presented in table 4.

Table 4. The results of tests for water absorption and swelling.

Processing temperature / processing time	Water absorption in 24 hours, %	Swelling in 24 hours, %
160/2	62,06	16,6
160/4	52,22	12,1
200/2	51,32	10,77
200/4	48,09	10,3

The maximum value of water absorption and swelling is observed for samples made of modified veneer at T = 160 ° C and τ = 2 h – 62,06 % and 16,6 %, respectively. The minimum value for samples made of modified veneer at T = 200° C and τ = 4 h is 48,09 % and 10,3 %, respectively. The water absorption of thermally modified samples is reduced by 5,4 to 33,5 % compared with general-purpose plywood. Swelling is reduced by a value of from 8,8 to 43,4 %. Thus, plywood based on thermally modified veneer acquires higher water resistance, which significantly expands the scope of application of urea-formaldehyde-based plywood.

High-temperature processing of wood is one of the ways to improve the physicomechanical properties of general purpose plywood through the use of thermally modified veneer for bonding.

The implementation of the PFE allowed to establish that the strength of plywood depends on the heat treatment mode of the veneer. The highest indicators of strength in static bending (78,6 MPa) and when shearing over the adhesive layer (2,1 MPa) have samples of plywood made of modified veneer with the following parameters: T = 160 °C and τ = 4 h; T = 200 °C and τ = 2 h.

Heat treatment of veneer, not leading to a decrease in the strength of plywood, reduces the water absorption and swelling by an amount from 20,4 to 33,5 %.

References

[1] ThermoWood® handbook. Guide to heat treatment of wood, 2003. available at: www.thermowood.fi.

[2] Akhmetova D A, Safina A V and Stepanova T O 2015 Review of research on thermal modification *Woodworking industry* No 4 pp 28-35

[3] Syunyaev R Z, Safieva R Z and Safin R R 2000 The influence of internal structure and dispersity to structural-mechanical properties of oil systems *Journal of Petroleum Science and Engineerin* 26 No 1-4 pp 31-39

[4] Calonego W F, Severo E T and Furtado E L 2010 Decay resistance of thermally-modified Eucalyptus grandis wood at 140 degrees C, 160 degrees C, 180 degrees C, 200 degrees C and 220 degrees C *Bioresour. Technol.* 2010. vol 101 (23) pp 9391-9394

[5] Boonstra M J, Tjeerdsma B F and Groeneveld H A 1998 Thermal modification of non-durable wood species //IRG/WP/98-40123 part 1 The PLATO technology-thermal modification of wood P 13

[6] Evans P D and Banks W B 1990 Degradation of wood surfaces by water *Holz als Roh-und Werkstoff* No 48 issue 5 pp 159-163

[7] Kollmann F Gefuge und Figenschaft 1969 Sandesangen im Holz durch mechanische und thermische Beanspruchung *Holz als Roh – und Werkstoff* No 1 pp 407–425

[8] Korkut S and Budak M 2009 Effect of high-emperature treatment on the mechanical properties of rowan (sorbus aucuparia L.) wood *Drying technology: an international journal* pp 1240-1247

[9] Giebeler E 1983 Dimensional stabilization of wood by moisture-heatpressur *Holz als Roh - und Werkstoff* **41** pp 87–94

[10] Bekhta P and Niemz P 2003 Effect of high temperature on the change in color, dimensional stability and mechanical properties of spruce wood *Holzforschung* No 57 (5) pp 539–546

Improving the environmental performance of laminated wood products

A I Krivorotova and O A Usoltsev

Reshetnev Siberian State University of Science and Technology, 31 Krasnoyarsky Rabochy Av., Krasnoyarsk, 660037, Russia

E-mail:tkmkai@mail.ru

Abstract. Modification of the formaldehyde resins with an extracts of softwood bark makes it possible to reduce a free formaldehyde emission from laminated wood products, in particular plywood, to increase the strength in chipping and static bending of plywood, to reduce the cost price of adhesive composition.

The laminated wood panels based on formaldehyde resin have many advantages with significant weakness of the high toxicity, whichwas caused by content of phenol and formaldehyde in the resins used. Production of laminated panels, in particular plywood,based on phenol-formaldehyde resins is environmentally hazardous for both environment and human health. One way ofthe toxicity reducing of the laminated products is modification of the phenol-formaldehyde resins with an aqueous extracts of softwood bark [1, 2].

Phenol-formaldehyde resin and extracts of softwood barkused for the modification of phenol-formaldehyde resin. Extracts obtained from bark of larch and fir with the use of inorganic and organic solvent, such as water, sodium hydroxide in water and isopropyl alcohol, were used as modifiers. The physico-chemical properties of extracts and their modified adhesives were initially defined [3].

The organoleptic properties, viscosity, pH ranges, refractive index and content of polyphenol substances are the basic properties of extractives influencing the properties of their modified adhesive compositions and bonded materials. The tables 1 and 2 show the basic physico-chemical properties of extracts investigated.

Table 1. The physico-chemical properties of extracts from a fir bark

Names of the indicators	Extraction of firbark		
	coldwater	hotwater	isopropanol
Appearance	muddy liquid		liquid
Relative viscosity on VZ-4, c	11,01	10,96	9,91
Color	light-yellow	dark-yellow	red-brown
Smell	wood		solvent
pHranges	6,50	5,80	3,00

Table 2. The physico-chemical properties of extracts from a larch bark

Names of the indicators	Extraction of larch bark		
	coldwater	hotwater	isopropanol
Appearance	muddy liquid		liquid
Relative viscosity on B3-4, c	9,81	12,60	10,01
Color	light-yellow	dark-yellow	red-brown
Smell	wood		solvent
pHranges	6,38	5,750	2,90

One of the most importantproperties of extracts, on which properties of modified adhesive depends, is quantity of polyphenol substances in an extract [4]. Figure 1 presents extractives emissions according to wood and concentration of solvent (sodium hydroxide in water).With increasing of solvent concentration, there has been an increase in quantitative emission of polyphenol extractives. Larch has quantitative emission of polyphenol extractives more than a fir.

Changing the concentration of extractives changes the viscosity of the extracts. High molecular weight phenolic components and intermolecular interactions, increasing with increasing concentration, increase the viscosity. The increase in concentration by 50 % leads to an increase in the viscosity of the aqueous extract by 34%, isopropanol extract by 25 %.

The obtained extracts were introduced into the adhesive composition based on SFZ-3013. Analysis of the physicochemical properties of the compositions obtained shows that with an increase in the concentration of polyphenolic substances in the extract in the modified composition, the mass fraction of alkali increases and the amount of brominated substances decreases, the time of gelatinization decreases. The remaining properties are unchanged [5].

To determine the physicomechanical properties of glued wood materials based on the modified glue composition in laboratory conditions, samples of FSF plywood were made. Testing of the mechanical properties of the obtained plywood samples was carried out in accordance with the requirements of the current.

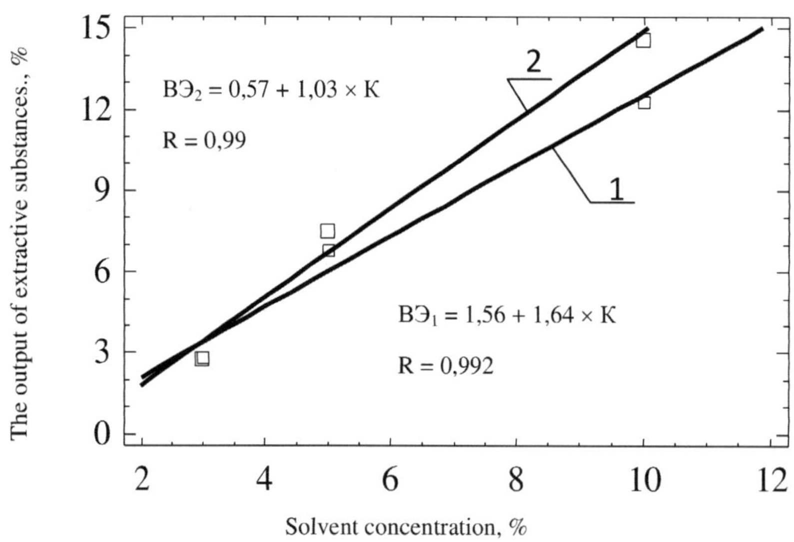

Figure 1. The dependence of the yield of extractives the concentration of the solvent (Wood species: 1 - fir, 2 – larch)

As a result of mathematical processing of the experimental results, regression equations were obtained, expressing the dependence of the mechanical properties of plywood on the modified binder from the pressing mode. As an example, we present the equation of dependence of the shear strength:

$$\sigma_{sh} = 1{,}48 + 0{,}106v + 0{,}098P + 0{,}032T - 0{,}09v^2 - 0{,}035vP + 0{,}0vT - - 0{,}07P^2 + 0{,}025PT - 0{,}08T^2$$

The influence of the studied factors on the shear strength of the adhesive layer was evaluated using a graphical interpretation of the regression equation and graphs of the effects of the factors and their interactions.

Analysis of the nature of the joint influence of factors on the strength of cleavage showed that the amount of extractives has a greater effect on the strength of the lower level of variation of the pressing pressure (strength change is 0,14 MPa). The change in strength at the upper level of pressure variation is only 0,08 MPa, and the dependence is extreme with an inflection point at 25 % of extractives. The change in strength when changing the amount of extractive substances at the minimum and maximum pressing temperature is the same, the difference between the indicators is from 0,02 to 0,03 MPa, which indicates a slight influence of temperature on the strength of the adhesive compounds on the modified adhesives. A similar character is observed when the pressure and the pressing temperature change. When the pressing pressure is at the lower level (1,2 MPa), the pressing temperature does not affect the strength of the adhesive joint. When the pressing pressure is increased to 1,8 MPa, the difference in strength between the samples compressed at 120 and 200 °C is 0,05 MPa.

In accordance with the accepted regimes, birch plywood samples were made to determine the effect of the amount of added extract on the tensile strength of the cleavage layer and the tensile strength of the static bending.

As a result, the greatest strength among the samples, using modified adhesives, have samples made with glue containing 25 % of extract. Samples of plywood made with glue containing 15 % of extract, the maximum tensile shear fracture do not meet the requirements of the standard (when using birch veneer ultimate strength in shearing adhesive layer after boiling for 1 h is for PSF 1,5 MPa), the tensile strength at a static bend is located at the lower boundary. This is due to the insufficient amount of the introduced polyphenolic substances in the adhesive composition for the polycondensation reaction, while reducing the viscosity, pH and increasing the gelatinization time of the resin. The depth of the polycondensation reaction depends on the amount of free reactive phenol, which, when modified by extracts, will depend on the content of polyphenolic substances. The amount of isolated formaldehyde (gas-analytical method) from the tested samples ranged from 1,5 to 3,6 $mg/m^3 \cdot h$, which corresponds to the E1 emission class [6].

As a result of researches possibility of use for modification of phenol-formaldehyde resins of extracts of coniferous breeds of the trees received with use of inorganic and organic solvents (water, water solutions of sodium hydroxide and isopropyl alcohol is established. Introduction to the adhesive composition of the above extracts can reduce the toxicity of plywood based on phenol-formaldehyde resins by reducing the concentration of phenol and formaldehyde in the adhesive composition, improve the physical and mechanical properties of plywood. The cost of modified adhesive composition is 35 % of the cost of phenol-formaldehyde resin.

References

[1] The use of spruce bark extracts in adhesives for the plywood / Plate and plywood: Express information, foreign experience1986 (NIPIEIlesprom) No 5 pp 2-10

[2] The properties of the extractives of the bark of the pine species Pinus halepensis / Plate and plywood: Express information, foreign experience 1986 (NIPIEIlesprom) No 5 pp 10-14

[3] Petrovskaya J A 2009 The Use of fir bark and fir in the production of composite materials *Young scientists in solving urgent problems of science* (Krasnoyarsk) pp 359 – 362

[4] Sokolova Y M and Krivorotova A I 2011 Modification of synthetic resins extracts of bark of conifers *Promising innovations in science, education, production and transport 2011* Vol. 2 (Odessa: Black Sea Coast) pp 100

[5] Krivorotova A I and Ushanova V M 2014 Improved the environmental properties and quality characteristics of plywood products, using adhesives modified with extracts of pine bark *Conifers of the boreal* V XXXII **1–2** pp 52–56

[6] Ushanova V V and Krivorotova A I Plywood: FGBOU VPO Siberian state technological University Patent RF №: 2440890 Appl. 05.07.2010 № 2010127766/13 publ. 27.01.2012 bul. No 3 IPC B27D 1/04 (20006/01)

IOP Conf. Series: Materials Science and Engineering **822** (2020) 012006 doi:10.1088/1757-899X/822/1/012006

Formation of project environmental thinking in the training of engineers

A A Lepeshev[1], Y Y Loginov[2], V V Kuimov[1], D A Tolstoy[1], A V Kozlov[1], T V Pogrebnaya[3] and O V Sidorkina[4]

[1] Siberian Federal University, Krasnoyarsk, 660041, Russia
[2] Reshetnev Siberian State University of Science and Technology, 31 Krasnoyarsky Rabochy Av., Krasnoyarsk, 660037, Russia
[3] Secondary School No 10, Krasnoyarsk, 660017, Russia
[4] Secondary School No 82, Krasnoyarsk, 660074, Russia

Abstract. At transition of civilization to the sixth technological way and the Industry 4.0 the speed of new technologies emergence, which influence on the environment can be different from former, increases. The created means of environmental protection can be insufficiently effective for new technologies. In these conditions the formation of project environmental thinking is important when training engineers to fully consider ecological requirements directly during the development of new technologies. In this work the possibilities of the theory of inventive problem solving (TIPS) use as effective means of modern engineering thinking for project ecological thinking formation are considered. The theoretical analysis and systemic synthesis of the TIPS methods and the didactic system of innovative education "TIPS pedagogics" are applied. The possibilities of TIPS in innovative thinking formation and its highest form, such as sustainable thinking, including ecological project thinking, modern ecological outlook and noosphere thinking, starting since school level, are shown. The possibility of a collaborative activity in interactive training, using the method of innovative projects, allows to find new problems solutions, including in ecological education.

1. Introduction

The transition of civilization to the sixth technological way as a set of associated production which core is formed by convergent NBICS (nano - bio - info - cognitive and social) technologies, to the Industry 4.0 significantly influences both environmental problems specificity and possibilities of their solution. The development of nanotechnologies, together with breakthrough achievements in various industrial products quality indicators, can lead to environmental pollution by nanoparticles the influence of which on living organisms is difficult to predict. The same can be said about biotechnologies wastes if released into the environment. The transition of production from subtractive to additive (for example, 3D - printing) reduces but does not fix completely a problem of wastes. More and more space technologies influence the environment: an increase in missile launches number influences, for example, agricultural workers health in adjacent to the spaceports territories; accumulating "space debris" subsequently forms extensive "garbage islands" in the oceans.

In general environmental protection problems during the transition to the sixth technological way (and especially to the subsequent ways which philosophers and sociologists already speak about) gain a significantly more common, systemic character. The main systemic problem is that the acceleration of new technologies creation leads to the lag in creation of protection means against new types of the harm from these new technologies. The solution of this common problem refers to the tasks of sustainable development (Sustainable Development).

Nowadays the relevance of sustainable development problems increases so that requires a scientific approach not only to specific areas of human activity, to specific territories and branches, but also to the concept of "sustainable development" in general. The need for a fast solution of the emerging environmental problems demands the corresponding methodology based on the common, fundamental ideas of development processes. Timely and high-quality training of future engineers in this methodology is not less important (as well as operating engineers in the form of further professional training).

2. Environmental and cognitive technologies

According to economists and sociologists forecasts, cognitive technologies will enter a core of the sixth technological way, along with new production technologies. The comprehension of cognitive technologies as technologies of computer model operation of a number of human brain functions [1] is the most widespread. Along with it, there is a comprehension of cognitive technologies as technologies of human brain functions improvement, in particular a creative function [2]. In both of these cases of comprehension, cognitive technologies are technologies of information creation (knowledge or project decisions). In this sense, cognitive technologies do not harm the environment. However, the use of cognitive technologies in the second comprehension if it is directed only to the projection of new, more economically efficient production technologies, without taking into account ecological requirements, can increase environmental damage..

The cognitive technologies having at the same time the property of creativity began to be created in ancient times: Socrates maieutics [3] and heuristics of Pappus Alexandria [4]. In the Middle Ages the philosopher Raymond Lully created "Lully's Circles" [5]. In the 20th century with the growing need for innovative ideas a number of new methods were created: brainstorming [6], lateral thinking [7], the method of focal objects [8], morphological analysis [9], sinektika [10]. The theory of inventive problem solving – TIPS [11] is considered to be the most modern and efficient, gaining more and more wide recognition in the world. TIPS has already been actively applied to innovative solutions generation by leading multinational corporations: Boeing, Kodak, Intel, Samsung [12] and, according to the growing demand for experts, is taught in a number of leading world universities of the USA, Europe, China, Southeast Asia, Australia.

The effectiveness of TIPS is caused by the fact that it is not just a set of divergent ("departing" from stereotypes) thinking methods, but it combines divergent and convergent ("merging" to the required decision) thinking. At the same time convergent thinking is based on the laws of technical systems development opened by the author of TIPS G.S. Altshuller and also on the principles, techniques, standards concretizing these laws. In turn, the named laws of systems development are a specification of philosophical science laws on dialectics development, therefore TIPS received one more name "applied dialectics", in particular in relation to its extension to not anthropogenic systems [13]. In this regard it is possible to predict that TIPS will become the main cognitive technology in the second comprehension (i.e. possessing creativity) which is applied in science, technique, economy of the sixth technological way [14].

Briefly the substance of TIPS can be expressed as follows:

1. Any solution of a problem task (the result of which is the innovative idea) is the act of some system development .
2. Development, according to dialectics, occurs through accumulation and overcoming (resolution, elimination) contradictions.
3. Therefore, if there is a problem, it is necessary to find a contradiction in it and to overcome it.

The external contradiction manifestation is usually the impossibility of system quality indicators improvement within its former structure. After contradiction overcoming a system performing some function better than the previous system in which there was a contradiction is formed. This new system has qualitatively different structure (qualitatively different principle of action).

To overcome contradictions in TIPS, in addition to classical dialectics, there are corresponding "intelligent tools": laws, principles, techniques, standards, algorithm. These "intelligent tools" reflect the regularities of systems development existing in the anthropogenic and not anthropogenic world.

Currently, a new based on TIPS class of computer programs – CAI (Computer Aided Invention) supplementing the complex CAD, CAM, CAE with design tools of fundamentally new structures of technical systems is being developed in the world.

The distribution of TIPS in the world has naturally led to the application of TIPS in the solution of ecological tasks [15, 16] as well. So far there is an experience of TIPS application for the solution of specific objectives of harm decrease or prevention to the environment from specific technologies. Based on philosophical, systemic character of TIPS ("system" – one of the basic concepts of TIPS), it is important to consider the possibilities of application of this science for the solution of the above-named systemic problem: possible lag in creation of protection means from new types of the harm caused by new technologies.

3. Concept "Degree of system ideality" as tool of environmental thinking

The first law of systems development in TIPS – the law of ideality degree increase: *"In the process of all systems development their degree of ideality increases, striving for infinity, but never reaching it"*. (In the literature on TIPS there are also other formulations which differ in separate words and their alternation, but not differing essentially) [11, 17].

The concept "ideality degree" is usually expressed with the formula which has not a strictly mathematical, but a methodical character:

$$И = \frac{\Sigma Ф_П}{\Sigma Ф_P} \to \infty ,$$

Where $И$ – ideality degree, $Ф_П$ – useful factors, $Ф_P$ – reckoning factors.

The degree of system ideality is the relation of its useful factors sum to the factors of reckoning sum (with what we "pay" for the useful factors). The concept "ideality degree", in essence, is the concept "purpose function"expansion which engineers use improving the composite technical system within its former structure, especially by computer optimization. The function of purpose has a precise mathematical expression. In TIPS the concept "ideality degree" is applied to the creation of fundamentally new systems, i.e. systems with a new structure. At the same time the degree of ideality increases so significantly that it is obvious without accurate mathematical formulation.

The given above definition of ideality degree refers to anthropogenic, including technical, systems. For not anthropogenic and social systems, as G.S. Altshuller followers researches show, it has specificities. However the questions of ecologically safe technical systems creation are considered here, therefore we will further use this definition.

From the point of view of the ideality degree the question of environmental protection comes down to whether such factor of reckoning as environmental harm is considered during new equipment and technologies designing. Judging by the fact that so far no noticeable acceleration of various environmental problems solutions has been observed in the world, the corporations applying TIPS in innovation consider this factor insufficiently so far. The named factor consideration leads to the emergence of additional contradictions which it is necessary to overcome, however potent "intellectual instruments" of TIPS, including application of CAI class programs, allow to make it.

Thus, TIPS methodology possession becomes the most important new part of engineers ecological culture while this methodology is applied to such increase in the new projected systems degree of ideality when the reckoning factor in the form of environmental harm is considered in it.

4. Innovative thinking and environmental thinking

Due to the accelerating process of innovations creation such term as "innovative thinking" is more often used. As the analysis shows, practically all formulations have common character and are expressed as an ability to do something (provide innovative activity; search, open, create a new; overcome stereotypes of public consciousness; call into a question effectiveness and success of traditional concepts and approaches. For actual formation of innovative thinking, in an education system as well, such formulations are not enough. Creation and distribution of TIPS engineering projection in practice allows to define the concept "innovative thinking" as the project thinking directed to various systems of the environmental anthropogenic and not anthropogenic world improvement by searching and overcoming contradictions in these systems. This definition does not reject all above formulated, but supplements them to such a degree when creation and use of the appropriate learning tools is possible.

At the same time the concept "environmental thinking" (it is possible to take "ecological thinking" for a synonym) should be also specified as the thinking acquiring a project character and directed to overcome contradictions between economic efficiency and ecological safety of the created innovative techniques and technologies.

Working on the theme of sustainable development which includes environmental protection as the major part and also considers other factors, we formulated the concept "sustainable thinking" as the project thinking directed to overcome contradictions between economic efficiency innovative technique and technologies and preservation of resources for future generations.

Respectively, the concept "ecological outlook" so far formulated as deep awareness of vital need of the environmental life, common for all mankind, preservation [18], it is expedient to add and formulate as vision of the world around, as a system consisting of various level subsystems, developing under the laws which can be learned and used for the life environment, common for all mankind, improvement.

Finally, the concept "noosphere thinking" on the basis of which the concept of sustainable development as the global thinking in its communication and interaction with a cognitive envelope of the earth [19] assuming the ability to project and predict consequences of the interaction with the world around [20] is built, it is possible to complement with ability to see and overcome the arising development contradictions between economic and ecological expediency. Since the noosphere existence opening [21], by V.I. Vernadsky, the noosphere developing as well as all systems of the world, has undergone a certain evolution and, according to the authors, entered a quantum leap stage in the development of the first part of the word "noosphere" – a noo, in Greek - mind. While the mind of people did not rely on ecological outlook in its offered modern comprehension, the world development was a resultant of various, often opposite, aspirations of people, and kept a substantially spontaneous character. The ecological outlook in the modern comprehension gives much more conscious character to the development of the world directed to its stability, preservation of resources for future generations.

5. Formation of environmental thinking in educational system

Formation of environmental (ecological) thinking is considered to be a problem of education for sustainable development (ESD) which develops in the world according to the United Nations decision by efforts of UNESCO. During 2005 - 2014 there was a Decade of education for sustainable development throughout which the Global action program (GAP) in the field of ESD started to be carried out. So far ESD was generally understood as the reflection of sustainable development questions in the form of separate training courses, separate subjects, lectures at various stages of education [22], though the task of integration of ESD into all types and stages of education was set. Creation and distribution of TIPS in the world, and together with the systems integrated with other subjects of TIPS study – the didactic TIPS- pedagogics system [23], caused consideration of TIPS-pedagogics use opportunities as an integrated system of ESD. First of all, it was necessary to extend the TIPS- pedagogics system to all stages of an educational process as earlier there was one method of creative tasks [24] in it. A method of knowledge invention (for a stage of new material studying) and a method of innovative projects (for a project activity and scientific-technical creativity including youth intensive schools) were developed by the experts who are the authors of the present article [25-27]. In this regard it meant that before improving the development processes, making a sustainable development, it is important to have a general representation of development mechanisms. This representation is formed by the method of knowledge invention according to which each studied system on any subject, any educational program is "re-invented" by TIPS methods as a result of contradictions in predecessor system overcoming. At the same time this promotes better assimilation of the studied material and formation of the major competences.

The method of innovative projects is, in essence, the association of Problem Based Learning (PBL)Problem Based Learning (PBL)[28] and TIPS. The method includes an option for youth intensive schools and other situations when a small number of teachers who know TIPS have to manage the creation of dozens of youth innovative projects at the same time This option provides double or group supervision of intensive school participants scientific works. One of heads – a methodologist of creativity who knows TIPS and raises questions to all school participants, according to TIPS, and other heads – scientists and experts in scientific and technical areas corresponding to the subject of the created projects. They provide answers to the questions of the first head.

Exactly with the method of innovative projects it is expedient to create such projects of new technique and technologies in which the contradiction between economic efficiency and ecological safety is resolved.

As the approbation of the method of innovative projects which was carried out more than 15 years during its development, pupils and students created dozens of innovative projects – winners and prize-winners of All-Russian forums, patents for inventions were obtained, the Big scientific Russian Cup was won. Among them, a number of innovative projects on environmental protection is created: a high-capacity tanker not collapsing on storm waves (the patent of the Russian Federation), mutual neutralization of various harmful wastage, economic technology of utilization of "garbage islands" (the victory in the competition "SHUSTRIK" of innovative regions of Russia Association), the economic "cleaner of space debris", etc.

The possibility of TIPS assimilation and creation of projects on the basis of TIPS by school pupils of various stages checked by authors and their colleagues from various cities and countries in practice allowed the authors to create the concept "New polytechnical school" offering a new system of pre-university engineering and engineering-ecological education: in specialized classes, studios of supplementary education, in youth intensive schools, in Children's science and technology parks. The concept assumes application of the method of knowledge invention when studying new material and application of the method of innovative projects in project activity, scientific and technical creativity.

Recently, methods of interactive learning have been widely developed. Interactive learning is the learning with well-organized feedback of subjects and objects of learning, with a two-way exchange of information between them.

Interactive training allows you to form the active-cognitive and intellectual activity of students, increase the motivation to study the subject, develop in various forms the communicative competencies of students [29, 30]. The possibility of joint activities in interactive training, using the innovative projects method, allows to find new solutions to problems, including in environmental education, and this contributes to the development of education for sustainable development (Education for Sustainable Development) in accordance with the plans of UNESCO.

The main requirement of educational programs in the field of "Engineering, technology and technical sciences" becomes a practice-oriented as the ability to form a certain set of professional and over professional competencies that allow the graduate to quickly join the production chain. In the undergraduate program, the focus is on the implementation of applied engineering competencies and basic soft skills, work in team engineering projects in competition formats and the introduction of the mandatory end-to-end module "Project Activities". Master programs focus on specialized engineering competence and soft skills in project management, end-to-end qualifying work in the form of a real engineering project. Engineering master programs aimed at training engineering teams in order business and the implementation of real engineering projects [31].

6. Conclusion

The possibilities of the theory of inventive problem solving (TIPS) use as an effective tool of modern engineering thinking for project ecological thinking formation are shown. The acceleration of scientific and technical progress rate causes new environmental problems, but due to the presence of cognitive technologies at the same time contains efficient means of these problems.solutions It is important to make full use of these means. The use of interactive technologies in the organization of training opens the opportunity to consider the issues of individual educational strategies defining the direction vector and content of future engineers education, to improve the quality of their training and creates opportunities for self-realization.

Acknowledgment
The reported study was funded by Russian Foundation for Basic Research, Government of Krasnoyarsk Territory, Krasnoyarsk Region Science and Technology Support Fund to the research project No 16-16-24023/17-OGON.

References

[1] Dror I E 2007 Gold Mines and Land Mines in Cognitive Technology. *Cognitive Technologies and Pragmatic of Cognition. Edited by Itiel E. Dror.* University of Southampton. (Amsterdam / Philadelfia: John Benjamins Publishing Company) 1-7
[2] *Finansovyy slovar* [Financial dictionary] 2017 *Cognitive Technologies* Access 06.12.2017, available at: https://dic.academic.ru/dic.nsf/fin_enc/23874
[3] IEP – Internet Encyclopedia of Philosophy (2017). *Socrates.* Access 05.12.2017, available at: http://www.iep.utm.edu/socrates/#SH3b
[4] Encyclopaedia Britannica 2017 *Pappus of Alexandria.* Access 05.12.2017, available at: https://www.britannica.com/biography/Pappus-of-Alexandria.
[5] Stanford Encyclopedia of Philosophy 2017 *Ramon Llull* Access 07.12.2017, available at: https://plato.stanford.edu/entries/llull/.
[6] Osborn A F 1963 Applied Imagination: Principles and Procedures of Creative Problem-Solving 3rd Edition (N.Y.: C. Scribner)
[7] De Bono Edward 1970 Lateral thinking: creativity step by step (N.Y.: Harper & Row)
[8] Whiting C H S 1958 Creative Thinking (N.Y.: Reinhold)
[9] Zwicky F 1957 Morphological astronomy. Springer-Verlag
[10] Gordon W J J 1961 Synectics: the development of creative capacity (N.Y.: Harper & Row)
[11] Altshuller G S 2007 The Innovation Algorithm: TRIZ Systematic Innovation and Technical Creativity Worcester (MA: Technical Innovation Center)
[12] Shaughnessy H 2013 What Makes Samsung Such An Innovative Company? *Forbes* Access 07.12.2017, available at: https://www.forbes.com/sites/haydnshaughnessy/2013/03/07/why-is-samsung-such-an-innovative-company/2/#7c0e4c0c7f96.
[13] Pogrebnaya T V, Kozlov A V, Sidorkina O V and Pogrebnaia T V 2017 TRIZ and applied dialectics, available at: http://www.metodolog.ru/01108/01108.html (access: 06.12.2017)
[14] Kozlov A V 2016 Thought Process of Engineering “Elite Force” Russian Development Technologies. *Engineering Education* 19 14-18 Access 07.12.2017, available at: http://www.aeer.ru/filesen/io/m19/art_2.pdf.
[15] Yang M and Guo M 2011 Application and analysis of TRIZ in man-made environment. In Proceedings of 2011 IEEE 18th International Conference on Industrial Engineering and Engineering Management (Changchun, China)
[16] Serban D, Man E, Lonescu N and Roche T A 2003 TRIZ Approach to Design for Environment *Product Engineering* pp 89-100
[17] Orloff M A 2003 Inventive Thinking Through TRIZ: A Practical Guide
[18] Ecology. Directory 2017 Access 07.12.2017, available at: http://ru-ecology.info/term/13769
[19] Noospheric Thinking 2017 *National encyclopedic service* Access 07.12.2017, available at: http://didacts.ru/termin/noosfernoe-myshlenie.html.
[20] The concept of celebrating the 150th anniversary of the birth of V.I. Vernadsky "Noospheric Thinking - Thinking XXI" 2013 Access 07.12.2017, available at: http://www.vernadsky.ru/files/o_vernadskom/150/Koncepciya%20150-letiya.pdf
[21] Vernadsky V I 1989 Biosphere and noosphere (Moscow: Nauka) p 261
[22] Hopkins C 2013 The Future of ESD. In Proceedings of International conference on education for sustainable development **18** (Khanty-Mansiysk, UNESCO Vestnik) pp 358 – 363

[23] Vikentiev I L, Gin A A and Kozlov A V 1998 TRIZ-pedagogy. In the Book of Creative Problems in Biology, Ecology (St. Petersburg: Akcident) pp 162–165
[24] Gin A A and Andrzheyevskaya I Yu 2010 150 tvorcheskikh zadach o tom, chto nas okruzhayet [150 creative tasks about what surrounds us] (Moscow: Vita-Press)
[25] Lepeshev A A, Podlesnyi S A, Pogrebnaya T V, Kozlov A V and Sidorkina O V 2013 TRIZ-based Engineering Education for Sustainable Development. In Proceedings of IEEE International Conference on Interactive Collaborative Learning (ICL) (Kazan, Russia) pp 489-493
[26] Lepeshev A A, Podlesnyi S A, Pogrebnaya T V, Kozlov A V and Sidorkina O V 2013 Development of creativity in engineering education using TRIZ. In Proceedings of IEEE Interdisciplinary Engineering Design Education Conference (IEDEC) (Santa Clara, USA) pp 6-9
[27] Kozlov A V, Pogrebnaya T V and Sidorkina O V 2013 ESD in UNESCO Associated schools. Didactics of sustainable development. In Proceedings of International conference on education for sustainable development **18** (Khanty-Mansiysk, Russia, UNESCO Vestnik) pp 228-237
[28] Boud D and Feletti G 1998 The Challenge of Problem-based Learning. Milton: Routledge
[29] Loginova I O and Chupina V B 2011 Technology of interactive learning in the university (Krasnoyarsk: KrasGMU)
[30] Yakovleva N O and Yakovlev E V 2014 Interactive teaching methods in contemporary higher education. Pacific Science Review **16** 75–80
[31] Kovalev I, Loginov Y and Zelenkov P 2015 An integrated system of training engineers for aerospace industry in Siberia using innovative technology of the student project-and-team work. Procedia – Social and Behavioral Sciences 174 pp 537-543

IOP Conf. Series: Materials Science and Engineering **822** (2020) 012007 doi:10.1088/1757-899X/822/1/012007

An integrated approach to environmental management

E A Zhirnova, I V Trifanov, O A Sukhanova and V I Trifanov

Reshetnev Siberian State University of Science and Technology, 31 Krasnoyarsky Rabochy Av., Krasnoyarsk, 660037, Russia

E-mail: sibgau-uks@mail.ru

Abstract. The issues of improving the system of environmental management on the basis of knowledge management, risk-based approach, environmental monitoring are covered in the article. As a result of the carried out researches the algorithm of improvement of system of ecological management is developed. The model of improvement of ecological monitoring is offered. Practical recommendations for improving environmental safety and placing points of observation of the state of the environment are presented.

Currently, an integrated approach based on the balance between the environment, society and economy to management is generally accepted in the world practice. The increasing impact of pollution on the environment, inefficient use of resources, improper waste management, climate change, degradation of ecosystems and loss of biological diversity make it necessary for enterprises to adopt an integrated approach to environmental management through the introduction of environmental management systems.

GOST R ISO 14001-2016 offers organizations an approach to protect the environment and respond to changing environmental conditions in balance with socio-economic needs. It emphasizes, however, that the application of this standard does not in itself guarantee optimal environmental results.

In accordance with the requirements of the legislation by legal and physical persons that operate businesses are obliged to comply with the approved technology and requirements in the field of environmental protection and must ensure compliance with the standards of environmental quality through the application of technical means and technologies of deactivation and safe disposal of production and consumption, clean emissions and discharges of pollutants and other best available technologies to comply with the requirements in the field of environmental protection [1].

The system of environmental management of the enterprise (hereinafter– EMS) is a voluntary system designed to achieve the most optimal allocation of resources in the organization to achieve its goals while reducing the negative impact on the environment. ISO 14001 in contrast to the environmental standards contain no specific quantitative standards (emissions, concentrations, etc.) [2]. The standard is based on the basic principle of achieving the requirements: prevention of environmental impact is more efficient and economical than the correction of the consequences, as well as the PDCA cycle that is widely used in management. The PDCA model represents a cyclical process that your organization uses to achieve continuous improvement (figure 1). It is based on the concept of "Plan-Do-Check-Act" (Plan, Do, Check and Act).

Therefore, the standard states that the two organizations carrying out similar activities may meet the requirements of this standard, but at the same time have different obligations set out in environmental policy, different environmental technologies and objectives, and objectively have a completely different impact on the environment. However, in modern conditions, even at certified enterprises, the environmental management system often remains in its infancy and does not function properly, which undermines the confidence of stakeholders in environmental management, in voluntary certification systems, as a conformity assessment procedure. The survey showed the public distrust toward the certificates of conformity [3].

So the problem of improving the environmental management system as a tool to improve work in the field of resource conservation and pollution prevention is relevant.

The main advantages of using EMS are improved planning and allocation of resources, reducing the possibility of man-made disasters (which can subsequently affect the state of the environment and lead to financial costs for their elimination), strengthening work in the field of resource conservation and prevention of environmental pollution at the production stage.

This paper proposes an integrated approach to the improvement of environmental management of the enterprise, based on the application of knowledge management, risk-based approach, integrated environmental monitoring and interested parties interaction.

We briefly describe the main definitions of the environmental management system. In accordance with GOST R ISO 9001-2016, the environmental management system is a part of the management system used to manage environmental aspects, fulfill obligations and take into account risks and opportunities.

Environmental policy means the intentions and direction of the organization with the regard to environmental performance, formalized by its senior management (figure 1).

An environmental aspect is an element of an organization's activities, products or services that interacts or can interact with the environment. Environmental impact is a change in the environment of a negative or positive nature, in whole or in part, which is the result of the environmental aspects of the organization.

Pollution prevention is the use of processes, practices, technical solutions, materials, products, services or energy to avoid, reduce or control (separately or in combination) the formation, release or discharge of any type of pollutant or waste in order to reduce adverse environmental impacts.

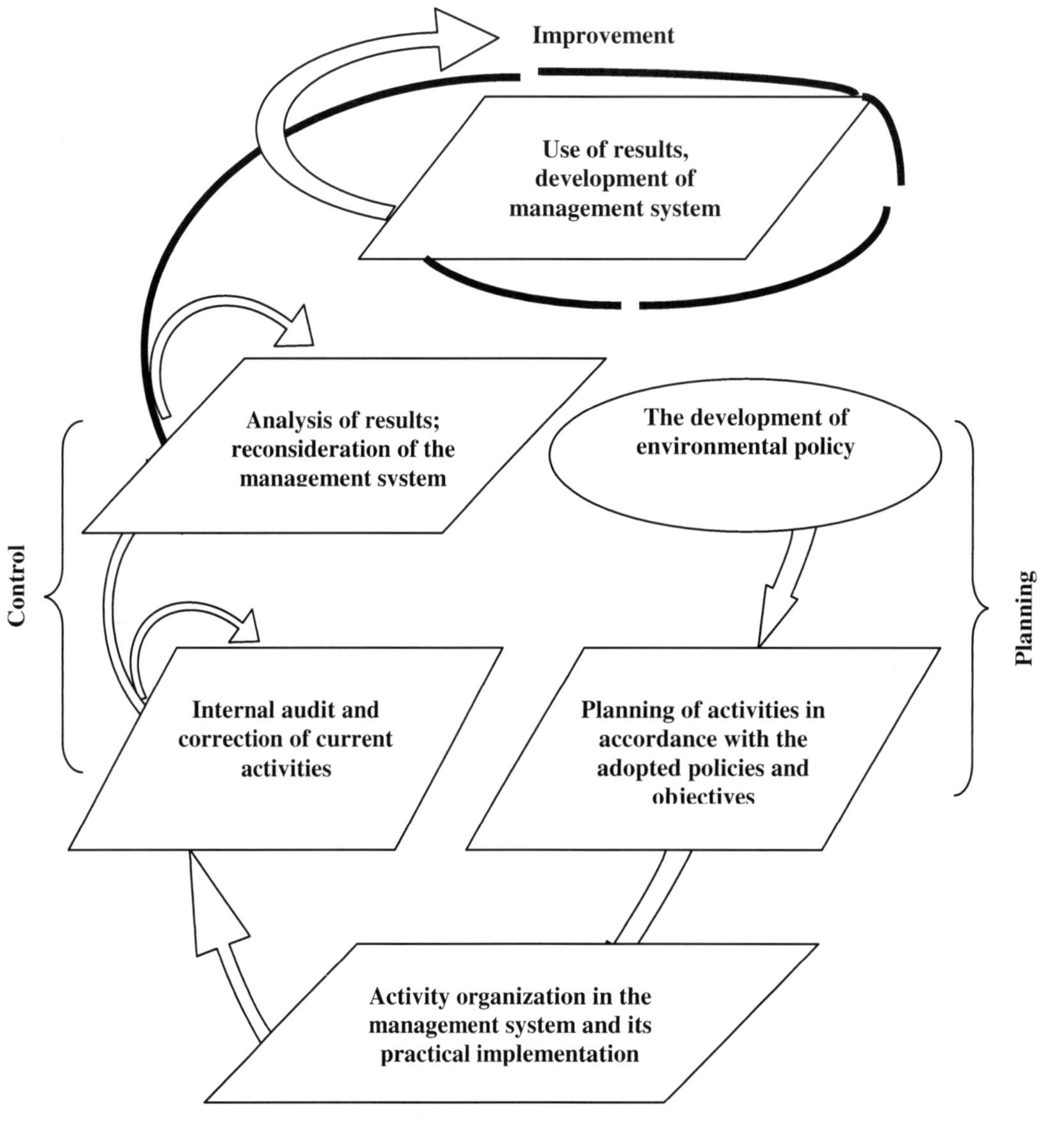

Figure 1. Implementation of PDCA cycle in the creation and development of environmental management system (EMS)

An integral part of the management of any organization that combines the actions associated with the formation of knowledge, their distribution and use, as well as the development of innovation and training is the management of knowledge. The knowledge management system creates a common information atmosphere in the organization, new and improved production processes that ensure the exchange of knowledge between specialists in different disciplines, information exchange and cooperation. There are management tools in knowledge management: an organized brainstorm, "help on the job" training, "storytelling", aid in the workplace, "knowledge cafe", the creation of clusters of knowledge methodology APO, document libraries, wikis, databases, social networking, locator expertise [4-7]. The algorithm for improving the environmental management system based on knowledge management and risk-based approach is shown in figure 2.

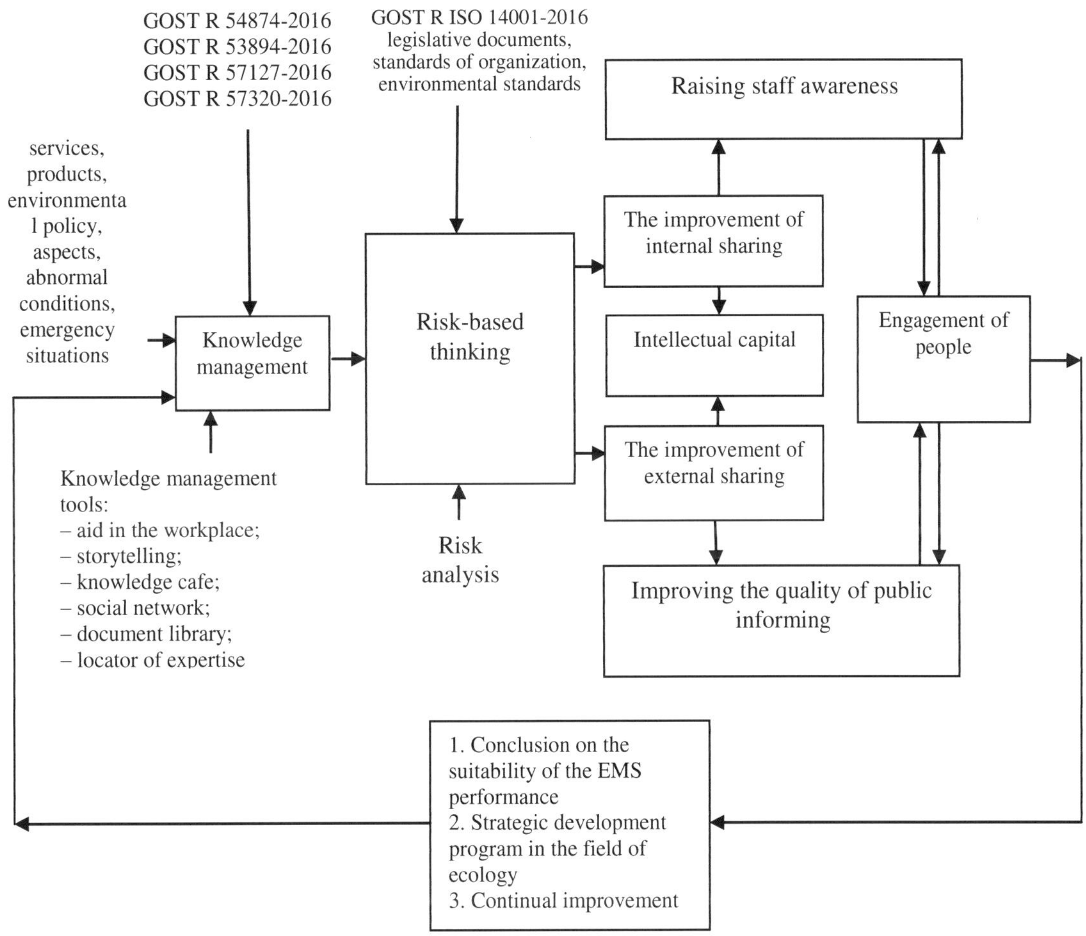

Figure 2. Improvement algorithm of ecological management system on the basis of knowledge management.

The tool of ecological management system improvement is storytelling and knowledge cafe. In the first case, to inform interested parties, especially the public about the mechanism of functioning of the EMS, the solution of environmental problems, the dynamics of changes in environmental indicators. Cafe of knowledge (figure 2) is a group of people who creates implements and disseminates new knowledge that will accumulate, analyze environmental information, as well as develop measures for self-improvement on its basis. The social network can be a powerful tool for the exchange of environmental information, providing its members with access to the most important knowledge, connections and advice.

An important aspect is the quality of informing the population about environmental problems. Environmental audit is an independent, comprehensive, documented assessment of compliance by a legal entity or an individual entrepreneur with the requirements, including standards and regulations, federal rules and regulations in the field of environmental protection, requirements of international standards and preparation of recommendations to improve such activities.

GOST R ISO 14001-2016 establishes that the organization must monitor (figure 3), measure, analyze and evaluate its environmental performance. The assessment includes an audit program, compliance assessment, and management review. The organization shall develop, update and apply the documented information in a manner sufficient to provide a suitable, adequate and effective environmental management system.

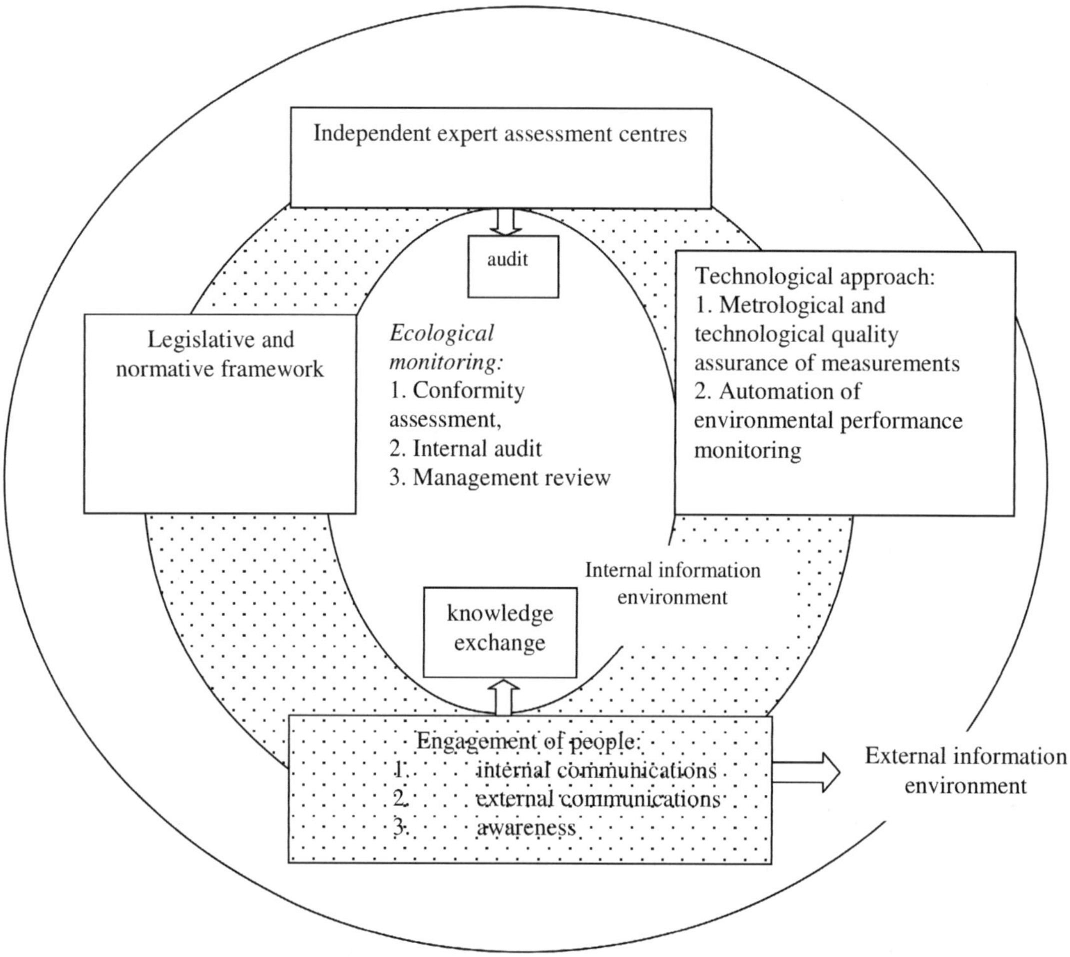

Figure 3. Model of improvement of environmental monitoring at factory

Metrological and technological quality assurance of measurements is aimed at fulfilling the requirements for measurements in accordance with the legislation. So for the measurements relating to the sphere of state regulation the measuring instruments of the approved type which passed checking shall be applied (item 1, Art. 9 of the Federal law "on ensuring unity of measurements No. 102-FZ" of 26.06.2008).

Measurements of pollutant concentrations in the air and industrial emissions are of particular importance for the environment. The government introduced in April 2018 in the state Duma a bill on the creation of systems for automatic control of emissions of pollutants into the air and discharges into water. In recent years, gas analyzers using the principle of direct measurements are becoming more common, since the use of these measuring instruments allows automating the measurement process, minimizing human involvement. Analysis of the metrological characteristics and measurement techniques underlying in the work allows drawing some conclusions [8-11]. To measure the concentrations of pollutants in the atmospheric air is better to use gas analyzers principle of operation, which is based on gas-phase chemiluminescence or fluorometry. For the measurement of organic compounds possible to use a chromatographic analyzer in the stationary post. Measurement of carbon oxide is advantageously carried out using electrochemical gas analyzers, as TLV -C of carbon oxide in the air on average is 10 times more than other major pollutants.

Low-cost electrochemical gas analyzers are suitable for measurements of industrial emissions pollutants. The problem of selectivity is partly solved by the fact that the composition of the exhaust gases is relatively constant and varies in time not so much.

When analyzing the air in the working area to measure the concentrations of ammonia, oxide and nitrogen dioxide, it is better to use gas – phase chemiluminescence, to measure the concentrations of sulfur dioxide and hydrogen sulfide – fluorimetric gas analyzer, measurement of other substances can be performed with electrochemical gas analyzer, taking into account the interfering effect of non-measurable components.

Improvement of environmental management system (figure 3) ensures the efficiency of the system of environmental control of emissions at enterprises, obtaining reliable information on environmental pollution by supervisory authorities and, ultimately, reducing the negative impact on the environment of the country.

The following recommendations can be used to improve environmental safety. In order to optimize the observation points for the state of the environment, it is necessary to place the monitoring equipment on the principle of choosing the priority zones for sources of air pollution, as well as for the aquatic environment. The criteria for selecting priority zones can be based on statistical analysis of observations. At the same time, it is advisable to create a network of environmental monitoring stations equipped with modern control devices created using controllers, digital and analog devices that operate simultaneously in real time and transmit the results of monitoring to the central control office of the city, where there is a simulation, a strategic analysis of environmental pollution processes, risk assessment based on an evolutionary strategy. The influence of geophysical characteristics of the territory where observation posts are placed should be taken into account. Mobile devices (quadrocopters, helicopters, etc.) can be used to determine the places of pollution. To monitor the state of the environment of large areas can be used spacecrafts that can monitor climate change, radiation, thermal and chemical pollution. Based on the results of environmental monitoring, knowledge and management, as well as a risk-based approach, it is necessary to develop an action plan for environmental protection of the population, animals, plants, through the introduction of environmentally friendly technologies, the use of filters and treatment facilities, removal of harmful industries outside the city.

Conclusion

Thus, on the basis of completed theoretical studies:

- an algorithm for improving the environmental management system, based on the introduction of knowledge management tools and risk management to improve environmental culture and awareness of the staff, public awareness and the development of intellectual capital of the organization is worked out;
- a model for improving environmental monitoring, based on the integrated management tools of information interaction of all interested parties, conformity assessment, internal and external audit conducted by independent expert organizations, as well as metrological and technological quality assurance measurements, automation control of environmental indicators is suggested;
- practical recommendations for improving environmental safety and placing points of observation of the state of the environment are presented.

References

[1] Russian Federation. *The Federal Law N7-FZ* of January 10, 2002, available at: http:// http://base.garant.ru/ (accessed 10.02.2018)

[2] *State Standard ISO 14001-2016 Environmental management system. Requirements and application guide* 2017 (Moscow, Standartinform Publ.) p 41 (In Russian)

[3] *The system Roskachestvo.* (In Russian) available at: http:// https://roskachestvo.gov.ru/ (accessed 19.03.2018)

[4] *State Standard 53894-2016 Knowledge management. Terms and definitions* 2017 (Moscow, Standartinform Publ.) p 35 (In Russian)
[5] *State Standard 57127-2016 Knowledge management. Guide to best practices* 2017 (Moscow, Standartinform Publ.) p 37 (In Russian)
[6] *State Standard 57320-2016 Knowledge management. Application of process-oriented knowledge management in small and medium – sized enterprises* 2017 (Moscow, Standartinform Publ.) p 29 (In Russian)
[7] State Standard 54874-2016 Knowledge management. Guide to best practices for the public sector 2017 (Moscow, Standartinform Publ.) p 32 (In Russian).
[8] *Description of the type "Multi-component gas analyzers "MONOLIT" 64349* (In Russian). available at: http:// https://all-pribors.ru// (accessed 25.03.2018)
[9] *Description of the type "Multicomponent gas analyzers "Polar" 64805/1* (In Russian). available at: http:// https://all-pribors.ru// (accessed 30.03.2018)
[10] *Description of type "Gas analyzers" Chromatotec GC 866" 55421* (In Russian). available at: http:// https://all-pribors.ru// (accessed 10.04.2018)
[11] *Description of the type of "Automatic gas chromatographic ACA-LIGA" 61315* (In Russian). available at: http:// https://all-pribors.ru// (accessed 12.04.2018)

IOP Conf. Series: Materials Science and Engineering **822** (2020) 012008 doi:10.1088/1757-899X/822/1/012008

Green plants in environmental monitoring of the environment of the large industrial city

E V Avdeeva and V F Poletaikin

Reshetnev Siberian State University of Science and Technology, 31 Krasnoyarsky Rabochy Av., Krasnoyarsk, 660037, Russia

E-mail: e.v.avdeeva@gmail.com

Abstract. The urgency of the problem is determined by the unfavorable ecological situation in the large industrial centers of Siberia, formed under the influence of man-made loads that determine the modification of the environment. This leads to changes in the properties of individual biotic components and, as a result, the quality of life of the population. An important element of the natural framework of urban ecosystems and a means to improve its urban planning qualities is the system of gardening. Currently, the existing methods of creating a system of greening the city of Krasnoyarsk and other cities of Siberia, the landscape-aesthetic principles of its formation, as well as the regulatory and technical design base do not provide the comfort of the urban environment. The basis for the development of directions and principles for the reconstruction of the greening system should be the results of environmental studies of landscape processes, the identification of the main sources of environmental stress, the study of the integrated impact of man-made stressors.

1. Introduction

The priority problem of optimizing the quality of life in the urban environment is a systematic approach to the study of woody plants as the main link in the city's ecological framework. This will allow them to evaluate their productivity, classify and model the transformation of green spaces, to reveal the features of their functioning in conditions of anthropogenic processes and landscape dynamics. Analysis of works studying the performance of the functional role of green plantations, taking into account the conditions of their growth, allowed to substantiate the methods of research and the system of parameters characterizing the processes of vital activity of woody plants under conditions of urban environment, to establish the laws of growth, the performance of ecological functions, the aesthetic perception of woody plants, territories [1-5]. The analysis of published works confirms the relevance of studies of growth processes, as well as the practical contribution of woody plants to improving the quality of life of the population. The results of the research are used as basic tools for monitoring the urban environment and for the scientific substantiation of regional concepts of landscape gardening of large industrial cities.

2 Results and Discussion

A retrospective analysis of the development and current state of greening in Krasnoyarsk has shown that the green plantations of the city are not a system of gardening, but a set of park, line, courtyard, sanitary protection areas that do not provide favorable environmental conditions in various zones of the city. An important indicator of the state of comfort of urbanized is the provision of residents with green areas of all categories, thanks to which an integral environmental and aesthetic effect develops. Analysis of the studies and the results of practical approbation showed that reaching the desired level of comfort in industrial cities of Siberia is possible with a total area of green plantations of at least 30% of the area of urban land [5]. Currently, the greenery of the city is 60% of the above-mentioned optimal ratio of built-up area and "green" areas. The lack of landscaping of public use makes up 67% of existing and 75% of promising standards. It is required to carry out works on the expansion of green common areas for two thousand hectares.

The main composition of woody plants in urban gardening is currently represented by a limited number of species: *Populus balsamifera* is 42; *Malus baccata* - 11, *Ulmus* - 8, *Larix sibirica* - 11, *Betula pendula* - 8, *Acer negundo* - 6, *Picea obovata* and *Picea abies* -5, *Tilia cordata* -5 and other species - 4%. At the same time, the analysis of the species composition of green plantations showed that, taking into account the collections of botanical gardens and arboretums, the range of woody plants in the city's gardening is quite diverse - more than 400 species and varieties [5].

The results of the analysis of the risk of air and soil pollution in the main street of the streets showed that at all sites there is an excess of concentrations of harmful substances for plants from 2 to 14. The data obtained allowed zoning of streets and highways of the city, determining the area of areas where the plants are under active impact of vehicles. Four levels of impact of the road-transport system on green spaces in the main spaces have been established. As the evaluation criteria, biological norms of maximum permissible concentrations (MPC) were used. It is established that only 30% of the plantings along the city's roads grow in satisfactory conditions, where excess pollution standards for individual impurities from the impact of motor vehicles are up to three MPCs, to which background concentrations of harmful substances are added [6-9].

To assess the conditions for the growth of green plantations, the authors developed a methodology that allows to classify territories as urban types of growth conditions, that is, to allocate territories with a certain natural resource potential and density of man-made loads that determine the growth and development of urban green spaces. Urban types of growth conditions reflect the intensity of anthropogenic impact on landscaping objects and allow predicting the productivity and transformation of green plantations under given growing conditions. Examination of plantations using this technique allowed us to distinguish four types of tree growth conditions in Krasnoyarsk - satisfactory (I), tense (II), conflict (III) and critical (IV). The developed methodology is a tool for monitoring the assessment of the state of green spaces and the quality of life of the population in the urban environment.

In the urban environment, it is actual to conduct a diagnosis of its ecological state by the reaction of changes in the biometric indicators of vegetation. Morphological features of woody plants are quite informative for determining the degree of stability of growth of individual plant species under the conditions of the industrial press and the state of the urban environment. Woody plants that grow in urban environments experience stress throughout their lives, that is, they are constantly in a state where additional costs are needed to compensate for stressful effects. Achieving threshold values of limiting anthropogenic factors violates the established balance, which leads to loss of stability and premature death of plants. To describe the relationship between the mean values of tree heights with age within the height growth series, we used the formula of Micherlich [10, 11]:

$$H = b_1 \times (1 - exp(-1(A / b_2)))^{b3}, \tag{1}$$

where A - is the age, b_1, b_2, and b_3 are the coefficients of the equation.

The adequacy of the equations to the experimental material was estimated by the determination coefficient (R^2) and the Fisher criterion (F). Analysis of the results showed that tree species react differently to anthropogenic loads of the environment and structural features of the plantations. The results of the analysis made it possible to distinguish four types of growth of *Larix siberica* in height, which corresponds to the number of types of plant growth conditions in Krasnoyarsk. The coefficients and parameters of the *Larix sibirica* growth models are represented by formulas (2).

$$\begin{aligned} Y_{(I)} &= 20{,}82(1-(exp(-1(X/26{,}90)))^{1{,}69}, R^2 = 0{,}990 \\ Y_{(II)} &= 15{,}07(1-(exp(-1(X/18{,}02)))^{2{,}23}, R^2 = 0{,}993 \\ Y_{(III)} &= 12{,}18(1-(exp(-1(X/21{,}81)))^{1{,}65}, R^2 = 0{,}996 \\ Y_{(VI)} &= 9{,}85(1-(exp(-1(X/27{,}02)))^{1{,}41}, R^2 = 0{,}976 \end{aligned} \tag{2}$$

Analysis of ecological growth models of Larix sibirica in height showed:

1. The growth of *Larix sibirica* plantations in height in an urbanized environment is reliably described by the Micherlich equation (R^2 varies from 0.98 to 0.99 for all dependencies).

2. Beginning at the age of 15, there is a tendency for trees to differ in height due to changes in ecological loads in green areas. For plantings of *Larix sibirica*, 4 types are distinguished, which corresponds to the number of types of conditions of growth in the city of Krasnoyarsk. The method of variance analysis established the reliability of their differences. The difference in mean height of trees growing in types I and II of growth conditions at the age of 45 years is 2.3 m, in II and III - 2.7 m, in III and IV - 2.6 m.

3. With increasing density of the complex of negative factors in the environmental conditions of the environment, a decrease in the plantation height is observed, reaching 19% by the age of 45 in growth type II, 34% in III, and 43% in IV compared to the satisfactory type of tree growth. Thus, the character of growth in height is determined by the cumulative influence of growing conditions. The type of growth in height is a clear indicator of this effect.

4. The spread of values relative to the growth series is 0.2 to 0.9 m in height. The greatest variance of values is observed for plantations growing in the conflict type of growth conditions, which indicates the adaptation of plants to environmental conditions. In the IV (critical) type - the range of variation of values is reduced, which reflects a significant suppression of the mechanisms of plant growth and development.

5. *Larix sibirica* is an effective indicator-indicator in environmental monitoring of large industrial cities.

3. Conclusion

Analysis of the conformity of the height and age of the plantation allows us to determine the type of growth conditions in a given area and, consequently, to predict the nature of plantation growth, coordinate the spatial structure and technology of care in order to improve the condition of plantations, and establish the level of ecological comfort for the population living in the urban environment. *Larix sibirica* has the greatest plasticity, reflects the stressful effects of the urban environment and is an effective kind of indicator. Dynamics of growth in tree height and trunk diameter are the main signs-markers in monitoring the environment and assessing the level of ecological comfort in urban areas.

References

[1] Tselnicker Yu, Korzukhin M and Zeide B 1997Morphological and physiological studies of the tree crown (Moscow: Urania)
[2] Ufimtseva M and Terekhina N 2005Phytoindication of the ecological state of urbogeosystems of St. Petersburg (St. Petersburg: Nauka)
[3] Shelukho V 2001 Bioindication of Chronic Industrial Impact of Alkaline Type on the Components of Coniferous Forests (Bryansk: Nauka)
[4] Pshenichnikova L 2004Bioindication of forest ecosystems (Krasnoyarsk: Siberian State Technological University)
[5] Avdeeva E 2007 Growth and indicative role of woody plants in an urbanized environment (Krasnoyarsk: Siberian State Technological University)
[6] Nikolaevsky, V S 2002Ecological assessment of environmental pollution and the state of terrestrial ecosystems using phytoindication (Pushchino)
[7] Krushlinsky V 1986 The city and nature of Siberia: architectural and planning aspects of Krasnoyarsk: (Krasnoyarsk University)
[8] Mashinsky V 2000 The need for an integrated approach to the problem of greening streets and highways (Moscow: Ecology of the Big City) V. 5
[9] Frolov A 1998 The environment of a large city and the life of plants in it (St. Petersburg: Science)
[10] Kuzmichev V 1977 Regularities in the growth of forest stands (Novosibirsk: Nauka)
[11] Kofman G 1986 Growth and form of trees (Novosibirsk: Nauka)

IOP Conf. Series: Materials Science and Engineering **822** (2020) 012009 doi:10.1088/1757-899X/822/1/012009

Improvement of environmental efficiency of plants of green zones by creation of subordinate crops of Siberian stone pine

Eh V Kolosovskij, R N Matveeva, O F Butorova, N P Bratilova, Yu E Shcherba, A M Pastuhova

Reshetnev Siberian State University of Science and Technology, 31 Krasnoyarsky Rabochy Av., Krasnoyarsk, 660037, Russia

E-mail: bratilovanp@sibsau.ru

Abstract. It is topical to increase the ecological efficiency of plantings of green zones of settlements by creating Siberian stone pine crops. The subordinate crops of this type, which were created in 1966 by quadrennial seedlings of different geographical origin at thick planting, were observed (9 pieces in a site measuring 0.7 x 0.7 m, the distance between the centers of the sites is 4 m). It was found that the canopy closure of the stand, consisting of birch and pine, has a significant effect on the growth rates of subordinate crops of Siberian stone pine. So at canopy closure 0,1-0,1 52 years old crops, despite the different geographical origin, had the highest rate of height, stem diameter, crown, number and diameter of lateral branches on the lower live node in comparison with the closure of 0.6-1.0 Average increment of branchlets at 32 year period was by 49,8% more in trees, growing at a canopy closure of 0,2 against 0,7-0,8. Trees growing on the open area characterized by early seeding, were selected and propagated by engrafting.

1. Introduction

Siberian stone pine (*Pinus sibirica* Du Tour) is one of the main timber species which is characterized by ecological efficiency, as evidenced by studies reflected in numerous literary sources [1-5].

There is an experience of growing cedar plantations both under the canopy of other forest-forming species, and in open areas. Studies of the growth of crops of woody plants under different intensity of illumination have shown that the species of woody plants are subdivided into long and short-days [6-10, etc.]

It was found that Siberian stone pine is a long-day plant. The best growth of plants of this species was noted under additional illumination [11, 12]. This probably explains the most intensive growth, the early seeding of Siberian stone pine trees in open areas, than under the forest canopy [13, 14, etc.].

It is noted that the growth of Siberian stone pine under the trees canopy can be adjusted, reducing the canopy closure by carrying out cuttings and selecting of planting material, which is grown on the basis of selection of fast-growing specimens growing under these conditions.

The purpose of our research was to establish the peculiarity of the growth of 56-year-old Siberian stone pine, which grows in subordinate crops, created in the spring of 1966 using four-year seedlings of different geographical origin (Altaic, Biryusin, Buryat, Kemerovo, Tomsk, Chita) Gorniy-2 ", located in the green zone of Krasnoyarsk.

2. Materials and methods

To create subordinate crops of Siberian stone pine seeds were harvested in populations growing under various growth conditions: from 50 ° 16 'to 56 ° 30'N, from 84 ° 48' to 108 ° 43 'E., from 100 to 1000 m above sea level, II-IV classes of bonitet, IV-VII classes of age, with Siberian stone pine in the composition of the stand from 3 to 9 units, forest types – mixed herbs, moss and cowberry Siberian pine forests.

Crops were created at a thickened planting. In each site 0.7 x 0.7 m under Kolesov' planting iron 9 seedlings were planted. The distance between the centers of the sites was 4 m. Each variant consisted of 80 sites, including 720 plants. After 52 years, 1 to 7 trees were preserved in the sites. Statistically, such indicators as height, increments of branchlets, diameter, stem volume, crowns, etc. were measured and processed in the leading trees in the sites according to existing methods.

3. Results and discussion

Research has shown that canopy closure has a significant effect on all biometric indicators of trees. The best growth and formation of the crown were observed in trees growing at a lower canopy closure of the stand, the lag in growth was observed with its increase. It was noted that trees grown from Buryat seeds (the progeny of the population 50 ° 16'N, 106 ° 30'E, 1000 m above sea level, III class of origin) were more affected by shading.

With the canopy closure of the stand 0.1-0.5, the highest values in the subordinate crops of the leading trees are observed according to the following biometric indicators: height, shoot growths over the last four years, diameter of the trunk, crown, number and diameter of the lateral branches on the lower node in comparison with cultures that grow at a canopy closure of 0.6-1.0. There are no differences in the angle of attachment of the lateral branches of the lower living node to the trunk, which fluctuated depending on a canopy closure of stand from 92.3 to 92.9 °.

A study of the increment of shoots over a 32-year period at different canopy closure showed significant differences in increment of shoots: when the canopy is closed, the average periodic increment is 49.8% higher in comparison with 0.7-0.8 (figure 1)

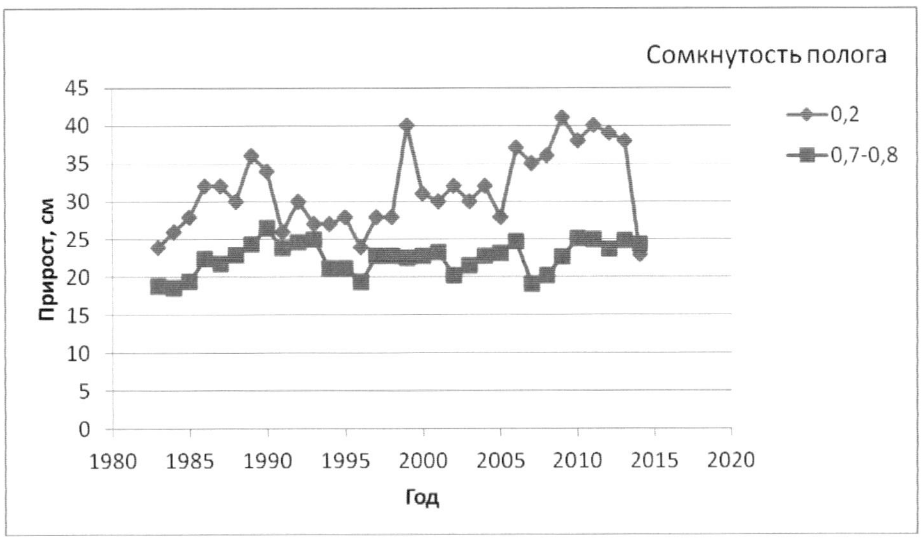

Figure 1. Increment of shoot of model trees over a 32-year period at a different canopy closure

At the same canopy cloure (0.5-0.6), differences in the intensity of growth of trees are observed depending on the geographical origin. The highest indicators for height, the current increment of shoots, the diameter of the trunk were in the trees of Altai, Chita origin. The trees of Chita origin have the best indices for the current increment of a shoot, the diameter of the trunk, the length, the diameter and the volume of the crown (the place of growth of the mother plantations: 50 ° 22`N, 108 ° 43`E, 700 m above sea level, class IV of bonitet).

Ranking of Siberian stone pine on the sum of the points characterizing intensity of growth and formation of a crown, has shown, that on the first places there are trees of the Chita and Altaic origin. Place of growth of the Altai maternal population: 51 ° 50`N, 86 ° 54`E, 500 m above sea level, III class of bonitet.

Comparative analysis of the growth rate of trees depending on the place of collection of seeds with the same canopy closure showed that the place of growth of parent plantations when creating subordinate crops affects biometric indicators of seed progeny. Significantly large indicators characterizing the intensity of growth and the formation of a crown, were in the seed progeny of trees from the plantations of mountain taiga and mountain-black forest seeded subareas.

There is a correlation link in variants of different geographical origin in subordinate crops between trunk diameter and height (r) from 0.350 of Chita origin to 0.872 of Kemerovo origin; the diameter of the trunk and the volume of the crown - from 0.543 of Altaic to 0.871 of Tomsk origin; crown diameter and crown volume - from 0.775 of Buryat origin to 0.942 of Chita origin.

High variability of biometric indicators among leading trees in geographical crops is noted, which indicates the manifestation of individual variability and the possibility of selecting valuable specimens regardless of their geographical origin.

Among the progeny of the maternity populations, the trees were selected, which are characterized by the most intensive growth and increased ecological efficiency for the purpose of growing varietal planting material for the creation of subordinate crops of Siberian stone pine in the green zone of Krasnoyarsk. Trees characterized by early and abundant seeding in the area where they grow in an open area are selected. The selected trees were propagated by engrafting using 7-year-old seedlings of Siberian stone pine as a stock. The best ramets are recommended to use for creating cedar plantations, which will have not only ecological but also economic importance, supplying the population with pine nuts.

4. Conclusions

The creation of Siberian stone pine crops in order to improve the ecological efficiency of existing plantations and in open areas should be justified by the use of planting material selected by economic features. It is necessary to carry out cuttings to reduce canopy closure to 0.5 in the created subordinate crops.

References

[1] Protopopov V V 1975 Sredoobrazuyushchaya rol' temnohvojnogo lesa [The mediating role of dark coniferous forest] (Novosibirsk: Nauka) p 328.
[2] Baboch N A, Sokolov N N and Bahtin A A 1996 Bescennyj dar tajgi [Priceless gift of the taiga] (Arhangelsk: ASTU) p 224
[3] Matvejeva R N, Butorova O F and Bratilova N P 2003 Poleznye svojstva i metody razmnozheniya kedra sibirskogo [Useful properties and methods of reproduction of Siberian cedar] (Krasnoyarsk: SibSTU) p 154
[4] Titov E V 2007 Car' sibirskoj tajgi [Tsar of the Siberian taiga] (M.: Kolos) 152 p

[5] Beh I A, Krivets S A and Bisirova A M 2009 Kedr - zhemchuzhina Sibiri [Cedar - the pearl of Siberia] (Tomsk: Pechatnaia manufaktura) p 50
[6] Wareing P F 1956 Photoperiodism in woody plants *Ann. Rev. Plants Phys.* 5
[7] Irgens-Moller H 1962 Genotypic variation in photoperiodic response of Douglas fir seedlings *For. Science* **8** 4
[8] Watt R F, Davis Mc and Gregor W H 1963 Growth of four Northern conifers under long natural photoperiods in Florida, Wisconsin *Forest science* 9 pp 115-128
[9] Sesaki S and Mori T 1981 Growth responses of Dipterocarp seedlings to light *Makays. Forest.* 44 No 2-3 pp 319-345
[10] Fembach E and Mohr H 1990 Coaction of blue/ultraviolet-A light absorbed by phytocrome in controlling growth of pine (*Pinus sylvestris* L.) seedlings *Planta* 180 No 2 pp 212-216
[11] Leman V M 1950 Ob uskorenii razvitiya seyancev hvojnyh porod pri nepreryvnom osveshchenii [On accelerating the development of seedlings of coniferous species under continuous illumination] (M: Proceedings of the RAS USSR) vol. 71 pp 163-166
[12] Matveeva R N and Butorova O F 2001 Uskorennoe vyrashchivanie seyancev i kul'tur kedra sibirskogo v Vostochnoj Sibiri [Accelerated cultivation of seedlings and crops of Siberian cedar in Eastern Siberia] (Krasnoyrsk: SibSTU) p 254
[13] Gabeev V N 1961 Rost kul'tur kedra pri razlichnoj osveshchennosti v lesostepnom Priob'e [Growth of cedar crops under different illumination in the forest-steppe Ob area] (Proceedings of the RAS Siberian branch USSR) No 7 pp 71-74
[14] Beh I A and Danchenko A M 1990 Rost odnovozrastnyh kul'tur kedra pod pologom lesa i na otkrytom uchastke. Problemy kedra [Growth of single-aged cedar crops under the canopy of the forest and on the open area. Problems of cedar] (Tomsk) 4 pp 65-70

Description of wildfires spreading and extinguishing with the aid of agent-based models

G A Dorrer and S V Yarovoy

Reshetnev Siberian State University of Science and Technology, 31 Krasnoyarsky Rabochy Av., Krasnoyarsk, 660037, Russia

E-mail: g_a_dorrer@mail.ru, ach_bask@mail.ru

Abstract. The article describes a new method of describing wildfires spreading and extinguishing based on agent-based models. The agent approach in conjunction with geoinformation systems (GIS) permits to efficiently describe the interaction of a big number of participants of wildfire fighting: spreading fire, fire brigades, means of mechanization, air means, etc. In the current paper, this approach is illustrated by an example of two types of agents denoted as A and B. The burning fire line is seen as a chain of type A agents, each of which imitates burning of an elementary area of vegetable fuel. The fire line movement is described by Hamilton–Jacobi equation and is calculated based on the moving grid algorithm. Agents of another type—B-agents—describe fire extinguishing, they approach the agents of A type and affect them decreasing fire intensity to zero. This model is presented by a GIS-oriented software system which may be enhanced and be useful both for fire control management and training personnel in wildfire fighting tactics.

1. Introduction

A wildfire is an uncontrollable fire spontaneously arising and spreading in the natural environment. This term implies forest, peat and steppe fires. Wildfire causes great economic and ecological damage to the countries and regions where they occur and result in casualties. By now a great number of information systems were developed to support efficient wildfire fighting. The systems permit to model and forecast fire behavior with various accuracy (systems Behave Plus [1], FARSITE [2], WFDS [3], etc.).

However, besides modeling of the wildfire line itself there is a necessity to model various external impacts on fire to localize and extinguish it. Use of such models may be actively applied in fire control, in particular, to improve efficiency of management of fire fighting forces and means. In practice the choice of the means and methods for wildfire localization depends on a big number of factors which are difficult to formalize. In such conditions, in our opinion, the agent-based modeling (ABM) is a quite efficient method.

The agent-based modeling, ABM, is a new approach to imitation modeling which is aimed at imitating behavior of complex dynamic systems through studying behavior and interaction of autonomous and independent subjects (agents) in a certain environment [4]. Agent-based models are used both in natural sciences (physics, biology) and in social sciences (sociology, economics). This type of models is the most effective in studying behavior of systems consisting of a big number of interacting subjects (from dozens to several thousands) being under the influence accidental impacts.

For modeling wildfire situations geoinformation systems (GIS) also are of special interest, since they have possibilities to store and display spatial information. The joint use of the agent-based modeling and GIS permits to combine the advantages of both directions, use advanced means for modeling processes running on in conditions of a real natural environment and landscape.

The idea of modeling wildfire line by the agent-based approach was considered in papers [4-9]. In these papers the agent-based models were created with the aid of specialized software systems (agent platforms) such as NetLogo, StarLogo, Repast Simphony, etc. This approach permits to simplify software implementation of an imitation model, but considerably complicates integration of agent-based models with GIS. Besides, papers [4-9] carry out modeling of only freely spreading fire without modeling the actions of fire-fighting forces and means.

Thus it is necessary to create imitation systems based on agent approach permitting to model fire situations on a real terrain. The current paper proposes a new method of creating such systems by presenting an example of an agent-based model, which imitates spreading of ground forest fire and a direct method of its extinguishing by non-mechanized firefighting brigades.

2. Description of fire front dynamics

The model of fire dynamics reviewed below is based on the following assumptions.

1. The area of modeling is a fragment of the map of a real territory covered with a layer of vegetable fuel. It permits to combine the data of GIS and agent models. The map contains the layers representing various elements of modeling environment: settlements, rivers, lakes, roads, ravines, forest ranges and other objects. The coordinates of each element are given. The so called model of thin layer of vegetable fuel [10] is used. This model does not consider the vertical structure of vegetable; and fuel materials are considered as a homogeneous layer on the top of the ground with average characteristics. Such characteristics are stored in the database of pyrologic descriptions of forest plots.
2. The process of fire spreading along the fuel layer and the process of fire spreading combating are considered as projection on the horizontal plane (on a map of corresponding scale). The coordinate system $X = \left[x^1, x^2\right]^T$ is mapped-referenced $X \in D$, where D – is the scenery area in review.
3. The micrometeorological, topographic parameters and characteristics of fuel materials in each map point are assumed to be known. It allows calculating the fire parameters in each map point with the use of known models of vegetable material burning (for example, by R. Rothermel's model for fire spread prediction [11]).

Let us regard $\varphi(X, t) = 0$ as the fire front equation (figure 1). This function describes the movement of fire line and satisfies the Hamilton–Jacoby equation,

$$\frac{\partial \varphi}{\partial t} + V \cdot grad\ \varphi = 0, \tag{1}$$

where $V = V(X) = \left[v^1, v^2\right]$ is the row vector of fire front speed.

The column vector of a gradient $grad\ \varphi = \left[\frac{\partial \varphi}{\partial x^1}, \frac{\partial \varphi}{\partial x^2}\right]^T = P = \left[p^1, p^2\right]^T$ is a normal to the fire front.

The equation (1) is considered in initial conditions,

$$X(0) = X_0(\alpha),\ P(0) = P_0(\alpha), \tag{2}$$

where X_0 and $P_0(\alpha)$ are a parametric representation of an initial fire line and multiple normals to it, U is an admissible set of values of the parameter α.

For the purpose of the numerical modeling of fire dynamics it is more convenient to use normal speed of a fire front $V_n = V \cdot P / |P|$, where $|P|$ is a length of a normal and $P/|P|$ is a unit vector of a normal to the front line.

And the equation (1) is transformed to the form

$$\frac{\partial \varphi}{\partial t} + V_n |P| = 0. \tag{3}$$

The equation (3) is also considered in initial conditions (2).

As it follows from the theory of burning processes and observations, the fire spreading speed depends on a set of external environment parameters, in the first run, on wind speed and a surface slope, where burning takes place. Therefore, normal speed of fire front movement may be presented as a function of several arguments,

$$V_n = V_n(X, t, W, S, \beta, \gamma), \tag{4}$$

where W is wind speed, S is a surface slope, β is an angle between a wind direction $\vec{W}$ and a direction of a normal to the front line P, γ an angle between a slope vector and a direction of normal to the front line. Normal speed of fire front movement at any point of a fuel materials layer may be presented as a product of two functions:

$$V_n(X, t, W, S, \beta, \gamma) = V_0(X, t, W, S)\chi_n(P, W, S), \tag{5}$$

where V_0 is a maximum fire spreading speed, when the direction of the normal to the line coincides with the wind direction and the surface slope direction.

The multiplier $0 \le \chi_n(P, W, S) \le 1$ determines a degree of fire spreading speed reduction V_n in other directions when the direction of a normal to the line does not coincide with the direction of wind and a slope. This function determines local directions of fire front movement and is called an indicatrix of the front's normal speed or according to H. Minkowski and H. Rund [12], a figurotrisa of the spreading process. The properties of this function are reviewed in paper [13]. For simplicity of representation we further on will designate the figurotrisa as $\chi_n(P)$.

The use the figurotrisa permits to create an effective algorithm of numerical modeling of the fire front spreading. The idea of algorithm implies use of the methods of moving grids proposed by S. Godunov. The backbone of the method consist in that the computational grid of a task is not built a priory, but is determined by the current solution of the task. The grid moves and develops while the computations are being made.

Let us consider the construction of a computational grid. The fire line at the moment t is a closed line going through the points $C_1, \ldots C_N$ (figure 1). Each point $C_i = C_i(t)$ is determined by the vector $C_i(t) = \{X_i, L(i), R(i), t\}$, $i = 1, \ldots, N$, where $X_i = (x_i^1, x_i^2)$ are coordinates of a point on the map, $L(i)$ and $R(i)$ correspondingly are the numbers of adjacent points located left and right from C_i.

The algorithm includes the following stages:

1. Creation of the grid nodes at the following moment $t+1$ (first iteration). For each point $C_i(t)$ a new vector of coordinates is computed

$$X_i(t+1) = X_i(t) + V_0(X, t)\chi_n(P_i)\Delta t \tag{6}$$

Here P_i is a vector of the normal to the front at the point $C_i(t)$. This vector is determined as the normal at the point $C_i(t)$ to the line passing through points $C_{i-1}(t)$, $C_i(t)$, $C_{i+1}(t)$; Δt is the time step of the calculation.

2. Grid ordering

In modeling fire spreading across terrain the distances between adjacent angles change and the grid may become irregular. To support a grid's regular structure two actions are foreseen: introduction of a new node, if a distance between adjacent nodes exceeds a maximum value *lmax* and exclusion of one of the nodes, if such distance turns out to be less than a given value *lmin*. For example, if a distance between nodes *C*(*i*) and *CR*(*i*) is bigger than *lmax*, then a new node is introduced *C*(*j*) between them and references to the adjacent nodes are changed: $R(i) = j$, $L(R(i)) = j$, $R(j) = R(i)$, $L(j) = i$. Similarly, if a distance between $C(i)$ and $C(R(i))$ is less $lmin$, the node $C(i)$ is excluded from the list and references change: $C(R(i))=C(L(i))$, $C(L(I))=C(R(i))$.

Due to the list structure of a fire front description modeling of the movement of several separate fires, their separation or merging is possible.

Input data for calculating fires are typical for many models [10]. Vegetable fuel parameters are taken into consideration:

- type of fuel materials,
- stock of fuel, kg/m^2,
- fuel calorific efficiency Q, J/m^2.

Parameters of external environment:

- fire hazard class as per weather (from 1 to 5),
- speed and direction of wind W according to data of the nearest weather station, m/sec
- value and direction of a surface slope, grades

These parameters are the most important for calculating burning intensity on the fire line I, W/m, maximum spreading speed V_0 m/min, figurotrisa parameters $\chi_n(P)$, which permits to calculate the speed of front movement at its any point.

3. Agent-based model

The model contains two types of agents, which are A and B. Below structure and functioning of the agents are described.

The agents of the type A (A-agents) are utilized for modeling fire spreading as a combustion wave based on the Hamilton–Jacoby equation and the moving grid algorithm reviewed above. Each node $C(i)$ of the moving grid corresponds to one agent, A_i. An array of A-agents are a fire line (figure 1). The spatial coordinates of the agents of this type $X_i(t)$ are calculated at each step of modeling. Besides, as was described above within modeling under certain conditions agents may perish or au contraire the new ones may arise. The figure also shows the movement of an agent $A_i(t)$ at the following time step $A_i(t+1)$. Vector $P_i(t)$ is a normal to fire front line at the point $A_i(t)$; W is a wind vector; $V_0\chi_n(P_i)\Delta t$ is a distance for which an agent moves A_i for one step.

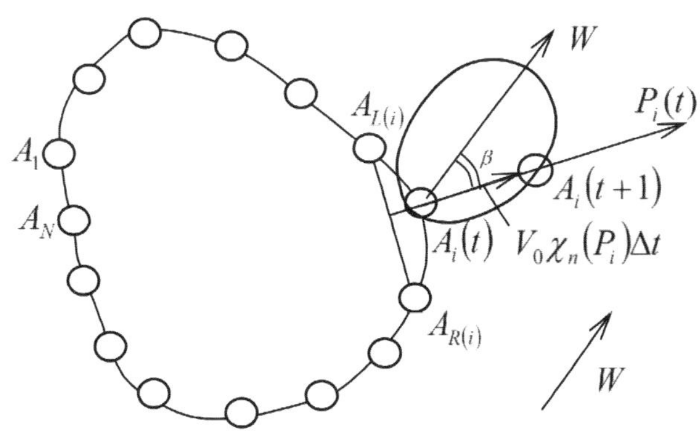

Figure 1. Presentation of fire line as a chain of A-agents, which move according to the moving grid algorithm.

The agents of the type A may be in an active and passive state. A-agent in an active state generates a heat current $I(t)$ and a smoke plume. Intensity of a heat current depends on burning conditions and may subside under the influence of the agents of type B. As value $I(t)$ reduces to zero A-agent becomes passive and no more takes part in the process of modeling.

The agents of type B model actions of fire fighting forces and influence the agents of type A. They pursue only one goal, which is to extinguish all nodes of burning, i.e. to make all A-agents passive. For this, *B*-agent moves across the modeling environment to the nearest *A*-agent and upon approaching reduces fire intensity $I_A(t)$:

$$I_A(t+1) = I_A(t) - \Delta I_B(t), \quad (7)$$

where $\Delta I_B(t)$ is reduction of fire intensity caused by *B*-agent within one step of modeling. Having *extinguished* a successive *A*-agent, *B*-agent moves to the nearest active *A*-agent and continues its actions.

Before the start of modeling number, location and characteristics of *A*- и *B*-agents are set. The characteristics of A-agents are set with burning parameters, speed of fire front movement and V_0 figurotrisa $\chi_n(P)$. Performance of *B*-agents is determined by wildfire fighting instructions [14, 15].

4. Implementation of an agent-based model

On the basis of the proposed multiagent model a software with client and server architecture and web interface was developed. There is a database on a server, which contains information on a modeling environment (real terrain). The program code for the server part is implemented in the PHP language. The system's user interface is implemented in JavaScript (JS) with the use of OpenLayers, an open source code library. This library is designed for creating maps based on the program interface (API) and permits to create a web interface for displaying cartographic documents. This library works with data of various cartographic servers, such as OpenStreetMap, Google, Yandex, etc.

Below an example of modeling with the aid of the developed system is given.

5. Example

Parameters of a modeling environment:

- type of vegetable fuel – lichen;
- fire hazard class – 2;
- wind speed under cover – 1 m/sec, wind direction – south-west.

The forest plot is located on a slope which is indicated by level lines. According to [14, 15] intensity of the modeled fire is determined as weak, therefore the scheme of fire extinguishing is along the whole perimeter. 3 fire brigades, each of 6 persons in number were engaged. The brigades were deployed around the fire line and got a task to go round fire counterclockwise. The area of the fire spot when detected is about 4 ha. Figure 2a represents the initial position of the fire brigades and the initial fire line.

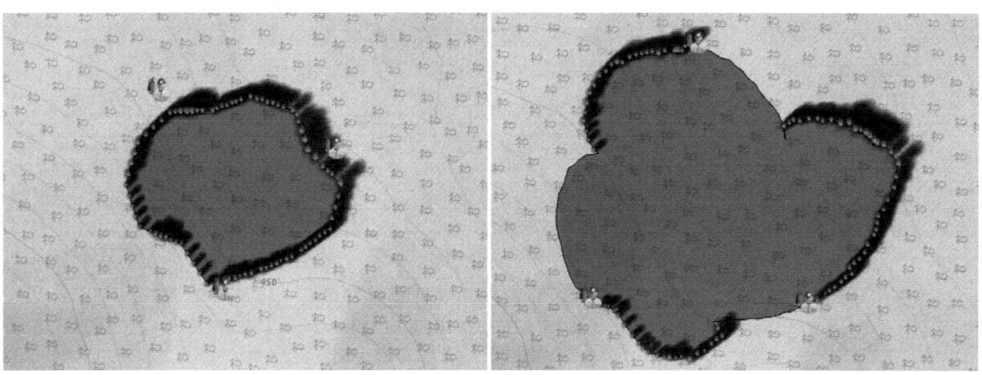

(a) *(b)*

Figure 2. Example of modeling the fire extinguishing.

Figure 2b represents the position of the brigades and the fire line 4 hours after the start of extinguishing (in 48 iterations of the modeling process). As it is seen from the figure after 4 hours the fire continues to be in progress, less than 50% of the active fire line is extinguished. Besides, the total length of the burning fire line at the 4th hour of extinction is approximately equal to the length of the initial fire line, which evidences that 3 fire brigades of 6 persons in number each fail to cope with this fire.

The given example of a negative result in planning a fire fighting event confirms, in our opinion, that the agent-based modeling is effective. The point is that the standards of using fire fighting forces [14, 15] give rather rough estimates, which with regard to terrain features may result in taking wrong decisions.

6. Conclusion

A new method of imitation modeling of wildfire spreading and extinguishing based on agent-based approach permitting simultaneous modeling wildfire spreading and combating therewith is proposed. The developed multiagent model for imitating a direct method of extinguishing a ground forest fire by non-mechanized firefighting brigades and the software developed on the basis of this model can be used for solving a wide range of problems. Firstly, the system may be useful for prompt determination of spreading parameters of ground forest fire in forest ranges, and for developing a complex of measures to prevent and combat fires. Secondly, this system may serve as a basis for instructional training systems designed for teaching students of forestry industry institutes and colleges, as well as forestry staff and Ministry of Emergency Situations in the basics of wildfire combating tactics in a form of a training-game. The possibilities of the proposed method may be enlarged by introducing new agents imitating behavior of other participants of wildfire combating, which will enhance modeling efficiency and reliability.

References

[1] Andrews P L, Bevins C D and Seli R C 2003 *BehavePlus fire modeling system, version 2.0: User's Guide*. USDA Forest Service Gen. Techn. Rep. RMRS-GTR-106WWW (Ogden) p 45

[2] Finney M A 1998 *FARSITE: Fire are simulator model, development and evaluation. USDA Forest Service Res* Paper RMRS-RP-4 (Ogden) p 47

[3] Mell W, Jenkins M A, Gould J and Cheney P 2007 *A physics-based approach to modeling grassland fires* International Journal of Wildland Fire (vol 16), p 1-22

[4] Aksenov K A and Goncharova N V 2006 *Dynamic modeling of the multiagents processes of the resources transformation* (Ekaterinburg: Ural Technical University) p 311 (on Russian)

[5] Kuhtna V B 2014 *Method of modeling a ground forest fire propagation in forest plantations with using an agent approach* (Moscow Forest University. Forestry Transactions, N°5) p 92-97 (in Russian)

[6] Dai D, Zhang Y 2013 *Simulating fire spread in a community using an agent-based model* Proceedings of the 12th International Conference on GeoComputation (LIESMARS Wuhan University, Wuhan, China) p 130-132

[7] Batty M, Jiang B *Multi-agent simulation: new approaches to exploring space-time dynamics within GIS*, Centre for Advanced Spatial Analysis Working Paper Series, Paper 10 (University College London, London)

[8] Yang F, Yang Q and Wang P 2015 *Modelling for Forest Fire Evolution Based on the Energy Accumulation and Release* (International Journal of Advanced Computer Science and Applications, vol 6) N°9 p 38-45

[9] Niazi M, Siddique Q, Hussain A and Kolberg M 2010 *Verification & Validation of an Agent-Based Forest Fire Simulation Model* Proceedings of the Agent Directed Simulation Symposium as part of the ACM SCS Spring Simulation Multiconference (Orlando, FL) p 142-149

[10] Rothermel R C 1972 *A mathematical model for fire spread predictions in wildland fuel* USDA Forest Service Research Paper INT-115 (Ogden) p 40 (Intermountain Forest and Range Exp. Stn.)

[11] Shatalov P S and Dorrer G A 2013 *Parallel computation of forest fire and its interaction with infrastructure objects* Fourth Fire Behave and Fuels Conference, conference abstracts (St. Petersburg) p 3
[12] Rund H 1959 The Differential Geometry of Finsler Spaces, Springer-Verlag.
[13] Dorrer G A 2008 *Dynamics of forest fires* (Novosibirsk, Ed. of Siberian Branch Russian Acad. of Sci.) p 404 (in Russian)
[14] Ivanov V A, Ivanova G A and Moscalchenko S A 2011 *Handbook on wildfires extinguishing* (Krasnoyarsk) p 130 (in Russian)
[15] Schetinski E A 2002 *Wildfires extinguishing: instruction for forest firefighters* (M) p 104 (in Russian)

From «Black sky mode» to ecologizing thought

S Y Piskorskaya, V A Kureshov

Reshetnev Siberian State University of Science and Technology, 31 Krasnoyarsky Rabochy Av., Krasnoyarsk, 660037, Russia

E-mail: piskorskaya1@rambler.ru

Abstract. The article introduces the environmental situation in Krasnoyarsky krai, it discovers the problem of "black sky mode", and indicates how to solve the problem adequately using additional landscaping and ecologizing people's thought. The research demonstrates the methodologies to form environmental thought and behavior. The article represents the social practices developed by Reshetnev Siberian State University of Science and Technologiesto reduce the air pollution.

1. Introduction

Krasnoyarsk krai is the second among the Russian Federation entities due to the area where the large amount of mineral resources are extracted, such as coal, nickel, platinum-group metals, siliceous sand, ores, and metanthracite. 70 % of the krai territory are timber resources concentrating in the area. The industrial wastes pollute the rivers and lakes, soil and air, the resources are extracted extensively; that results in environmental disasters and catastrophes. Therefore, Krasnoyarsk krai is among the three most polluted territories in Russia.

Industrial metallurgical and power plants' toxic emissions contribute to bad air pollution in the krai, though not only big enterprises but small businesses dramatically damage the environment. Experts characterize the air pollution as intensive, and it often happens that it is extremely high with maximum air concentration of such dangerous substances as phenol, benzapyrene, formaldehyde, ammonia, oxidide and sulfur oxide [1].

There were 22.4 thousand stationary polluting substance generators in 2016 according to the data of the Federal Service Agency of State Statistics (Krasnoyrskstat). The polluting substance amount produced by the stationary generators was 8653.2 thousand tons in 2016. The air was contaminated with 2363.3 thousand tons of polluting substances, 95,1% consisted of gaseous and liquid substances (2247.9 thousand tons) [2].

Both the industries and the yearly increasing transport contribute to the air pollution: the car drivers number is increasing annually; the city population is on the 12-th place in Russia, the krai occupies the third place in the country due to the number of cars, 2 cars per family and the transportation level is usual for the industrial krai [3]. The situation is becoming worse, as Krasnoyarsk is located between two hills, and polluted air is stagnated.

As a result, lately, the krai and city population have got accustomed to the so called "black sky mode". The mode is announced if there are industrial pollutants in the air; dangerous substances such as toxic fumes, (chemical substances), the fire smoke and wood fire smoke. There is heavy smog above the city districts and the krai for a long period in windless weather or higher humidity weather after the heavy and long rains, the air turns to smoke, people cannot breath easily.

2. Results and Discussion

As harmful contaminants head in the air, the krai authorities inform the citizens about the black sky mode or unfavorable weather conditions in politically correct language. Though the krai mass media has been far from being politically correct lately; as they suppose: the more information about the dramatic environmental conditions is in air, the greater the probability for the federal authorities to be aware of them.

Elena Penzina, Member of Legislative Assembly of Krasnoyarsk krai, speculating about 'the black sky mode', notices the main problem exists due to the lack of the complex solutions concerning 24 hour work of inspection agencies; she suggests, illegally functioning enterprises should be banned; public transportation should be transferred to EURO-4, big transport enterprises should follow environmental protection control rules [4].

Since organizing the environmental protection control is really difficult because of the political aspects in the governance; and the Krai industrial enterprises cannot be closed down, as it results in sufficient economic losses, the only possible solution can be additional territory gardening. Annually there should be more than 150 thousand new plants in the krai and ecologizing citizens' thought.

We need to underline the long period of research of environmental problems in their global aspect and ecologizing thought, consciousness and science in general, G. Reale and D. Antiseri mention ecologising science to be an absolutelly different approach to science, within its framework the issues of developing industrial civilisation turns to the issues of environmental protection and forecasting environmental consequences. Due to the approach, the main aspect becomes a certain social engaging and training new scientists to meet the criteria of responsibility for the biosphere state and the future of human beings. The environmental protection problems are certain to solve efficiently; the current technical development, technical tools to monitor the environment, and sufficient number of engineers-ecologists greatly contribute to the solution; it is necessary to set up the system environmental thought and a new ranking of social values [5].

It is apparent that we cannot solve the environmental protection problems without educating environmental thought and environmental culture; as a result, the up-to-date environmental recommendations are turning to the requirements to human beings, they need to minimize negative impact of the technology to nature, to redirect technological development to resource-conscious production technologies friendly to nature [6]. The fact is highlighted by many specialists; for example N.G. Mal'kevich writes a human being should live following the environmental norms preserving the nature, taking into account the rules preserving air and water purity, natural landscape, flora and fauna protection, decreasing personal energy and resource consumption [7].

Speculating the environmental protection challenges at the global level, V.N. Vasil'eva confirms that the future development of a society will mainly depend on the level of environmental culture and cautiousness of human beings. Forming the culture is closely connected with environmental thought and consciousness of citizens and their certain environmental focuses generating a basement of environmental behavior. The need in ecologizing thought is caused by the environmental challenges, tangible environmental crisis, the current state of environment; it reflects the sufficient demands of a society [8].

The system advocacy of environmental knowledge is one of the ways to ecology thinking in the society; it needs to provide scientifically excluding emotional and any personal prejudice. Mass media is the most powerful tool of advocacy; it can be used to inform the public about the levels of global and local environmental problems, as well as to promote environmental protection recommendations, norms and requirements systematically.

The important role in developing environmental thinking can belong to unofficial environmental institutions, for example, non-governmental 'Greenpeace' organizations, science technological and science education environmental protection centres, summer environmental protection camps for teenagers, environment-oriented exhibitions, excursions to national parks.

Currently, many cultural subjects (libraries, museums, community centres, youth centres) hold different discussions, disputes, conferences, lectures to promote environmental problem awareness, organize environmental exhibitions, contributing to development of ecologizing thinking of the population. Rather significant are other non-governmental institutions and establishments. One of the outstanding examples is the activity of 'The Consolidated National Front' (CNF). CNF has joined the movement 'LESEGAIS Volunteers' in Krasnoyarsk krai. Using the National Information System for Recording Wood Resource and Wood Business, the CNF enthusiasts will fight against illegal logging; continue involving citizens to landscaping, setting up parks. At present the enthusiasts are busy with forming 'the green shield' of the town including not only woods, but also recreational areas, rivers, and lakes, vegetation plantations inside the municipal area [9].

Providing ecological education as continuous movement to realize the environmental knowledge is a guarantee of successful formation and development of environmental thinking and behavior of citizens.

However, comprehensive efficient forming environmental thinking and behavior needs to involve the education system into the process; the education system is responsible for forming personality, their knowledge, skills, and behavior at different levels (at preschool, school, college, and university). The system of education should form a citizen able to greatly contribute to life of a society including its ecologizing; the system should train professional staff and professional managers for all spheres, who are environmentally responsible. In this regard, the system of education focuses on the following assignments: 1) environmental education; 2) activity approaches in connection with the natural environment; 3) transformation of environmental values, norms, and recommendations to concise, rationally determined performance; 4) development of responsibilities for environmental protection results of their performance.

The stated tasks are certain to simultaneously perform social missions of any institute. Primarily and mainly, it relates to the sphere of higher education, namely, realizing the social functions by the universities in the society. In this context, the university social responsibility does not mean the common thought reproduction, culture development, and behavior norms, but environmental education. The last is accomplished in different forms: implementing significant environmental disciplines into the curricular; integrating humanitarian, environmental, and special disciplines due to their contents integration; including students' community to the environmental projects, directed to solve local problems of the town and the region, expanding the range of volunteer movement, organizing and holding discussion platforms to the environmental protection problems. When the university participates in all events of the kind, it poses its social mission, on the one hand; on the other hand, the university forms the environmental thinking of education process participants.

We could demonstrate the social practices of Reshtnev Siberian State University of Science and technologies (SibSU). The Programme of Flagship University Development determines SibSU social mission in chapter 'Development of the local communities of the town and the regional environment' [10]; according to the document, the university needs to create and develop the Centre of Siberian gardening on the basis of Botanic garden named after V.M. Krutovsky. The Centre tasks are urban improvement and landscape gardening of the town territories, expert-advisory performance within the examining landscape projects, industrial technologies while implementing projects of landscape architecture, and designing the urban environment, gardening, and ornamental fruit-farming.

The Centre has been working for two years; it has proved its efficiency to form environmental culture, environmental thinking of the krai and town population. Moreover, there is cultural-educational centre on the territory of Botanic garden named after V.M. Krutovsky, working permanently; it is opened for all citizens and the tourists. The authority plans to develop an education demonstration zone of physic garden, where the herbs and medicinal herbs, garden crop will grow, cultivated as compositions.

The Centre of Siberian gardening advocates environmental protection by promoting environmental tourism, holding master-classes to fruit-trees and trees grafting, excursions to the sites designed by the Centre students' projects; the Centre proposes the landscape projects to design parks, beach promenades, Surikov historic house museum, it organizes events to improve the urban territories. The Centre volunteers actively beautify the town districts by planting apple trees, designing ornamental flower beds, planting broad-leaved trees due to the gardening project for 'Dendrosad' park.

In general, the Centre employees and volunteer have improved approximately ten hectare territory for two years, developed two new tourist environmental routes.

Developing a volunteer movement, managing the young people environmental initiative consistently, organizing exhibitions, discussion platforms are an efficient university social practice; developing environmental tourist routes for the citizens greatly contributes to forming environmental thinking and behavior, and environmental culture of the population in general.

The described practices cannot solve all environmental protection problems in the town and krai completely; neither do they reduce the pollution of metallurgical enterprises and power plants, nor liquidate deforestation, water body pollution, soil pollution, they are not able to eliminate the black sky mode.

However, the environmental situation in the town can improve due to regular environmental education, consistent work of non-governmental organizations and institutions, educational and cultural institutions, involving variously-aged citizens into realizing environmental protection issues, mutual search for their solutions and the found solutions (cleaning up the Yenissei banks, urban territories, gardening).

References

[1] Ekologicheskie problem Krasnoyarskogo-Kraya (Environmental problems of Krasnoyarky Krai) available at: https://ecoportal.info/ekologicheskie-problemy-krasnoyarskogo-kraya/. Access 12.04.2018

[2] Vybrosy v atmosferu ot statsionarnykh istochnikov v 2016 godu: pressvypusk Upravleniya federalnoy sluzhby gosudarstvennoy statistiki po Krasnoyarskomu Krayu; respublike Khakasiya I respublike Tyva (Krasnoyarskstat) (Air Contamination by Stationary Sources in 2016: press release by Federal Government Statistical Service in Krasnoyarsky Krai; Republic of Khakasiya and Republic of Tyva (Krasnoyarskstat) available at: http://krasstat.gks.ru/wps/wcm/connect/rosstat_ts/krasstat/ru/statistics/tuvStat/environment/ Access 17.04.2018

[3] Kamenchenko P Rezhim chernogo neba (Black Sky Mode.) available at: https://lenta.ru/articles/2018/04/12/blacksky/. Access 12.04.2018

[4] Temnov A. Elena Penzina: «Vse khotyat krovi» (Elena Penzina: «Everyone needs blood») available at: - http://babr24.com/kras/?IDE=167154. Access 14.04.2018

[5] Reale Dg and Antiseri D 2003 Zapadnaya filosofiya ot istokov do nashikh dney. Ot romantizma do nashikh dney (4) (Western Philosophy from the origin up to current days. From Romantic Period up to current days) (St. Petersburg: Pnevma) p 67

[6] Ivanova E M and Gorshkova O A 2012 Ekologizatsiya myshleniya sovremennogo cheloveka kak vazhnaya komponenta ego obrazovannosti i bytiya. (Environmentalising modern human being consciousness as a significant component of their education and being) *Sovremennye tekhnika i tekhnologii materialy XVIII mezhdunarodnoy nauchnoprakticheskoy konferentsii* p 421

[7] Mal'kevich N G 2015 Ekologizatsiya myshleniya i povedeniya lyudey. (Environmentalising people's thinking and behavior) *Nauka – obrazovaniyu, proizvodstvu, ekonomike. Materialy XIII mezhdunarodnoy nauchnotekhnicheskoy konferentsii* (Minsk: BNTU) vol 2 pp 467–468

[8] Vasil'eva V N 2003 Formirovanie ekologicheskogo myshleniya v protsesse obrazovaniya. (Forming environmental thinking while getting education) *Innovatsii i obrazovanie: Sbornik materialov konferentsii* (St. Petersburg) **29** p 273

[9] ONF derzhit na kontrole vopros sozdaniya «zelenogo-shchita» vokrug Krasnoyarska. (published in March, 27, 2018) (All-Russian National Front controls the issue to create "green shield" around Krasnoyarsk) available at: https://onf.ru/2018/03/27/onf-derzhit-na-kontrole-vopros-sozdaniya-zelenogo-shchita-vokrug-krasnoyarska/ Access 17.04.2018

[10] Programma-razvitiya (dorozhnaya karta) SIBGAU (The programme of development (road map) of SibSAU). available at: https://disk.sibsau.ru/index.php/s/8drYDIhECpyvvjA. Access 09.04.2018

Post-fire succession in scots pine forests of southern taiga central Siberia

V A Ivanov[1], G A Ivanova[2] and N M Kovaleva[2]

[1] Reshetnev Siberian State University of Science and Technology, 31 Krasnoyarsky Rabochy Av., Krasnoyarsk, 660037, Russia
[2] Sukachev Institute of Forest, Siberian Branch, Russian Academy of Sciences, Krasnoyarsk, Russia

E-mail: ivanovv53@yandex.ru

Abstract. Fire is one of the main disturbance factors in the boreal forests of Russia. Forests dominated by Scots pine (*Pinus sylvestris* L.) are widespread in Central Siberia, with large areas burned annually. We studied post fire succession of ground vegetation following experimental fires of various intensities on typical Scots pine forest sites. The greatest changes in ground vegetation and biomass of these pine forests were after fires of high intensity. After fires of low intensity ecosystem components recovered rather quickly to pre-fire composition. These results are important for determining the effects of fire behavior and intensity on rates and patterns of post-fire vegetation recovery in southern taiga Scots pine forests of Siberia.

1. Introduction

Forest fires can have a long-lasting and substantial influence on successional dynamics of boreal forests. Being an important evolutionary and ecological factor, wildfires influence forest structure and composition, affect surface vegetation and animal habitat, and regulate a wide range of ecosystem processes [13, 15].

Fire impact on vegetation shows itself variously both as a result of a direct influence on plant communities during wildfires and indirectly - through pyrogenic (post-fire) succession. The post-fire succession of a plant community depends on many factors: physical-geographical conditions, climate, and pre-fire structure of plant community [3, 7, 12].

Ecological heterogeneity of vegetation in the subcanopy and surface vegetation, which is often barely visible from above the tree canopy, is strongly evident in post-fire succession. The resulting partial or complete changes in surface vegetation can lead to diverse pathways of post-fire vegetation regrowth and recovery. Forest fires facilitate invasion of early successional species, although often the post-fire community is dominated by species that have survived the fire [18].

Forest fires change not only species diversity of surface vegetation within habitats but also diversity of ecotopes themselves. The nature of post-fire successions is determined by fire intensity [11, 16]. Fire intensity influences germination and amount of viable seeds in soil and also soil nutrients on which the post-fire vegetation succession depends [8]. Increased fire intensity impacted on species composition, percent cover and biomass of ground vegetation [5]. The post-fire regeneration of vegetation results in biomass accumulation. The accumulation of post-fire biomass can be impacted by many factors such as forest type, age, tree stand density and so on, as well as the type and intensity of a fire [6].

Pine forests of Central Siberia are exposed to fires of widely varying intensity dependent on fuel loads, fuel condition, and the weather conditions before and during the fire. As a result these forests contain a complicated landscape matrix of stands at different stages of post-fire regeneration [1]. The fire impact in one case may be limited to partial destruction of the lower vegetation layers or partial damage of a tree stand, in another case the fire may cause high levels of understory fuel consumption and canopy mortality, killing almost all living vegetation of the pre-fire plant communities. Fire impact on plant communities should be considered in respect to their structural elements: ground vegetation (shrubs, herbs, moss, and lichen), understory, tree regeneration and the overstory tree stand.

The objective of this paper is to describe changes in forest floor vegetation and above ground biomass in southern taiga pine stands of Central Siberia after surface fires of various intensities.

2. Materials and methods

Studies were carried out in pine stands, which are representative for Central Siberian Scots pine-dominated forests. Experimental plots (4 ha each) were set in the Angara river basin in Govorkovo (Plots 1, 2) (58°35'N, 98°55'E) and Khrebtovo (Plots 3, 4) (58°42'N, 98°25'E). Soils were alluvial-ferriferous podzols on experimental plots.

Plots 1 and 2 were in dry Scots pine forest with lichen and green moss in ground vegetation. This stand was on a flat step of a slope, and had last been burned in 1922, more than 80 years before experimental fires. The average tree age was 90 years. Species *Vaccinium vitis-idaea*, *Pleurozium schreberi* and *Cladonia rangiferina* dominated in the ground vegetation. Plots 3 and 4 were in a more mesic Scots pine forest where dominated herbs and green moss in ground layer. This stand occupied a gentle slope, last burned in a surface fire in 1948, more than 50 years before our experimental fires. The average tree age was 100–120 years. Understory (*Alnus* sp.) forms the layer height up to 3.5 m on the average. Species *Vaccinium vitis-idaea, Linnaea borealis* L., *Maianthemum bifolium* L. F. W. Schmidt, *Iris ruthenica* Ker-Gswl, *Pleurozium schreberi* dominated in ground vegetation.

In 2002, 2003 years were conducted prescribed fires to characterizing and modeling the fire behavior and ecosystem effects of fire burned under different weather on a series of plots in central and southern taiga stands [4, 10]. During these experiments we conducted prescribed fires to simulate the behavior of natural surface fires, which are typical for the Scots pine stands of Central Siberia. According to the intensity classification of forest fires [10] a high intensity of fire was on experimental plot 2 (4876 kW/m), moderate intensity – on the plot 1 (3195 kW/m), low intensity of fire – on plots 3 and 4 (878 and 924 kW/m).

To estimate the structure and biomass of forest floor vegetation, litter and duff the methods developed by D. J. McRae [9] were used on these experimental plots. The samples of living ground vegetation, litter and duff were collected before fire and every year after fire to monitor biomass. The biomass of down woody debris was estimated by the line-intersect approach [9, 19]. The tree stand biomass was assessed by inventory methods and by harvesting model trees by diameter classes. Based on obtained data on biomass of model trees and on distribution of trees by diameter classes the biomass of living trees and dead standing trees was calculated for the experimental plots.

Species diversity and percent cover were measured on 25 (1m x 1m) sub-plots on every plot before and after experimental fires [17]. Structure of ground vegetation was mapped on plots before and after fires (1, 3, 7, 13 years). The plant microgroup was the main mapped unit [14]. Schematic maps of vegetation were developed using the ESRI *Arc*Map program.

3. Results and discussions

We compared the initial stage of post-fire succession in southern taiga pine forests after fires of different intensity. Natural tree mortality in these forests stands before fire was from 2 to 15% of the total tree number. The main post-fire tree mortality – up to 90% from the total tree number died during 13 years occurred in the first two to three years after fires. By thirteen years past the high intensity of fire (Plot 2) the tree mortality was 71.1% and after fires of moderate and low intensity 18.3 (Plot 1) and 9.8% (Plot 3), 6.4% (Plots 4) of the total number of trees, respectively (table 1). On our obtained data for experimental plots in southern taiga and our earlier data about experimental fires in central taiga Scots pine forests [4. 6], the tree mortality correlate with fire line intensity, the correlation coefficient characterizing this relationship is 0.83 (figure 1).

Table 1. Dynamics of the tree mortality in pine forests after fires of different intensity (% of the number of trees before the fire)

Fire intensity	Plot	Time after fire, year						
		1	2	3	4	5	8	13
Low	3	3.3	5.4	5.4	7.6	8.7	9.8	9.8
	4	3.2	4.2	6.4	6.4	6.4	6.4	6.4
Moderate	1	11.8	16.1	16.1	16.1	16.1	17.2	18.3
High	2	61.8	70.1	70.1	70.1	70.1	71.1	71.1

Regardless of fire intensity the highest mortality occurred in trees of smaller diameters. The relation of mortality with tree diameters was strong (r = 0.88; $p < 0.05$). A study of causes of mortality on our earlier set of prescribed fires in central taiga Scots pine forests showed that mortality was significantly related to char height and degree of colonization of a tree by bark insects after the fires то есть Fires of high intensity stimulated weakening of trees and increased tree colonization by insects [2].

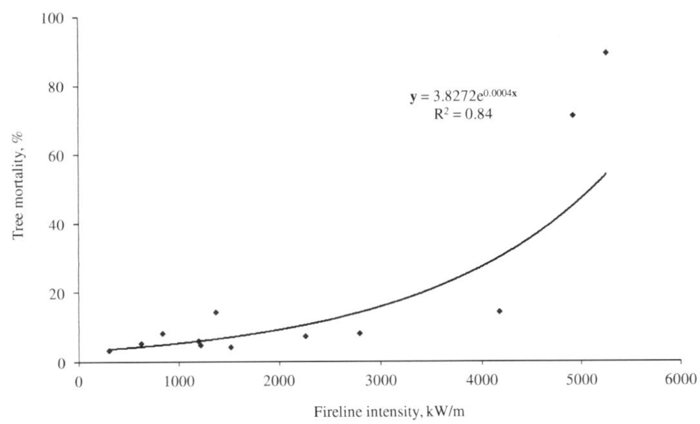

Figure 1. Relationship between tree mortality and fire line intensity in pine stands of Siberia

We analyzed species composition of ground layer before and after experimental fires. Pre- fire vegetation pine stand (Plots 1, 2) had low species diversity: shrubs and herbs (11 species), green mosses (3 species) and lichens (7 species). Plots 3 and 4 had great abundance of shrubs and grass (20 species) and moss-lichen layer (6 species).

The number of species increased up to 16 species (Plots 1 and 2) and up to 24 species (Plots 3 and 4) after experimental fires. All the species pre-fire also remained in the years following the fires. Such species as *Calamagrostis epigeios* (L.) Roth, *Chamaenerion angustifolium* L., *Rubus saxatilis* L. appeared in ground layer after fires of moderate and high intensity (Plots 1 and 2). Species *Hieracium umbellatum* L. and *Chelidonium majus* L. appeared in ground layer after the low intensity of fire (Plot 4) (figure 2).

Percent cover of herbs and shrubs decreased after fires but the cover of ground layer increased with time since fires. Regeneration of herbs and shrubs proceeds by vegetative reproduction of surviving plants or their seeds. The percent cover of the dominant shrub *Vaccinium vitis-idaea* reached its pre fire levels in during 7–8 years.

The percent cover of moss layer decreased after fires depending on intensity. Lichen layer burned completely in fires of moderate and high intensity. Scattered lichen patches (< 1% percent cover)

remained among mosses in unburned areas on the plots after low intensity fires (Plot 4). Sites where forest floor burned to the mineral layer was colonized by mosses (*Polytrichum commune* Hedw., *P. strictum* Brid.), which reached about 10% cover by the 13 year after fires of moderate and high intensity. Percent cover of pre-fire mosses (*Pleurozium schreberi*, *Dicranum polysetum*) was only 1–10% by 13 years post fire (before fire – 30–80%).

After experimental fires the herbs and shrubs dominated the plant community on all plots, independent of fire intensity. *Vaccinium vitis-idaea* microgroup dominated in the ground vegetation five years after the prescribed fires (figure 2). Initial species such as *Calamagrostis arundinacea* (L.) Roth, *Chamerion angustifolium* and *Polytrichum* dominated in sites with the greatest depth of burn in the litter and humus layers. However, in sites with the lowest depth of burn the role of such herbs as *Maianthemum bifolium*, *Lycopodium complanatum* L intensified. After fire of low intensity regeneration of pre-fire mosses (*Pleurozium schreberi*, *Dicranum polysetum*) led to expanded cover of these species in the microgroup (*Vaccinium vitis-idaea–Pleurozium schreberi*) relative to before fire.

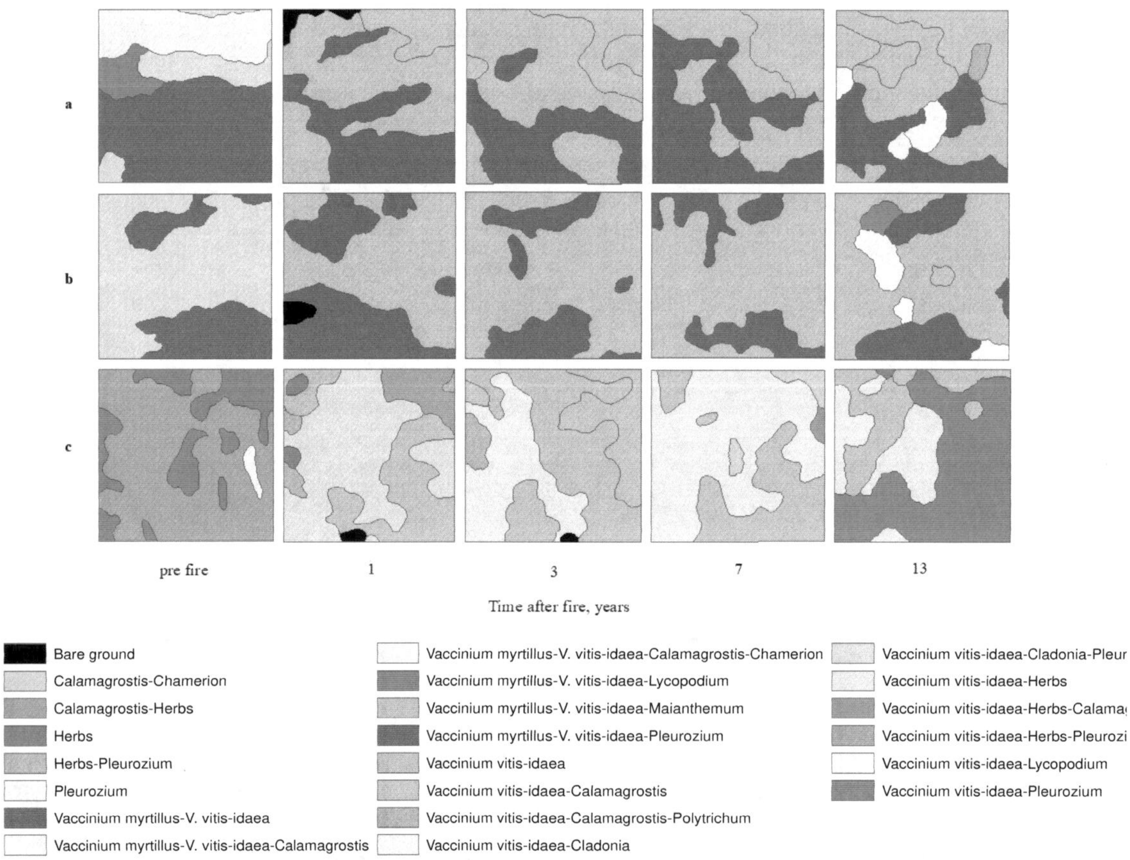

Figure 2. Structure of plant microgroups before and after the experimental fires of a) high (Plot 2), b) moderate (Plot 1), c) low intensity (Plot 3).

Biomass of forest ecosystems is the fuels that support wildfire spread. It is extremely rare for this biomass to burn completely. We evaluated the effects of fire on major components of the Scots pine ecosystems studied here. These components include the trees, tree regeneration, understory, and ground soil vegetation. These different components of the forest ecosystem contribute differently to available fuels in a fire and experience different impacts from fires as well. The amount of burned biomass is determined by the intensity of the fire. Mosses and lichens were fully destroyed by fire. All fires resulted in reduction of biomass lichen, moss, herb, small bushes, small branches and windfall. On all plots fire resulted in death of all young tree regeneration.

Over the years, dead organic matter accumulates at all sites after fires as a result of the cast of needles, bark and branches from trees damaged by fire and the fall of deadwood trunks. By 13 years post fire the biomass of living trees in the pine forests on plot 1–4 decreased 16–62% by five years post fire depending on fire intensity.

Accumulation of above ground biomass after fires in these southern taiga pine forests greatly depended on burn intensity. After fires the live biomass killed but not consumed in the fire is re-distributed to mort mass

At the same time forest floor vegetation gradually increased after fire. Fire intensity affected depth of burn of forest litter, which undoubtedly had an effect on the success of seedling regeneration. By 13 years post fire the number of seedlings after the moderate intensity fire in pine forest with lichen and green mosses (Plot 1) reached 68 thousand per ha.

Total above ground biomass on experimental plots changed slightly after fires of low and moderate intensity. A greater initial decrease of living ground vegetation and duff loads is typical after moderate intensity fires due to higher fuel consumption and tree mortality. Living biomass decreased an average of 10% after the low intensity fires by 15% after the moderate fire intensity. The greatest changes in the structure of aboveground organic matter distribution were after the high intensity fire, when the aboveground living biomass decreased by 85% (figure 3).

Biomass of species *Vaccinium vitis-idaea* also increased after fires on all plots. By four to six years after high intensity fire biomass of *Calamagrostis arundinacea* increased up to 24% (compared to 3% before fire). Lichens were not regenerated in the first eight to nine years after fires. Biomass of mosses was less than 1% of the pre fire value 13 years after fires of high and moderate intensity. Fires of low intensity resulted in reduction of moss biomass by 68% (1 year after the fire). In the following post-fire years forest floor biomass gradually increased.

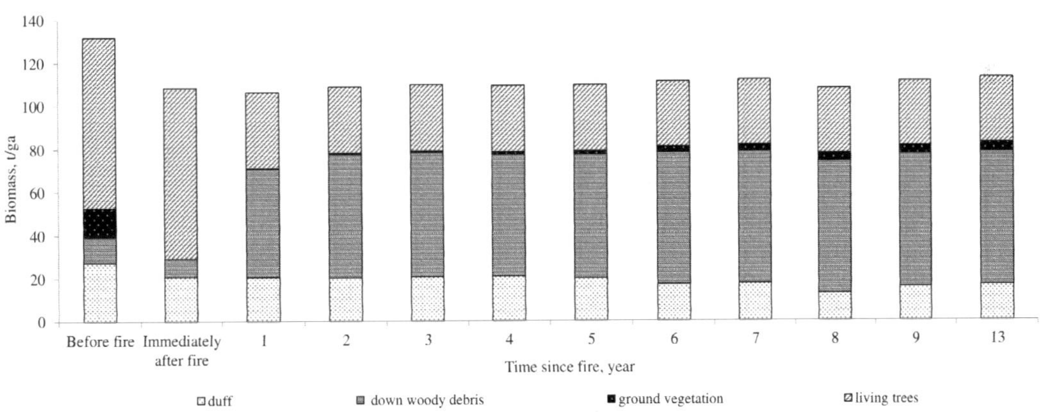

Figure 3. Forest above-ground biomass dynamics in southern taiga lichen pine stands following fires of high intensity (Plot 2). The living biomass includes the tree stand, regeneration, lichen-moss and grass-small shrub layers; dead biomass consists of dead standing trees, down woody debris, litter and duff.

4. Conclusions

Post fire ground vegetation at the initial stage of fire succession depends on a complex of factors (fire intensity, depth of duff consumption, and pre-fire forest type). In the course of studies the experimental data on impact of different intensity fires on dynamics of the above ground biomass of southern taiga pine forests of Central Siberia were obtained.

Fires, regardless of their intensity, decreased percent cover and biomass of ground vegetation. The greatest changes in ground layer were in pine forests with dominated lichen and green moss after fires of high and moderate intensity. In these situations where most or all of the lichen-green moss cover was burned, *Vaccinium vitis-idaea* recovered fairly rapidly, but there was also considerable invasion of grass and herbaceous species that had been largely absent pre fire.

The post-fire accumulation of biomass was strongly related to fire intensity. In the first several years after fire the increase of dead biomass correlates well with fire intensity. Later in succession, the impact of fire intensity on vegetation recovery and on dead biomass accumulation decreases. After fires the dead biomass continues to accumulate, especially following high intensity fires, as dead branches and standing dead tree fall to the forest floor. The high intensity fire leads to high levels of dead biomass, which becomes an increasing proportion of the total biomass.

References

[1] Buzikin A I 1975 Impact of surface fires on pine forests of Central Angara region *Protection of forest resources of Siberia* pp 141–153 (In Russian)

[2] Conard S G, Tsvetkov P, Ivanova G A and McRae D J 2004 Impacts of Fire Severity and Fire Behavior on Mortality of *Pinus sylvestris* in Pine Forests of Central Siberia *Climate Disturbance Interactions in Boreal Forest Ecosystems IBFRA 12-th Annual Scientific Conference 3-6 May 2004, Fairbanks, Alaska, USA Program and Abstracts* p 54

[3] De Luis M, Garc´ıa-Cano M F, Cortina J, Ravent´os J, Gonz´alez-Hidalgo J C and S´anchez J R 2001 Climatic trends, disturbances and short-term vegetation dynamics in a Mediterranean shrubland. *Forest Ecology and Management* vol 147 pp 25–37

[4] Ivanova G A, Conard S G, Kukavskaya E A and McRae D J 2011 Fire impact on carbon storage in light conifer forests of the Lower Angara region, Siberia *Environmental Research Letters* vol 6 **4**

[5] Komarova T A 1992 Post-fire successions in the forests of Southern Sihote-Alinya (Vladivostok: DVNC) p 222 (In Russian)

[6] Kukavskaya E A , Ivanova G A, Conard S G, McRae D J and Ivanov V A 2014 Biomass dynamics of central Siberian Scots pine forests following surface fires of varying severity *International Journal of Wildland Fire* vol 23 **6** pp 872–876

[7] Lloret F andVil`a M 2003 Diversity patterns of plant functional types in relation to fire regime and previous land use in Mediterranean woodlands *Journal of Vegetation Science* vol 14 pp 387–398

[8] Malanson G P and Trabaud L 1988 Vigor of post-fire resprouting by *Quercus Coccifera* L. *Journal of* Ecology vol 76 **2** pp 351–365

[9] McRae D J and Alexander M E, Stocks B J 1979 Measurement and description of fuels and fire behavior on prescribed burns: A handbook. Information Report O-X-287 (Sault Ste. Marie: Environment Canada, Canadian Forest Service, Great Lakes Forestry Research Centre) p 44

[10] McRae D, Conard S, Ivanova G, Sukhinin A, Baker S, Samsonov Y, Blake T, Ivanov V, Ivanov A, Churkina T, Hao W, Koutzenogij K and Kovaleva N 2006 Variability of fire behavior, fire effects, and emissions in Scotch pine forests of Central Siberia. *Mitigation and Adaptation Strategies for Global Change* vol 11 (1) pp 45–74

[11] Moreno J M and Oechel W C 1991 Fire intensity effects on germination of shrubs and herbs in Southern California Chaparral. *Journal of* Ecology vol 7 p 1993–2004

[12] Pausas G J and Vallejo V R 1999 The role of fire in European Mediterranean Ecosystems *Remote sensing of large wildfires in the European Mediterranean basin* p 3–16

[13] Popov L V 1982 *South taiga forests of central Siberia* (Irkutsk: Irkutsk University Publications) p 330 (In Russian)

[14] Ramensky L G 1971 Problems and methods of study of vegetation cover. Selected publications (Leningrad: Nauka) p 334 (In Russian)

[15] Sannikov S N 1973 Forest fires as evolutionary factor pine regeneration in Zayralye burning and fires in a forest (Krasnoyarsk: IF and W SB AS USSR) pp 236–277 (In Russian)

[16] Schimmel J and Granström A 1996 Fire severity and vegetation response in the boreal Swedish forest *Journal of Ecology* vol 77 pp 1436–1450

[17] Sukachev V N and Zonn S V (Eds) 1961 *Forest Type Study Methodologies* (Moscow: USSR Academy Publications) (In Russian)

[18] Turner M G, Romme W H, Gardner R H and Hargrove W W 1997 Effects of fire size and pattern on early succession in Yellowstone National Park. *Ecological* Monographs vol 67 pp 411–433
[19] Van Wagner C E 1968 The line intersect method in forest fuel sampling. *Forest Science* vol 1 pp 20–26

Contemporary analysis of Krasnoyarsk environmental problems

V I Litovchenko[1], I I Misuna [2] and N A Shumakova[1]

[1] Reshetnev Siberian State University of Science and Technology. 31, Krasnoyarsk Rabochy Av., Krasnoyarsk, Russia, 660037
[2] Krasnoyarsk Aluminum Smelter. 40, Pogranichnikov St., Krasnoyarsk, Russia, 660111

E-mail: litovchenkovi@yandex.ru

Abstract. The purpose of the article is to analyze modern environmental situation in Krasnoyarsk. Industrialization, which started during the war, didn't take into consideration ecological question. Krasnoyarsk Aluminum Smelter and the Krasnoyarsk Hydroelectric Power Station negatively affected the city ecology. Besides, the increasing number of transport also produces environmental discharge in great amount. According to the research of total index of atmospheric layer air pollution five top priority pollutants are at their critical level. Benzapyrene, formaldehyde, suspended substance, nitrogen dioxide and hydrogen nitride mainly influence city air. Urgent measures must be taken to solve Krasnoyarsk environmental problems. The industries have to recycle and dispose hazardous waste, bituminous coal plants should be converted to gas, motor transport should be transferred to environmentally friendly fuel, and a green belt round Krasnoyarsk should be created. A document of intentions to improve the ecological situation in Krasnoyarsk has already been adopted, and the process of fulfilling it has already started. Otherwise, if situation isn't changed, life-threatening consequences are inevitable.

According to the World Health Organization data published in 2019, air pollution is considered to be one of the most serious threats for health associated with the environment [1].

The sources of air pollution are diverse in Krasnoyarsk and the composition of emissions is complex.Krasnoyarsk is among the cities characterized by a super high level of 1st and 2nd class hazard of air pollutants.

According to the research of total index of atmospheric layer air pollution five top priority pollutants are at their critical level. The priority substances creating a high level of air pollution are: benzapyrene, formaldehyde, suspended substance, nitrogen dioxide and oxide [2].

According to the Environmental Safety Report issued by the regional Ministry of Ecology in 2017, the level of air pollution in Krasnoyarsk is characterized as extremely high. The complex index of air pollution (IAP) is 5> 14, standard index (SI) is 20,1 % (benzapyrene level) and the level of formaldehyde is 23,1%. Benzapyrene, formaldehyde, suspended substance, nitrogen dioxide and hydrogen nitride mainly influence air pollution [3].

The consequences of living within the conditions of poor ecology are as follows [1]:

- in areas with high nitrogen dioxide pollution, lung function of both children and adults decreases (a review of current studies prepared by Greenpeace in 2017);
- when nitrogen dioxide exposure increases by 10 μg / m3, the risk of asthma increases by 15% (a review of current studies prepared by Greenpeace in 2017);
- formaldehyde causes severe irritation of the mucous membranes of the respiratory tract and is a carcinogenic substance and increases allergy to house dust (study of Perm Medical Academy);
- with prolonged exposure, an increase in the concentration of PM2.5 fine particles by 10 μg / m3 results in a decrease in language learning ability and general cognitive abilities (research conducted by American scientists, 2016).

This situation has been shaping for many years. Krasnoyarsk was an ecologically clean city until the post-war years. According to its parameters, it had fallen in line with resort cities, as it had a lot of sunny days, clear air, stable weather and landscape.

Industrialization had started during the war and ecology was not taken into consideration. A lot of enterprises had been removed from the western part of Russia and located on the right bank of the Yenisei River as the left bank had a difficult terrain. That is why the ecological situation here was better.

In 1960s, Krasnoyarsk began to develop as an industrial center. The period when the dam of the Krasnoyarsk Hydroelectric Power Station and Krasnoyarsk Aluminum Smelter were built was of high importance. 70% of the hydroelectric power should go to Krasnoyarsk Aluminum Smelter. It was planned to build Aluminum Smelter 60 km further from the city center, downstream of the Yenisei River, its capacity should be less than today. Secretary General N.S. Khrushchev received a proposal from the head of the region to build a plant near the village of Industrialny (15 km from the city center). With the money saved, it was proposed to double aluminum production, and the volume of production of commercial aluminum was about 1 million tons.

In 1964, the world's largest plant for the production of primary aluminum in the suburbs of the developing city of Krasnoyarsk was launched. In the world there is no practice of local concentration of aluminum production enterprise using outdated technologies. After the launch of the last facility, emissions to the atmosphere amounted 120 thousand tons, which was unacceptable. At the same time, the set of harmful components exceeded the permissible rate by several times. It is primarily fluorine and its compounds like benzapyrene. Today, after simple renovations and upgrades, emissions from the plant to the atmosphere amount to 18% of the citywide ones. Many researchers and environmentalists dispute this figure.

After starting working at full capacity Aluminum Smelter took the major share of city's electricity, so it was necessary to build a bituminous coal plant. Nowadays there are several bituminous coal plants in the city. Coal burned at them is low-grade, that is harmful for the city's environment. Total emissions from bituminous coal plants today are about 40,000 tons per year.

The Krasnoyarsk Hydroelectric Power Station continues to influence local ecology negatively. Water enters the turbines from the downstream of the cold layers of the dam lake. Water falling from a great height forms an ice-free ice lane 300-350 km from the dam. For the ecology of Krasnoyarsk, this situation is similar to a catastrophe. Above the riverbed of the Yenisei there is a fog. All hazardous substances accumulated by bituminous coal plants, Aluminum Smelter, road transport and other industries are accumulated here. This poisonous mixture, enters the lungs of Krasnoyarsk, causes various bronco-pulmonary diseases. It is simply impossible to swim in the summer due to the low temperature of the water in the Yenisei.

Different scientists had proposed several ecology improving projects but their financing in early 1990s was stopped.

In London there was a similar environmental problem when London fog began to accumulate harmful substances during the urbanization and smog fell on the city. But they found recourses to transfer all the enterprises and boiler rooms to gas and the problem was solved [4].

It should be noted that there are a number of developments to create smokeless coal, there are even examples of an applicable project based on the boiler house of the former Television Factory. About

30 patents were issued for this, but the massive introduction is hampered by the lack of funding. Meanwhile, the Mongolian government purchased technology from Krasnoyarsk scientists for Ulaanbaatar, converted incineration at its boiler houses, and solved the smog problem over the city.

A definite contribution to the unfavorable ecological situation is brought up by the issue of municipal and industrial wastes. After nearly 20 years of the 21st century in Krasnoyarsk, there is no sane strategy, policy on waste management. Until now, there is no waste processing enterprise in the city. Waste during collection is not sorted; hazardous, long-decaying chemical waste and household waste are mixed. The level of processing and return to production at enterprises is very low. In the vicinity of Krasnoyarsk there are no certified landfills for waste disposal. Environmental protection legislation is violated while administration, pro-environmental supervision and Federal Service on Surveillance for Consumer rights protection and human well- being are tolerant to the problem. Industrial waste is buried near the cemetery of Badalyk which is in the city of Krasnoyarsk.

Federal Law No. 89 prohibits waste disposal in places of settlement [5]. Neighborhood Solnechny, Badalyk and part of the neighborhood Severniy are affected by the landfill a lot. There are a lot of unauthorized landfills around the city, which negatively influences the environment.

The issue of the Aluminum Smelter waste, concentrated on the mud storage area of the plant, is of great concern and importance today. Sludges on artificial excavation in the east of the plant has high content of fluorine. Currently, about a million tons of hazardous waste is stored there. Research shows that the territory of the Sovietskiy district, part of the Central and the right bank of Berezovka are infected with fluorine. Fluorine is carried from the storage area with the wind in the form of a dust-like fraction. This dangerous situation leads to thyroid disorders, arthridis, cancer. Nowadays it is necessary to apply preventive measures: Rusal has to recycle and dispose hazardous waste.

Despite a huge amount of unsolved environmental problems, managers at different levels continue to be interested only in the profits of enterprises and industrial zones. In 2016 at the economic forum in St. Petersburg RUSAL, the Aluminum Association and, at that time, Krasnoyarsk governor Viktor Tolokonsky signed a "protocol of intent on creating an innovative industrial territory "Aluminum Valley", which was immediately called the "obituary" of Krasnoyarsk [6].

In fact, most environmental problems are directly related to environmental engineering. Therefore, it is possible to calculate the time needed before people will have cancer 4-5 times more often when an aluminum smelter is built in the city, as it happened in Krasnoyarsk.

High percentage of environmental discharge, along with industrial enterprises, is produced by motor transport.

According to the official data of the Krasnoyarsk City Hall, 43% of environmental discharge goes from motor transport. The share of public transport in the total amount is only 0.4%. Today there are about 410,000 cars in the city, a large amount of which has a worked out resource. The most dangerous cars emission is benz(a)pyrene, which is the strongest carcinogen [7].

The issue of parks and green zones is of great importance for Krasnoyarsk. The city is developing quickly today but park zones are not provided. The only natural green space is the Tatyshev Island which is not enough for such a large territory. The island can become an attractive park zone in future. Krasnoyarsk takes the last place in terms of the number of green spaces compared to even cities with a smaller population.

In 1997, the administration of the Krasnoyarsk Territory adopted the resolution No. 644-P, which not only defined the boundaries of the green zone of Krasnoyarsk, but also regulated the functional division, as well as the mode of use and the list of activities. However, in 2009, Decree No. 481-p. was signed, which recognized Decree No. 644-P as invalid, and a year later a document appeared that abolished the mandatory environmental impact assessment of construction sites on the territory of reserves. To improve the ecological status of Krasnoyarsk, it is necessary, at least, to reestablish the by-law act that established green zones. It is also essential to create sanitary protection zones around potentially dangerous objects [2].

The existing green plantings are cared of the way, which often leads to their degradation and death. Green areas in the sanitary protection zones of large enterprises in Krasnoyarsk are not planted. A good example is the Krasnoyarsk Aluminum Smelter sanitary protection zone, where today practically all the trees that were planted before have been cut down, and construction is developing instead of a green zone. During the construction of the market in the area of Gaydashovka in the sanitary protection zone of Aluminum Smelter, trees were cut down by 4.5 million rubles. The environmental prosecutor's office and the district administration don't take any measures offering RUSAL to sue the car market.

Deputies of the Regional Legislature approved the initiative of activists to create a forest green belt around Krasnoyarsk in 2018. An interdepartmental working group was formed to develop proposals. A list of additional parts to include in the green belt zone was sent to the Ministry.

According to the scheme prepared by the Regional Ministry of Environment, the area of the forest green belt will be about 250,000 hectares. It should include the urban forests of Krasnoyarsk, the Stolby reserve, partly the territory of Divnogorsk, as well as the lands of several regions. As part of the regional program "Environmental Protection and Reproduction of Natural Resources", about 9 million rubles of budget funds were allocated for carrying out the necessary land management works, the formation of cartographic materials and determining the coordinates [8].

In the year of ecology, the Environmental Charter was adopted at the Krasnoyarsk Economic Forum. The time had come to take measures on the ecological situation in our city and to leave the "honorable" place in the top ten of the most polluted cities in the world. Otherwise, as a well-known environmental scientist R.G. Khlebopros remarks, this city will be uninhabitable in 15 years [9].

The Charter is a document of intentions of the administration and the large business community to improve the ecological situation in the Krasnoyarsk Territory, it declares and takes responsibility before the present and future generations for providing a favorable environment.

We hope this document will be more effective than the previous ones. The authorities should realize that life with the "black sky" will lead to the irreversible environmental consequences in a couple of decades.

Within the next few years scientists need to work out economic and ecological state of Krasnoyarsk. First of all they should calculate the cost of converting bituminous coal plants - 1, 2, 3 to gas.

As part of a large-scale modernization programme, carried out in 2004-2009, the Aluminum smelter was converted to «dry» anode technology and equipped with dry gas scrubbing stations and automated alumina point feeders. The programme led to a considerable cut in emissions: hydrogen fluoride was reduced by 1.5 times, tar substances by 2.7 times, and benzo(a)pyrene by 2.5 times.

The plant is currently undertaking a second stage of modernization, which was scheduled to be completed by 2018. It includes converting the potrooms to the environmentally friendly technology "Clean Soederberg" and improving the efficiency of the SEI calcining systems which are used in anode paste production. A complex environmental program in 2015 led to pollution reduction by 1.3%.

The second phase of the modernization plan is expected to result in a 39% reduction of pollutant emissions [10].

It is also necessary to continue to support measures to transfer motor transport to environmentally friendly fuel, gas, and, in the long run, to electricity. It is also important to develop tram and trolleybus parks in the city, to expand and introduce new routes for electric transport. The idea of construction of subway is of great interest again. It will lead to the reduction of the environmental burden on the city.

The Minister of Ecology and Environmental Management of the Krasnoyarsk Krai, Vladimir Chasovitin, called the beginning of the modernization of the oldest bituminous coal plant in Krasnoyarsk the main environmental event in 2018. He hopes that this year the government will finally approve an environmental recovery plan in Krasnoyarsk, which includes 39 events, each of which will bring real benefits for the residents of Krasnoyarsk [11].

Using the right to a favorable environment and reliable information, relying on Article 42 of the Constitution of the Russian Federation, residents of Krasnoyarsk must demand from the government and the business community to radically change the ecological situation in the city.

References

[1] To breathe is bad, to move is useless, available at: https://greenpeace.ru/blogs/2019/03/06/dyshat-vredno-pereezzhat-bespolezno/
[2] Krasnoyarsk: the million city needs millions for ecology, available at: https://greenologia.ru/eko-problemy/goroda/krasnoyarsk.html
[3] State report on the state and environmental protection in the Krasnoyarsk Territory in 2017, available at: http://www.mpr.krskstate.ru/envir/page5849
[4] Khlebopros R, Dykhno Yu 27.06.2016 "Aluminum Valley": an environmental sentence for Krasnoyarsk, available at: https://salt.zone/radio/2035
[5] Federal law 24.06.1998 N 89-FZ, available at: https://rg.ru/1998/06/30/utilizaciya-dok.html
[6] Tarasov A 22.06.2016 Aluminum desert, available at: https://www.novayagazeta.ru
[7] Sources of air pollution, available at: http://www.admkrsk.ru/citytoday/ecology/Pages/IstochZagrVozd.aspx
[8] The Krasnoyarsk mayor is asked not to delay the creation of a "green shield", available at: https://regnum.ru/news/polit/2598681.html
[9] Khlebopros R 21.05.2008 Yenisey is donated to one owner *Vecherniy Krasnoyarsk* **18** pp 33-4
[10] Krasnoyarsk Aluminum Smelter https://rusal.ru/en/about/41/index.php
[11] The environmental results of 2018 were summed up in Krasnoyarsk, available at: https://krsk.sibnovosti.ru/ekologiya/371294-v-krasnoyarske-podveli-ekologicheskie-itogi-2018-goda

IOP Conf. Series: Materials Science and Engineering **822** (2020) 012014 doi:10.1088/1757-899X/822/1/012014

Influence of antipiren on the properties of small density wood plates

V N Ermolin, A V Namyatov, M A Bayandin and V A Prokopenko

Reshetnev Siberian State University of Science and Technology, 31 Krasnoyarsky Rabochy Av., Krasnoyarsk, 660037, Russia

E-mail: mihailbayandin@yandex.ru

Abstract. In this work a new method of small density wood plates obtaining without adhesives use based on preliminary hydrodynamic processing of wood is offered. It has been established that this method allows to receive the plates with indexes of physicomechanical properties and water resistance surpassing the known analogs made with the use of synthetic adhesives.

1. Introduction

Wood plates are widely applied as constructional and heat-insulating materials. At the same time in most cases synthetic adhesives which in turn predetermine toxicity of plate materials are used for production. Due to severe restrictions on this index, especially in living premises, a large number of studies directed to this question solution are carried out. They are generally aimed at finding new adhesives or modification of the known. In the works [4, 6] it is established that obtaining materials without binding agents with high operational properties use is possible. In most cases it requires wood pre-treatment. Thus in the work [5] it is claimed that at particle sizes of lignotsellulosic raw materials decrease physicomechanical properties of plates without glue use increase. In a number of works it is noted [4, 6] that physicomechanical properties of plate materials significantly depend on fractional composition of initial wood pulp. Our previous studies [2] point to the effectiveness of preliminary hydrodynamic activation of wood particles for obtaining plates of average density without adhesives use. Such processing allows to obtain wood particles with a big specific surface area which depends on the form and the size of wood particles. At increase in an external surface of wood pulp particles more dense and uniform stacking is reached, the particles contact surface increases that causes the increase in the intermolecular interactions number.

The plates of average density (from 650 to 850 kg/m^3) are used in construction and wood processing in a large number. At the same time obtaining of small density plates (from 200 to 350 kg/m^3) is of great interest. Such plates have a low heat and sound conductivity that is very important, for example, at manufacturing of enclosure structures in construction. Such plates obtaining is a serious problem. Due to low density contact area between wood particles in a plate decreases dramatically what complicates its structure development. Thus, to obtain low density plates without binding agents from wood particles with high physicomechanical properties it is necessary for mechanical wood pulp to have a high specific surface area since binding formation (autohesion interaction) between wood particles during plates obtaining depends on it.

The most available raw material for plates manufacturing is the crushed woodworking waste – sawdust. The possibility of obtaining competitive production from this wastage is extremely limited.

Proceeding from this, we conducted the research directed to the studying of obtaining low density wood plates possibility from the sawdust tentatively subjected to hydrodynamic activation.

2. Materials and research techniques.

In the research works Siberian larch (Lárix sibírica) sawdust received on the timber sawing machine was used. Activation of sawdust was made by means of a laboratory hydrodynamic disperser of a rotary–pulsating type. The processing was carried out at the hydromodule 1:10. As a criterion for estimation of mechanical wood pulp processing degree we accepted an index of water-retaining ability – WRV which was determined according to Dzhayma technique [3]. The mechanical wood pulp obtained as a result of hydrodynamic processing was filled in a form. Mechanical squeezing in a cold press was made for mechanical wood pulp dehydration and consolidation. The density of ready plates depends on the value of the pressure. During the research squeezing was carried out with the pressure of 0,5 MPa. The given thickness of plates was 50 mm. After, the plates were placed in a convective drying chamber. The drying was carried out at 105 °C and a circulation time of the drying agent of 2,5 m/s to plates humidity of 4%. At this humidity the plates density was ≈ 250 kg/m^3. Drying duration was 24 hours. After drying the plates were kept in the laboratory conditions, and then the samples for determination of tensile strength at bending, stretching perpendicularly of the face and swelling sizes after soaking in water within 24 hours were made from them.

3. Results and Discussion

As the results of research works showed (figure 1) low density plates properties significantly depend on an initial mechanical wood pulp WRV index. Tensile strength at bending (figure 1a) and at stretching perpendicularly the face (figure 1b) with increase in an WRV index from 100% to 250% respectively increase in 12,5 and 5 times. Sharp increase in plates strength indexes is observed after increase in WRV more than 200%. The obtained results indicate that considering strength characteristics the studied material surpasses the low density plates with synthetic adhesives use, for example Glunz, produced according to the standard [1]

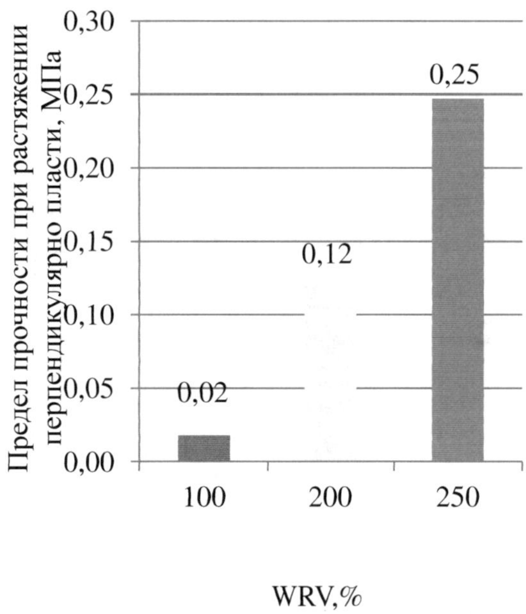

(a)

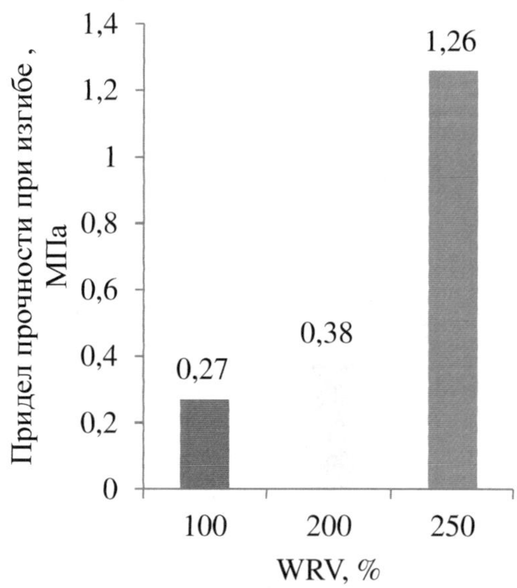

(b)

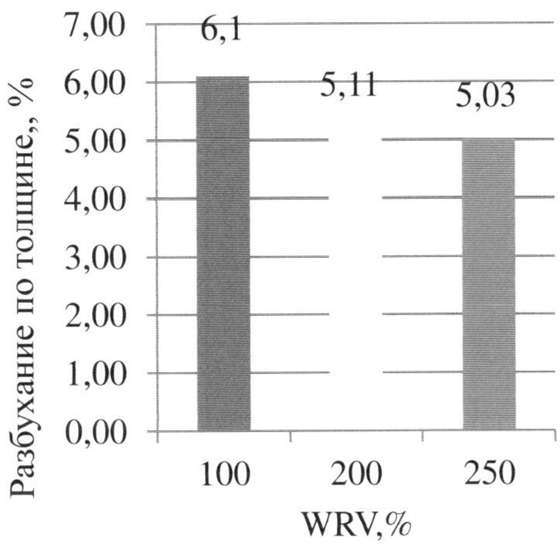

(c)

Figure 1. Physicomechanical properties of small density plates: (a) Tensile strength perpendicularly to face, MPa; (b) Uensile strength at bending, MPa; (c) Swelling on thickness, %.

The size of plates swelling at soaking in water within 24 hours does not significantly depend on mechanical wood pulp processing degree. At the same time this index values are much lower than it is allowed by the standard for similar wood plates [1]. The peculiarity of the obtained plates is that after the wetted plates drying their sizes return to their initial values. And this is observed at multiple repetition of soaking and drying. It is necessary to highlight that after drying the plates indicators of durability are from 80% to 100% of initial values. The obtained results can be explained as follows: at the increase in a mechanical wood pulp processing degree the quantity of particles with a big specific surface area increases. As we noted earlier, it allows to receive the greater number of contacts between particles that causes the increase in values of plates physicomechanical properties. All similar wood plates which are made with synthetic adhesives use completely lose mechanical characteristics even at single humidification and drying. High water resistance of the received plates allows to broaden the sphere of their application considerably. They can be used in premises with high air humidity, in the conditions of contact with an atmospheric precipitation, as fixed timbering and etc.

4. Conclusion

1. Preliminary hydrodynamic processing of sawdust allows to obtain plates of low density without adhesives use.

2. Mechanical characteristics of plates depend on the degree of hydrodynamic wood processing. With increase in WRV index the plates strength at bending and at stretching perpendicularly to face increases. The most significant increase in strength occurs after WRV rise more than 200%.

3. Indexes of low density plates strength obtained without adhesives use do not concede to similar plates produced with adhesives use. The obtained plates have high water resistance that broadens the sphere of their application.

Acknowledgment

The reported study was funded by Russian Foundation for Basic Research, Government of Krasnoyarsk Territory, Krasnoyarsk Region Science and Technology Support Fund to the research project №17-48-240553

References

[1] DIN EN 622-4-2010
[2] Ermolin V N, Bayandin M A and Kazitsin S N 2016 Mechanical Activation of Wood for Adhesive-free board Production *IOP Conf. Series: Materials Science and Engineering* 155 012038 doi:10.1088/1757-899X/155/1/012038
[3] Jayme G 1961 "Zellstoff und Papier" 11 pp 432–438
[4] Kelvin M Chapman 2006 Wood-based panels: particleboard, fibreboards and oriented strand board Primary Wood Processing pp 425-475
[5] Xu J, Widyorini R, Yamauchi H and Kawai S 2004 Development of binderless fiberboard from kenaf core *Journal of wood science* 50:53-61
[6] Kawasaki T, Zhang M and Kawai S 1998 Manufacture and properties of ultra-low-density fiberboard The Japan Wood Research Society 44:354-3

Influence of sweep on the efficiency of sawing

V V Ogurtsov, D N Derevyannykh, E V Kargina and I S Matveyeva

Reshetnev Siberian State University of Science and Technology, 31 Krasnoyarsky Rabochy Av., Krasnoyarsk, 660037, Russia

E-mail: vogurtsov@mail.ru

Abstract. The influence of the sweep of logs on the volume yield of sawn timber and the profitability of its production has been investigated by the method of simulation modeling. It is established that when the average sweep varies from 0 to 0.25%, the volume yield is reduced by 1.2 ... 2.9%, the profitability is reduced by 2.7 ... 5.8%. The contribution of the sweep of logs to the deviation of the real volume yield of sawn timber from its calculated value can reach 70 ... 80%.

1. Introduction

The efficiency of a saw mill is greatly influenced by the group sorting of logs according to their diameters [1, 2]. All subsequent technological processes, namely, cutting saw logs, sorting sawn sections, drying sawn timber, processing sawn timber after drying and, as a consequence, harmoniousness and efficiency of the entire production process of a saw mill depend on it [3, 4] Therefore, the authors [5] defined group sorting of logs according to their thickness as the most important controlled variable for maximizing the efficiency of sawn timber production. The profitability of sawmill production was chosen as the criterion of its optimization [6]. The volume yield of sawn timber is used as a competing optimization criterion and as an indicator of great independent value for evaluating the efficiency of the sawing process. Mathematical models [5] connecting the characteristics of logs and the processes of their cutting with the volume yield and profitability of a saw mill are proposed. Algorithms and programs of simulation studies of the process of sawn timber production with reproduction of real (random) variability of the sizes and shapes of sawn logs, as well as random displacement of logs relative to the center of saws have been developed [6]. Dependences of the volume yield of sawn timber and the profitability of its production on the amplitude of variation of the log thicknesses, without taking into account the variation in the characteristics of the shape of sawn logs, as well as the random displacement of logs relative to the center of the saws, are investigated [7].

The article continues the series of articles written by the same authors and devoted to the investigation of the influence of probabilistic characteristics of the shape of a log and its orientation relative to saw position on the volume yield of sawn timber and the profitability of its production which was not taken into account in previous works. The subject of the study is the sweep of the logs.

2. Results and Discussion

To establish the general patterns of changes in the efficiency of sawmills in relation to the main technological controllable parameters, that is, group sorting of logs according to thickness, the number of simultaneously produced thicknesses of boards, grading of lengths of boards, the real (casual) variability of the sizes and shapes of sawed logs as well as their random displacement relative to the center of the saws are taken into account. To do this, we use the method of simulation modeling with multiple reproductions of a set of different values of the original data on the computer. It is believed that ellipticity, sweep and bias relative to the center of saws obey the normal law, and the thickness of the logs obeys an equiprobable distribution law. In the machine experiment, in each sawing schedule "1000 logs are sawn".

The research is carried out on logs of thickness and length of 22 cm / 5.5 m using the sawing schedule number 4: 25 - 150 - 25; 25 - 50/3 - 25, providing a theoretical maximum volume yield of 60.21%; sawing schedule number 5: 25/2 - 100 - 25/2; 25 - 50/3 - 25 with a theoretical maximum volume yield of 58.51%; sawing schedule number 6: 25/2 - 100 - 25/2; 25/3 - 50 - 25/3 with a theoretical maximum volume yield of 57.04% with a length of boards of 1.5 ... 6.3 m with a gradation of 0.3 m, with a share of raw materials costs RMC = 0.8, with a constant basic cost and with a basic profitability of 25%.

The results of the research of the influence of log sweep on the volume yield of sawn timber and the profitability of its production are presented in tables 1 ... 4 and in figures 1 ... 8.

In figures 1 ... 8, the average arithmetic values of logs sweep are plotted along the abscissa. The mean square deviations of the sweep are 1/3 of the arithmetic mean. Thus, the point "0.10" on the abscissa corresponds to a variation from zero to 0.20%, the point "0.20" - from zero to 0.40%, etc. according to the normal sweep law. Sweep refers to a simple sweep (one curve), which is defined as the percentage of the bilge to the length of the log.

Table 1. Volume yield of sawn timber and profitability of production with variation amplitude (ddmax) of log thicknesses of 2.5 mm.

Sweep, % at ddmax=2,5 MM	Profitability, %			Volumetric yield, %		
	Sawing schedule 4	Sawing schedule 5	Sawing schedule 6	Sawing schedule 4	Sawing schedule 5	Sawing schedule 6
0,00	2,8	0,4	-2	60,138	58,68	57,299
0,05	2,1	-0,4	-2,8	59,7	58,25	56,86
0,10	1,4	-1,1	-4,3	59,27	57,82	55,97
0,15	0,3	-2,2	-5	58,63	57,204	55,579
0,20	-0,7	-2,9	-5,1	58,08	56,76	55,48
0,25	-2,1	-3,4	-5,2	57,234	56,46	55,45

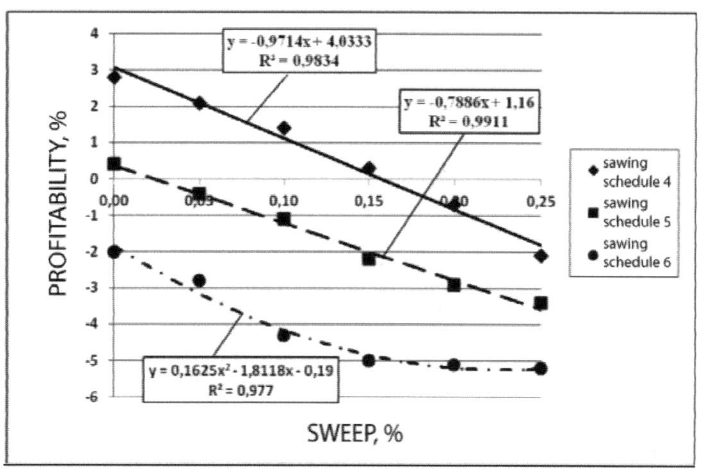

Figure 1. Dependence of profitability on the sweep of logs with variation amplitude of their thicknesses of 2.5 mm

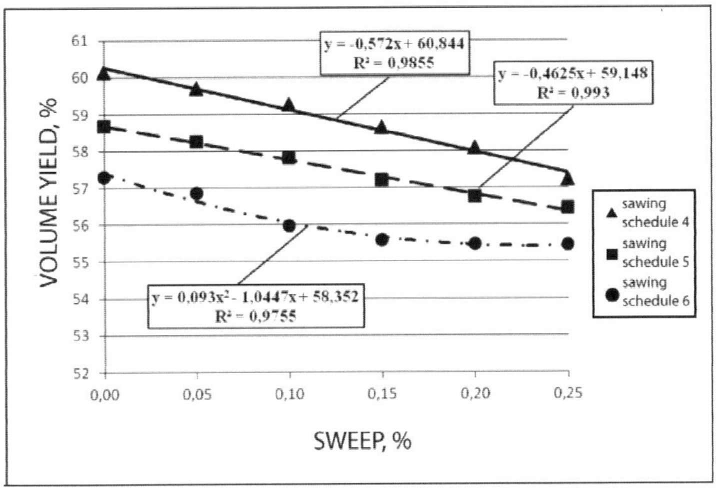

Figure 2. Dependence of the volume yield of sawn timber on the sweep of logs with variation amplitude of their thicknesses of 2.5 mm.

Table 2. Volume yield of sawn timber and profitability of production at variation amplitude of log thicknesses of 10 mm.

Sweep, % at ddmax=10 mm	Profitability, %			Volume yield, %		
	Sawing schedule 4	Sawing schedule 5	Sawing schedule 6	Sawing schedule 4	Sawing schedule 5	Sawing schedule 6
0	22,7	20,8	17	58,94	58,09	56,37
0,05	21,8	19,9	15,8	58,49	57,65	55,75
0,1	20,5	18,7	14,9	57,9	57,09	55,35
0,15	19,4	17,6	14,4	57,37	56,56	55,09
0,2	18,4	17,0	13,9	56,89	56,26	54,86
0,25	17,2	16,5	13,5	56,32	56,05	54,68

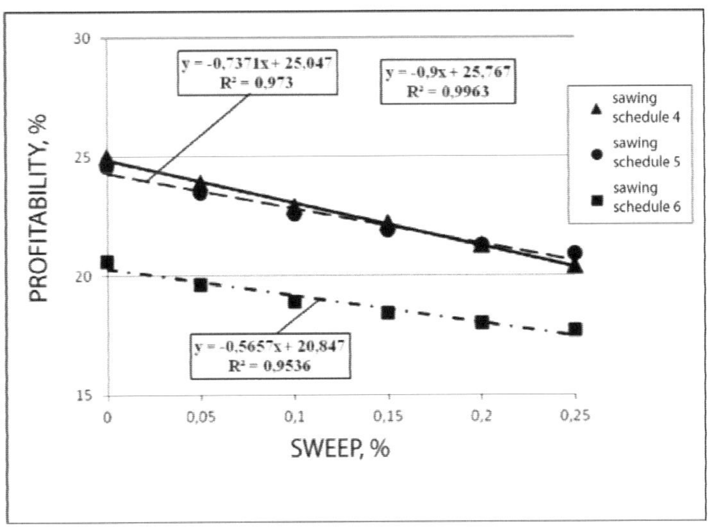

Figure 3. Dependence of profitability on the sweep of logs with variation amplitude of their thicknesses of 20 mm.

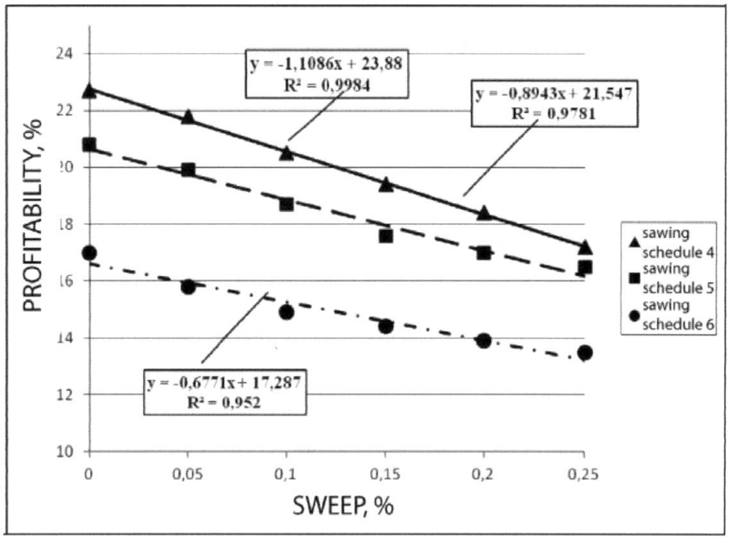

Figure 4. Dependence of profitability on the sweep of logs with variation amplitude of their thicknesses of 10 mm

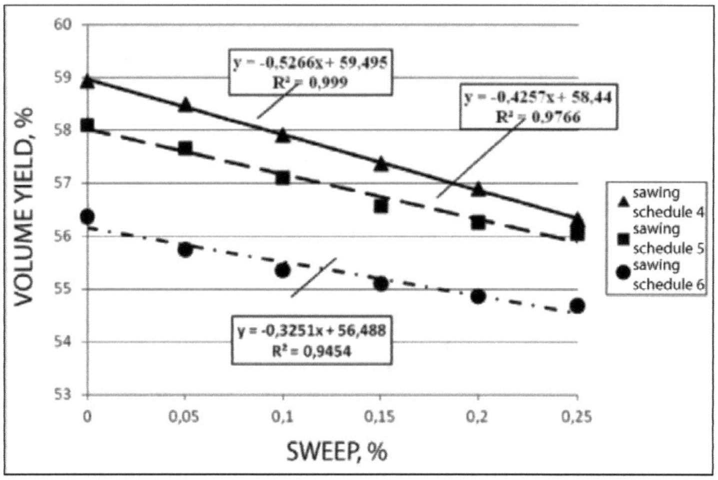

Figure 5. Dependence of the volume yield of sawn timber on the sweep of logs with variation amplitude of their thicknesses of 10 mm.

Table 3. Volume yield of sawn timber and profitability of production with variation amplitude of log thicknesses of 20 mm.

Sweep, % at ddmax=20 mm	Profitability, %			Volume yield, %		
	Sawing schedule 4	Sawing schedule 5	Sawing schedule 6	Sawing schedule 4	Sawing schedule 5	Sawing schedule 6
0	25	24,6	20,6	58,00	57,67	55,98
0,05	23,9	23,5	19,6	57,50	57,19	55,53
0,1	22,9	22,6	18,9	57,04	56,74	55,22
0,15	22,2	21,9	18,4	56,69	56,41	54,98
0,2	21,3	21,3	18,0	56,30	56,17	54,79
0,25	20,4	20,9	17,7	55,88	55,95	54,63

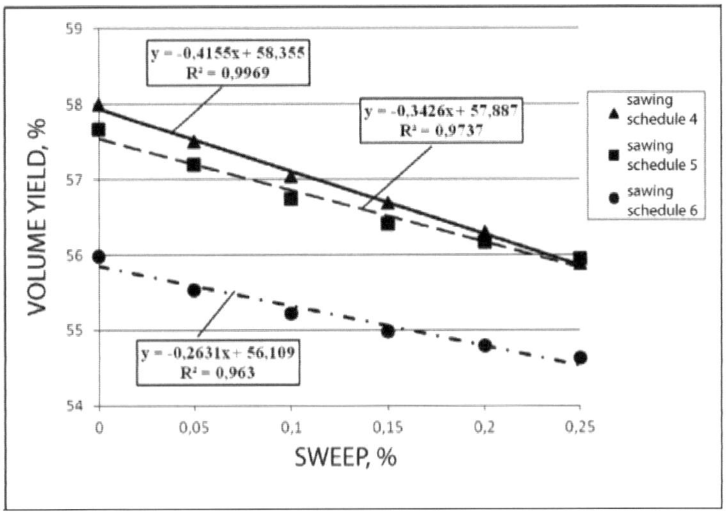

Figure 6. Dependence of the volume yield of sawn timber on the sweep of logs with variation amplitude of their thicknesses of 20 mm.

Table 4. Volume yield of sawn timber and profitability of production at variation amplitude of log thicknesses of 30 mm.

Sweep, % at ddmax=30 mm	Profitability, %			Volume yield, %		
	Sawing schedule 4	Sawing schedule 5	Sawing schedule 6	Sawing schedule 4	Sawing schedule 5	Sawing schedule 6
0	24,3	25,7	21,8	56,89	57,33	55,77
0,05	23,5	24,9	21	56,51	56,95	55,39
0,1	22,7	24,2	20,3	56,14	56,62	55,11
0,15	22,2	23,6	19,9	55,90	56,36	54,90
0,2	21,6	23,2	19,5	55,62	56,18	54,73
0,25	20,4	22,7	19,1	55,09	55,97	54,54

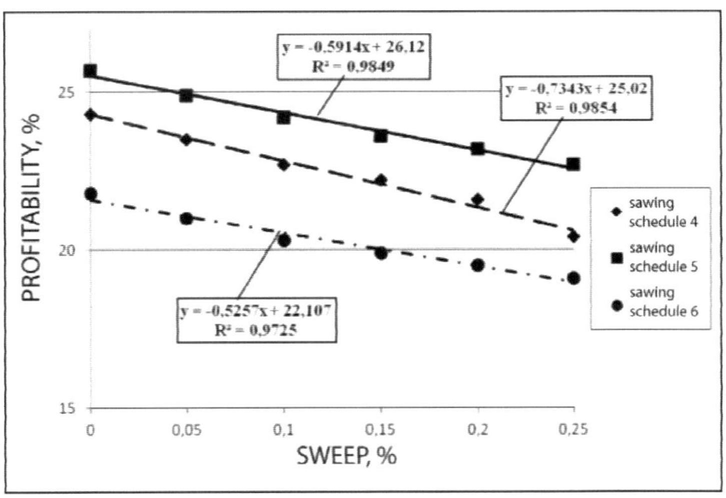

Figure 7. Dependence of profitability on the sweep of logs with variation amplitude of their thicknesses of 30 mm.

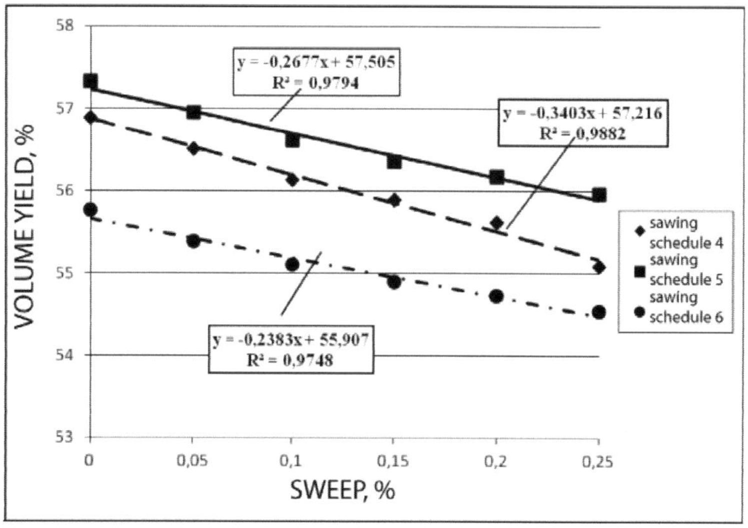

Figure 8. Dependence of the volume yield of sawn timber on the sweep of logs with variation amplitude of their thicknesses of 30 mm

From figures 2 ... 8 it can be seen that the volume yield of boards from logs of 22 cm / 5.5 m decreases with increasing sweep either by a linear or a quadratic law (with sufficiently high determination coefficients: from 0.945 to 0.999). The magnitude of the decrease in the volume yield depends on the amplitude of variation in the thickness of the logs and on the position of saws. When the average sweep value changes from 0 to 0.25% with a range of variation from 0 to 0.50%, the volume yield for № 4 sawing schedule is reduced by 2.9% at variation amplitude of the log thickness of 2.5 mm, for № 5 sawing schedule it is reduced by 2.2%, and for № 6 - by 1.8%. With the amplitude variation of log thicknesses of 30 mm, the decrease in the volume yield is 1.8: 1.4: 1.2%, respectively. It is easy to see that even a small sweep of logs (up to 0.25%) leads to a significant decrease in the volume yield of sawn timber (up to 2.9%). Moreover, the more efficient the sawing schedule and the higher the accuracy of log selection according to thickness are, the more the sweep reduces the volume yield.

It can be seen from figures 1 ... 7 that the profitability, as well as the volume output of boards from logs of 22 cm / 5.5 m, decreases with increasing sweep either by a linear or a quadratic law (with coefficients of determination from 0,952 to 0,999). Moreover, the quadratic dependence is observed with a very accurate group timber sorting according to thickness (up to ± 5 mm) and low quality of the sawing schedule (№ 6). The magnitude of the decline in profitability depends on the amplitude of variation in the thickness of logs in a sawing lot and on the position of saws. When the average sweep value varies from 0 to 0.25% with a range of variation from 0 to 0.50%, the profitability for № 4 sawing schedule decreases by 5.8% if the amplitude variation of the log thickness in the sawing lot is 5 mm; for sawing schedule № 5 it decreases by 4.2%, and for sawing schedule № 6 - by 3.5%. If an amplitude variation of log thicknesses is 30 mm, the profitability decrease is 3.9: 3.0: 2.7%, respectively. It should be noted that even a small sweep of logs (up to 0.25%) leads to a significant decrease in profitability of sawn timber production (up to 5.8%). Moreover, the more effective the sawing schedule and the higher the accuracy of log selection according to thickness are, the more the sweep reduces the profitability of timber production.

Comparing the patterns of decrease in the volume yield of sawn timber and the profitability of its production, caused by the sweep of the logs, we see that they are also similar, as well as due to ellipticity. This is explained by the fact that sweep as well as ellipticity reduces profitability only through a decrease in the volume yield of sawn timber.

3. Conclusion

The study of the influence of the sweep of logs on the volume yield of sawn timber and the profitability of its production makes it possible to draw the following conclusions:

1. Reducing the volume yield of sawn timber because of the sweep of logs measuring 22 cm / 5.5 m should be recognized as very significant. When the average sweep varies from 0 to 0.25% (with a range of variation from 0 to 0.50%), the volume yield is reduced by 1.2 ... 2.9%. If we take into account that the real sweep of the logs varies from 0 to 2%, then we can expect a decrease in volume yield by 5 ... 12%.

2. The expected contribution of the log sweep to the deviation of the real volume yield of sawn timber from its calculated value can reach 70 ... 80% (the contribution of ellipticity is about 10% [5]). The contribution of other causes will be described in the following papers of the authors.

3. Profitability of timber production is highly dependent on the sweep of logs measuring 22 cm / 5.5 m. When the average sweep changes from 0 to 0.25%, the profitability decreases by 2.7 ... 5.8% (ellipticity decreases by 0.8 ... 2, 3% [5]).

4. At the next stage, the influence of sweep on the dependence of the profitability of sawn wood production on the amplitude of variation in log thickness will be investigated, which is used as an objective function in the search for the optimal sorting of log by diameters.

References

[1] Kaliteyevskiy R E 2008 Lesopileniye v XXI veke. Tekhnologiya, oborudovaniye, menedzhment (Sawmill in the 21st century. Technology, equipment, management) the second edition, corrected and supplemented (S-Pb.: "PROFIX") p 496 (In Russ.)

[2] Pizhurin A A 1988 Kompleks matematicheskikh modeley zadachi optimizatsii raskroya khlystov i breven: nauchnyye trudy (A complex of mathematical models for the problem of optimization of cutting logs and full-length logs: scientific works) Center for Scientific Research, Institute of Mechanical Processing of Wood (Arkhangelsk) p 147 (In Russ.)

[3] Sobolev I V 1981 Upravleniye proizvodstvom pilomaterialov (Management of lumber production) (Moscow: Forest industry) p 184 (In Russ.)

[4] Turushev V G 1975 Tekhnologicheskiye osnovy avtomatizirovannogo proizvodstva pilomaterialov (Technological foundations of the automated production of sawn timber) (Moscow: Forest Industry) p 208 (In Russ.)

[5] Kargina E V, Matveeva I S and Ogurtsov V V 2011 Teoreticheskiye osnovy raschota postavov dlya raspilovki broven s porokami formy (Theoretical bases of calculation of the posts for sawing logs with defects of form) *Coniferous boreal zone* V 28 No **1-2** pp 141-145 (In Russ.)

[6] Kargina E V, Ridel L N, Matveyeva I S and Ogurtsov V V 2011 Algoritm imitatsionnykh issledovaniy ekonomicheskoy effektivnosti lesopil'nykh predpriyatiy (Algorithm of simulation studies of economic efficiency of sawmills) *Coniferous boreal zone* V 28 **1-2** pp 146-153 (In Russ.)

[7] Ogurtsov V V, Kargina E V and Matveyeva I S 2013 Zavisimost' ob"yomnogo vykhoda pilomaterialov ot drobnosti sortirovki broven po tolshchine (Dependence of the volume yield of sawn timber on sorting of logs according to thickness) *Coniferous boreal zone* V.31 No 5-6 pp 71-75 (In Russ.)

IOP Conf. Series: Materials Science and Engineering **822** (2020) 012016 doi:10.1088/1757-899X/822/1/012016

Dependence of saw timber volume output and production efficiency on logs ellipticity

V V Ogurtsov, D N Dereviannykh, E V Kargina and I S Matveeva

Reshetnev Siberian State University of Science and Technologies, Krasnoyarsk, 660037, Russian Federation

E-mail: vogurtsov@mail.ru

Abstract. We investigated the effect of the log ellipticity on the volume output of timber and production efficiency by means of simulation. The degree of ellipticity influence on the forecasting of timber production economic indicators accuracy is shown. We have found out that when the average ellipticity from 0 to 6 mm, so the volumetric output is reduced by about 1 %, and profitability decreases by 0.8...2.3 %.

1. Introduction

The authors have analyzed and explained in their papers [1-6] the possibility of improving the economic efficiency of sawmills by optimizing the fractional sorting of logs in thickness. We took saw timber volume output and production efficiency as competing criteria of optimality. Mathematical models that connect the characteristics of logs and the processes of their cutting with the saw timber volume output and the profitability of a sawmill have been developed. Algorithms and programs for imitation studies of the production process for saw timber have been developed. Dependencies of profitability and volume output of saw timber on the fractional sorting of sawlogs in thickness are determined without taking into account the random variation of the curvature and log ellipticity, and also without taking into account their spontaneous displacement relative to the saw mill.

This article is the first one in a cycle of articles concerning the influence of probabilistic characteristics of a log shape and log's orientation on the saw timber volume output and the production efficiency that were not taken into account in previous works.

2. Results and Discussion

In order to understand what affects the efficiency of the sawmill enterprises, we need to take into account the main technologically controlled parameters; this includes the fractional sorting of logs by thickness, the number of simultaneously produced thicknesses of boards and gradation of board lengths. The real (random) variability of sizes and shapes of sawed logs and their random displacement relative to the center of the saw mill depend on those parameters. For this simulation we used method with multiple playback on the computer with different values in the source data. It is supposed that the ellipticity, curvature and displacement of the saw mill from the center follow usual rules, and the log thickness follows the equally possible distributive law. In the experiment every saw mill has 1000 sawn logs.

Research is being conducted on the logs of a thickness and length at 22 cm / 5.5 m, saw mill used #4: 25 – 150 – 25; 25 – 50/3 – 25, providing the theoretical maximum volumetric output 60.21 %;

saw mill #5: 25/2-100-25/2; 50/3-25-25 with a theoretical maximum volume output 58.51%; saw mill #6: 25/2 – 100 – 25/2; 25/3 – 50 – 25/3 the theoretical maximum of volumetric output is 57.04 % when the length of the boards is 1.5...6.3 m with a step of 0.3 m, the proportion of commodity costs equals 0.8, while the permanent base cost and the underlying profitability of 25%.

The results of the study of the log ellipticity influence on the saw timber volume output and the production efficiency are presented in tables 1 to 3 and figures 1 to 6.

In the figures, the x-axis deposit the average value of log ellipticity. Average quadratic deviation of ellipticity is equal to 1/3 of the average values. Thus, the point "2" on the x-axis corresponds to the variation of the ellipticity according to the normal law of zero to 4 mm, and the point "4" corresponds from zero to 8 mm, etc. The ellipticity refers to the difference between the maximum and average diameter or polprasert between the maximum and minimum diameter of the apical end of a log

Table 1. Saw timber volume output and production efficiency in amplitude (ddmax) of varying log thicknesses in 2.5 mm.

Ellipticity, mm at ddmax=2.5 mm	Profitability, %			Volume output, %		
	Saw mill 4	Saw mill 5	Saw mill 6	Saw mill 4	Saw mill 5	Saw mill 6
0	2.8	0.4	-2	60.14	58.68	57.30
2	2.4	-0.1	-2.5	59.91	58.41	57.01
4	2	-0.6	-3	59.62	58.12	56.72
6	1.4	-1.1	-3.8	59.32	57.82	56.29
8	0.8	-1.7	-4.8	58.97	57.46	55.71
10	0	-2.5	-6	58.48	56.99	55.01

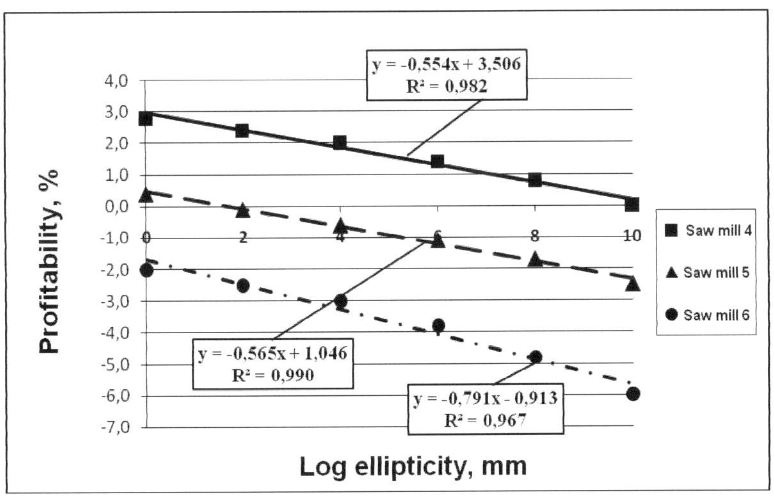

Figure 1. Dependence of the profitability on the log ellipticity in the amplitude of thicknesses variation of 2.5 mm.

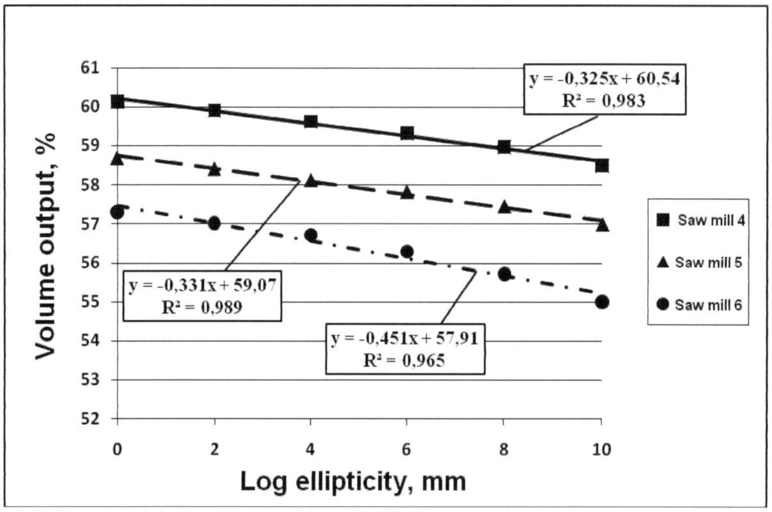

Figure 2. Dependence of the saw timber volume output from the logs ellipticity in the amplitude of thicknesses variation of 2.5 mm.

Table 2. Saw timber volume output and production efficiency in amplitude (ddmax) of varying log thicknesses in 2.5 mm.

Ellipticity, mm at ddmax=2.5 mm	Profitability, %			Volume output, %		
	Saw mill 4	Saw mill 5	Saw mill 6	Saw mill 4	Saw mill 5	Saw mill 6
0	22.7	20.8	17	58.94	58.09	56.37
2	22.1	20.1	16.3	58.67	57.76	56.03
4	21.3	19.4	15.5	58.35	57.40	55.64
6	20.5	18.5	14.7	57.99	56.96	55.22
8	19.7	17.4	13.6	57.57	56.45	54.69
10	18.7	16.1	12.4	57.09	55.84	54.14

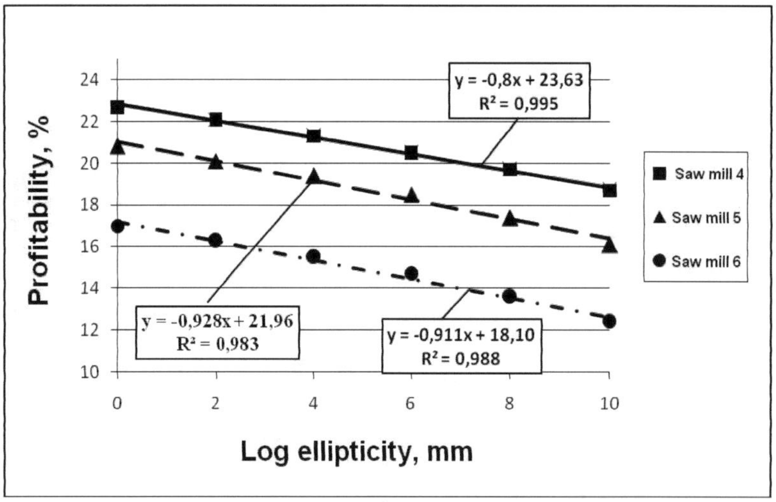

Figure 3. Dependence of the profitability on the log ellipticity in the amplitude of thicknesses variation of 10 mm

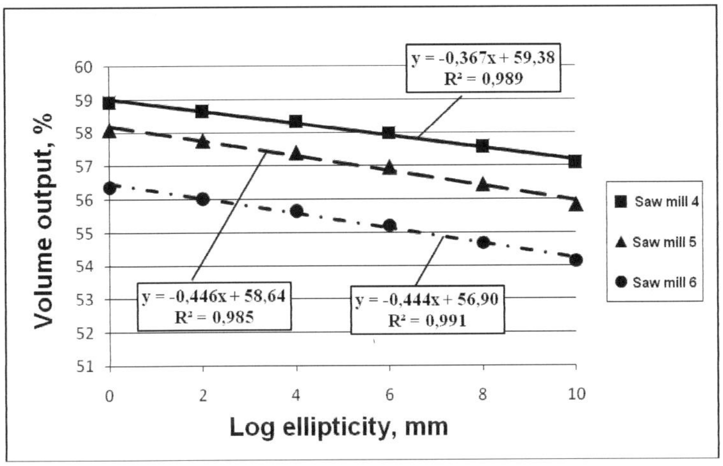

Figure 4. Dependence of the saw timber volume output on the ellipticity of logs in the amplitude of thicknesses variation of 10 mm

Table 3. Saw timber volume output and production efficiency in amplitude (ddmax) of varying log thicknesses in 30 mm.

Ellipticity, in mm at ddmax=30 mm	Profitability, %			Volume output, %		
	Saw mill 4	Saw mill 5	Saw mill 6	Saw mill 4	Saw mill 5	Saw mill 6
0	24.3	25.7	21.8	56.89	57.33	55.77
2	23.7	25.0	21.1	56.59	57.02	55.47
4	23	24.2	20.5	56.28	56.66	55.16
6	23.5	23.6	19.7	56.12	56.26	54.83
8	22.7	22.9	19	55.74	56.04	54.50
10	21.9	22.2	18.2	55.37	55.72	54.14

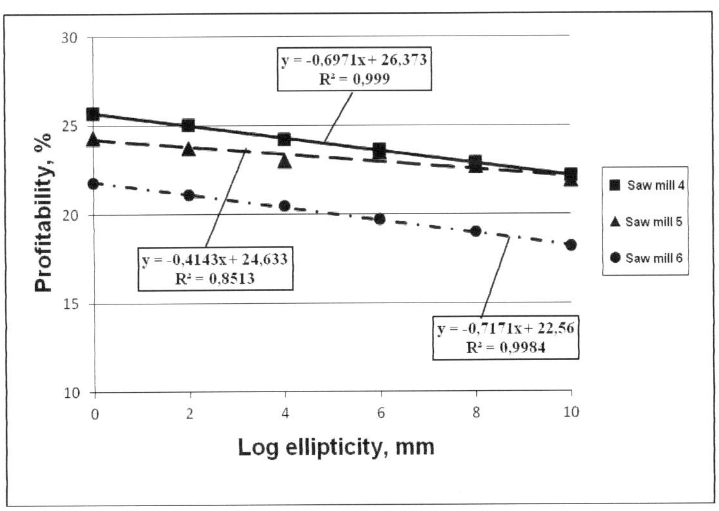

Figure 5. Dependence of the profitability on the log ellipticity in the amplitude of thicknesses variation of 30 mm.

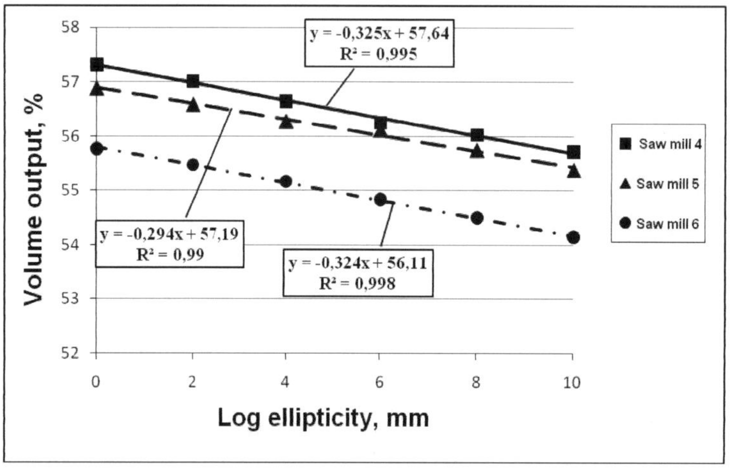

Figure 6. Dependence of the saw timber volume output on the ellipticity of logs in the amplitude of thicknesses variation of 30 mm.

Figures 2, 4 and 6 show that saw timber volume output from logs of 22 cm / 5.5 m decreases with increasing linear ellipticity (with sufficiently high determination coefficients: from 0.965 to 0.999). The reduction in volumetric output depends on the log thickness variation amplitude of in the sawing lot and saw mill. If you change the average value of ellipticity from 0 to 10 mm with a variation from 0 to 20 mm volumetric output is reduced by 1.5 to 2.3 percent. When the average ellipticity changes from 0 to 6 mm, most characteristic of logs with thickness of 22 cm, the volumetric output is reduced by 0.8...1.2 %. Such a reduction should be mentioned as significant.

Figures 1, 3 and 5 show that production efficiency and saw timber volume output from logs of 22 cm / 5.5 m decreases with increasing linear ellipticity (with determination coefficients from 0,915 to 0,999). Reducing volume of profitability approximately equally depends on the amplitude varying log thicknesses and saw mill. If you change the average value of ellipticity from 0 to 10 mm with a variation from 0 to 20 mm volumetric output is reduced by 2.4 to 4.7 %. When the average ellipticity changes from 0 to 6 mm then profitability is reduced by 0.8...2.3 %. This decline in profitability is quite significant and therefore a factor of log ellipticity must be considered when forecasting the economic indicators of a sawmill.

Comparing the pattern of reduction in saw timber volume output and production efficiency due to the log ellipticity, we see that they are very similar. The reason is that ellipticity reduces the profitability only through the reduction in saw timber volume output.

3. Conclusion

Investigation of the effect of log ellipticity on the saw timber volume output and the production efficiency allow to make the following conclusions:

1. The decrease in saw timber volume output from the log ellipticity with dimensions of 22 cm / 5.5 m should be recognized significant. When the average ellipticity changes from 0 to 6 mm then profitability is reduced by 1 %.
2. The influence of the log ellipticity on the deviation of the saw timber volume output from the calculation of its value is about 10 %. The contribution of other causes will be described in the following papers of the authors.
3. The production efficiency of sawn timber is highly dependent on the log ellipticity with dimensions of 22 cm / 5.5 m. When the average ellipticity changes from 0 to 6 mm then profitability is reduced by 0.8...2.3 %.
4. In the next step, the influence of ellipticity on the dependence of the production efficiency of sawn timber from the amplitude variation of the thickness of the logs, which is used as the objective function when searching for optimal external sorting logs by diameter.

References

[1] Kaliteyevskiy R E 2008 Lesopileniye v XXI veke. Tekhnologiya, oborudovaniye, menedzhment [Lumbering in the XXI century. Technology, equipment, management. Second edition, corrected and augmented] (Saint Petersburg: "PROFIX" 2008) **2** p 496

[2] Pizhurin A A 1988 Kompleks matematicheskikh modeley zadachi optimizatsii raskroya khlystov i breven: na-uchnyye trudy [Complex mathematical models of optimization of cutting whips and logs: scientific papers] (Arkhangelsk: Centre of scientific research Institute of mechanical wood processing) p 147

[3] Sobolev I V 1981 Upravleniye proizvodstvom pilomaterialov [Managing the production of lumber] (Moscow: The forest industry) p 184

[4] Turushev V G 1975 Tekhnologicheskiye osnovy avtomatizirovannogo proizvodstva pilomaterialov [The technological basis for automated production of timber] (Moscow: The forest industry) p 208

[5] Kargina E V, Matveyeva I S and Ogurtsov V V 2011 Teoreticheskiye osnovy raschëta postavov dlya raspilovki brëven s porokami formy [Theoretical basis of calculation for saw mill of logs with defects of the form] *Conifers of the boreal zone* **28** No1-2 pp 141-145

[6] Kargina E V, Riedel L N, Matveeva I S and Ogurtsov V V 2011 Algoritm imitatsionnykh issledovaniy ekonomicheskoy effektivnosti lesopil'nykh predpriyatiy [Algorithm simulation research of economic efficiency of sawmills] *Conifers of the boreal zone* **28** No 1-2 pp 146-153

IOP Conf. Series: Materials Science and Engineering **822** (2020) 012017 doi:10.1088/1757-899X/822/1/012017

On some group properties of the functional presentation of information

Anatolii M Kukartsev[1], Alexander A Kuznetsov[1], and Anna I Kuklina[1]

[1]Reshetnev Siberian State University of Science and Technology
31, Krasnoyarsky Rabochy Av., Krasnoyarsk, 660037, Russian Federation

E-mail: amkukarcev@yandex.ru

Abstract. The paper provides an overview of research results on processing the information represented as Boolean functions. Actions of the hyperoctahedral group (or the Jevons group) are set on these functions. Information objects transformed by such actions have special frequency (entropy) characteristics. Such actions can be used to generate information messages with equal frequency and / or entropy characteristics. Preservation of the frequency characteristics of informational messages also allows us to solve one of the problems of classifying Boolean functions with respect to the Jevons group action. This problem has existed since the Harvard catalog was created and it has not been solved until the results of the studies covered in the paper were applied. Additionally, the possibilities of optimizing the calculation of actions tens and hundreds of times due to the properties of the elements of the Jevons group and modern processors are highlighted. This optimization allows the construction of information processing algorithms using the Jevons transformations for real data (commensurate with 1 MiB).

1. Introduction

Digital information is represented mainly in two ways: a combinational one and a functional one. The first way considers information as a combination of symbols of some alphabet, and the second one – as a set of values of a certain discrete function. Combinational methods of presenting information and the corresponding algorithms for its processing began to develop in the early twentieth century. These ways of presenting are usually associated with the works of K. Shannon [1], R.V. Hamming [2] and many other researchers. Functional methods of information presentation began to appear at the end of the 20th century in connection with the development of information technologies and natural need for qualitatively new methods of information processing. A series of JPEG [3] graphic data compression algorithms is an example of such an information processing method. Digital information is mapped to some function of two arguments. Further, the function is transformed and its parameters (Fourier coefficients) are rounded, coded, and their codes are compressed using combination methods.

Boolean functions (hereinafter – BF) [4] are naturally suitable for the functional representation of digital binary information. It is possible to injectively associate any binary vector (hereinafter – BV) consisting of zeros and / or units with a BV of length that is a multiple of two. The latter can be mutually associated with some Boolean function. For this, it is important to unambiguously order the domain of the definition of BF. This can be done by fixing the order of its arguments and by arranging, for example, lexicographically, its definition area accordingly. The BF itself can be described by the functional elements implementing it or by the formula [5].

The property of reducing the volume of information when using algorithms for graphic compression of the JPEG series is not the only advantage of functional presentation methods. Some transformations of information presented in the form of BF have special entropic properties, and also allow solving a number of complex mathematical problems related to the classification of BF with respect to the action of algebraic groups.

Two special groups can be distinguished among the various groups of actions on the set of BF: the set of negations and the set of permutations of the BF arguments, forming the Jevons group [6] or the hyperectahedral group [7]. The specified action is remarkable because it is neutral, that is, it does not affect the connections between functional elements of BF and does not change its formula. The Jevons-equivalent BF will have the same CNF and DNF [5].

Studying the functional presentation of information allows us to develop qualitatively new information processing algorithms. This paper aims to review the works in researching the properties of transformations associated with functional presentation of information. It covers the following tasks:

1. constructive presentation of the Jevons group for the development of information processing algorithms;
2. definition of actions of the Jevons group on sets of BF and BV as well as the analysis of their properties;
3. the use of functional information representation for solving the problem of BF classification regarding the action of Jevons group elements.

2. Definitions and notation

Hereinafter let n be a nonnegative integer, $k = 2^n$, i, j are nonnegative integer indices (counters) not exceeding k.

We define $E = \{0, 1\}$ as a set consisting of two elements. We define the Cartesian product of this set by itself as $E^n = \underbrace{E \times \cdots \times E}_{n}$. BV will be an element of such a set. For its coordinates we will use the numeric L2R notation (left to right), i.e. larger coordinates are to the left. We denote an arbitrary element of the set as $x \in E^n, x = \{x_{n-1}, \ldots, x_i, \ldots, x_0\}$. For BF argument vectors we will also use the L2R notation.

The Boolean function of n arguments $f(x_{n-1}, x_{n-2}, \ldots, x_1, x_0)$ implements the mapping $E^n \to E$ [4]:

$$\begin{array}{c} y_f = \{f(11\ldots11), f(11\ldots10), \ldots, f(00\ldots01), f(00\ldots00)\} \\ (y_f)_j = f(x_{n-1}, x_{n-2}, \ldots, x_1, x_0), j = \sum_{i=n-1}^{0} x_i \cdot 2^i \end{array} \quad (1)$$

Let BF be given through the truth table [4]. If we unambiguously determine the order of the arguments, then each BF can be associated with a binary vector y_f (1) of length k (column of values).

We denote the variable inversion group (or the linear shift group [4]) by $E_n = (E^n, \oplus)$. Variable permutation group (the symmetric group) is denoted by S_n. The Jevons group is denoted as $D_n = E_n \leftthreetimes S_n$ [7]. Due to the specifics of the targets, we agree that the symmetric group acts on the set of numbers $[0; n-1]$.

The set of Boolean functions of n arguments will be denoted as $B(n)$.

3. Constructive representation of the Jevons group for actions on sets of BV and BF

When developing information processing algorithms using the actions of the Jevons group on BV and BF there appear two problems:

1) non-unique specification of the action of the symmetric group on the BV set;

2) the specification of the homomorphism of the external semidirect product of the Jevons group.

In [8] and [9] these problems are addressed and their solutions are suggested.

It is important to distinguish between the group operation in the symmetric group and its action on BV. It is empirically shown that for tasks of the Jevons group action on BF it is convenient to define the homomorphism of the external semi direct product of the Jevons group from the symmetric group to the group of automorphisms of the variables inversion group through the very action of the element on BF. Such an action may not be specified in the only way. It is empirically shown that two sets of actions can be distinguished: type A and type B. Both types of actions are expressed through each other. When the action of a Jevons group element takes place on BV, type A is preferable [8], but when it takes place on BF then type B is preferable [8]. It is therefore impossible to single out one type of action as the main one, and the second as reducible to it. At the same time the developers can choose the type of action at their discretion depending on the problem to be solved.

The operation of A type of an element of the group $\pi \in S_n$ on BV $x \in E^n$ will be called the calculation of the result $x' \in E^n$: $x' = x^\pi$ according to the following rule: "old" indexes are in the top substitution line, and "new" indexes are at the bottom one, or in a symbolic form:

$$x'_{\pi(i)} = x_i. \tag{1.A}$$

The operation of B type of an element of the group $\pi \in S_n$ on $x \in E^n$ will be called the calculation of the result $x' \in E^n$: $x' = x^\pi$ according to the following rule: "new" indexes are in the top substitution line, and "old" indexes are at the bottom one, or in a symbolic form:

$$x'_i = x_{\pi(i)}. \tag{1.B}$$

The second problem is the very way of setting the Jevons group. It is that the external semidirect product of groups can be set in many ways and as a result can give different (non-isomorphic, for example, direct product) groups [4]. In [8] it is proposed to isomorphically embed the Jevons group into a substitution subgroup $\beta_n < S_k$ acting on the set of rows of the truth table of BF. Such a group is decomposed into an inner semidirect product, whose homomorphism is set in a natural way by conjugations. From where we get two representations of Jevons group for $z_0, z_1 \in E_n$ and $\pi_0, \pi_1 \in S_n$:

$$(z_0\pi_0)(z_1\pi_1) = \left(z_0 z_1^{\pi_0^{-1}} \; \pi_0\pi_1\right) \tag{2.A}$$

$$(z_0\pi_0)(z_1\pi_1) = \left(z_0 z_1^{\pi_0^{-1}} \; \pi_1\pi_0\right) \tag{2.B}$$

All necessary evidence and examples are given in [8] and [9].

4. Frequency properties of the action of the Jevons group on BF

The frequency characteristics of BV specifically depend on the actions of the Jevons group elements on the corresponding Boolean functions. These properties can form the basis of the message generation methods possessing the same frequency and spectral distributions in all their allowable alphabets simultaneously. We present the necessary series of definitions from [10].

Definition 1. The alphabet A_i is a set constructed as a Cartesian product E^{2^i}.

Definition 2. An alphabet symbol is an element of the alphabet A_i encountered in the vector y_f.

BV y_f can be divided into characters in any of the alphabets A_i, where i runs through all the values of $[0; n]$. In this case, the division starts from the first value of the vector y_f and the characters do not overlap. The length of the BV y_f (the number of characters) in each of the alphabets is calculated as $l_i = 2^{n-i}$.

Definition 3. Symbol frequency is the number of times that a given symbol has been encountered from the A_i alphabet in the y_f vector.

Definition 4. The frequency distribution of the BV y_f (spectrum) over the alphabet A_i is the ratio $Q_i(y_f) \subset A_i \times [0;k]$, i.e. a set of symbol – frequency pairs.

Definition 5. The spectral distribution of the BV y_f over the alphabet A_i is the ratio $R_i(y_f) \subset [0;k] \times [0;k]$ (the derived set from $Q_i(y_f)$), the elements of which show how often the frequencies in the vector y_f are repeated. First the frequency from $Q_i(y_f)$ and then the number of various characters with this frequency in the vector y_f is indicated in the relationship element.

We define $c_{n-1}, \ldots, c_0 \in E^n$ – the generating set of the group E_n, and each c_i has the value 1 at the position of i, and the value 0 on the other positions.

Theorem 1. (On the invariance of frequency spectra under the action of E_n) *If the element c_i acts on the BF $f \in B(n)$, then the frequency distributions of the BV y_f are invariant with respect to this action for the alphabets $A_{i'}: i' \leqslant i, i \in [0; n-1]$, and the spectral distributions of the BV y_f are invariant with respect to the same action for the alphabets A_i.*

Theorem 2. (On the invariance of frequency spectra under the action of S_n) *If the transposition $(i,j) \in S_n, i,j \in [0;n-1] : i < j$ acts on $f \in B(n)$, the frequency distributions of BV y_f are invariant with respect to this action for alphabets $A_{i'}: i' \leqslant i$, and the spectral distributions of BV y_f are invariant with respect to the same action for alphabets $A_{i'}$ and $A_{j'}: j' > j$.*

Entropy is a function that depends only on frequencies, and not on frequency distribution symbols. The value of frequencies and their number are determined by the spectral distribution, so the following consequences are true.

Corollary 1. The entropy of BV y_f is invariant in all alphabets $A_i, i \in [0; n-1]$ with respect to the action of any element E_n over BF f.

Corollary 2. The entropy of BV y_f is invariant for alphabets of indices up to i inclusive and greater than j with respect to the action of the transposition $(i,j), i,j \in [0;n-1] : i < j$ of group S_n over BF f.

In [10] all the necessary descriptions and proofs of the theorems are given and specific decompositions are introduced: a monotone decomposition of an element of a substitution group and a canonical decomposition of an element of the Jevons group.

5. Solution of the problem of BF classifying with respect to the Jevons group action

Let$f, g \in B(n)$. We denote the negation of the arguments by $z \in E_n$ and the permutation of the arguments by $\pi \in S_n$. Both the negation and / or permutation of the arguments are denoted by $(z\pi) \in D_n$, then the basic equation of the action of a Jevons group element on a BF with respect to the unknown $(z\pi)$ is denoted as:

$$f^{(z\pi)} = g \tag{3}$$

Equation (3) describes the problem of classifying the BF with respect to the Jevons group action. The specified task is set in [11]. It admits a trivial solution with complexity $O(2^n \cdot n!)$, but such a solution is already inefficient for $n > 4$. The solution of this problem is closely related to the notion of an invariant of the group acting on the set BF [4]. S. Golomb proposed the full invariant [12], but the complexity of its calculation as well as the trivial solution is exponential. E.A. Yakubaytis also proposed the invariant [13], but it is not complete. As a result the formulated problem is difficult to solve. Due to lack of solutions, besides the trivial transformation of information by negation and / or permutations of BF arguments, it is used as a cryptographic primitive for encrypting information by algorithms based on controlled operations in which an element of the Jevons group is the encryption key [14].

Consideration of the specified in (3) BFs as information objects allows us to construct an algorithm for solving equation (2) in polynomial time $O\left(n^2+n\right)$. The correctness of the algorithm is proved by the following theorem, described in [15].

Theorem 3. **(On calculating the zero action)** *The set H_n coincides with the set of all solutions of the equation $f^{(z\pi)} = g$ and can be calculated as $H'_{i+1} = \left[\bigcup_{j_i=i}^{j_i<n} (e_E(i,j_i))\, H_i\right] \bigcup \left[\bigcup_{j_i=i}^{j_i<n} (c_i(i,j_i))\, H_i\right]$ in n steps in series for $0\leqslant i<n$ and $H_{i+1} = \left\{h \in H'_{i+1} \mid Q_{i+1}\left(f^h\right) = Q_{i+1}(g)\right\}$, starting with $H_0 = \{(e_E e_S)\}$.*

Modern PCs are mainly focused on processing computer words, rather than individual bits. Jevons transformations are remarkable in that their action can be localized in a machine word. In fact, this means speeding up the computational process equal to the number of bits in a machine word. So some extensions of Intel processors have 512 bit registers. This allows increasing optimization hundreds of times. In [16] such methods for optimizing calculations with the necessary evidence are described.

References

[1] Shannon K 1963 *Works on information theory and cybernetics: Trans. from English ed. R. L. Dobrushushin and O. B. Lupanova. From prev. A. N. Kolmogorov* (Moscow: Publishing house of foreign literature) – 829 p. (In Russ)

[2] Hamming R V 1983 *Coding Theory and Information Theory: Trans. from English* (Moscow: Radio and communication) – 176 p. (In Russ)

[3] Salomon D 2004 *Compression of data, images and sound. Trans. from English V.V. Chepyzhov* (Moscow: Technosphere) – 368 p. (In Russ)

[4] Logachev O A, Salnikov A A, Yaschenko V V 2004 *Boolean functions in coding theory and cryptology* (Moscow: ICMNO) – 470 p. (In Russ)

[5] Yablonsky S V 1986 *Introduction to discrete mathematics: textbook for universities. – 2 ed., correct. and add* (Moscow: Science. Ch. ed. Phys.-Mat. lit.) – 384 p. (In Russ)

[6] Jevons S 1881 *Fundamentals of Science* (Saint-Petersburg)

[7] Young A A 1930 Quantitative substitutional analysis *Proc. London Math. Soc.* **31** 273–388

[8] Kukartsev A M, Kuznetsov A A 2015 Constructive Representation of the Jevons Group for Engineering Solutions of Information Processing *Software Engineering* **11** (Moscow: New Technology) 25–33 (In Russ)

[9] Kukartsev A M, Kuznetsov A A 2016 On the actions of the Jevons group on sets of binary vectors and Boolean functions for engineering information processing solutions *Software Engineering* **7** (Moscow: New Technology) 29–36 (In Russ)

[10] Kukartsev A M 2016 On the frequency properties of actions of the Jevons group on Boolean functions *Software Engineering* **7** (Moscow: New Technology) 515–521 (In Russ)

[11] Glukhov M M, Remizov A. B., Shaposhnikov V. A. 1988 *Review on the theory of k-valued functions. Part 1. Reference book. Ed. N.P. Yemelyanov. Order number 163f* (Moscow: Typography of military unit 33965, 125040, Moscow A–40) – 153 p. (In Russ)

[12] Golomb S W 1959 On classification of Boolean functions *IRE, Trans.circuit theory. Spec.Suppl.* **6** 176–186

[13] Yaubaytis E A 1974 Subclasses and classes of Boolean functions *Automation and computing* **1** (Riga: Zinatne) 1–8 (In Russ)

[14] Shniperov A N 2008 Synthesis and analysis of high-speed symmetric cryptosystems based on managed operations *Information Technology* **137** (Moscow: Information Technology) 36–41 (In Russ)

[15] Kukartsev A M, Kuznetsov A A 2017 On the effective algorithm for solving the equation of the Jevons group action on Boolean functions *Software Engineering* **8** (Moscow: New Technology) 76–87 (In Russ)

[16] Kukartsev A M, Kuznetsov A A 2017 On the Equimorphic Computation of the Jevons Group Element Action over Boolean Functions, *Software Engineering* **8** (Moscow: New Technology) 300–309 (In Russ)

IOP Conf. Series: Materials Science and Engineering **822** (2020) 012018 doi:10.1088/1757-899X/822/1/012018

Impact of brush fire and pinetum horizontal structure on the tree condition

A A Vais, S L Shevelev and V N Nemich

Reshetnev Siberian State University of Science and Technology, 31 Krasnoyarsky Rabochy Av., Krasnoyarsk, 660037, Russia

E-mail: shevelev@sibsau.ru

Abstract. The research focuses on pinetum in the South of Central Siberia. Studying was performed in the territory of low-mountain suburban forests. Sample plots were set up in the sites where wood fires of different strength happened; the pinetum plantations withstanding to burning were in the observational focus. It was found that post-fire condition of trees depends on the tree size and plant population. The fire damage to trees (scorch height) depends on a tree size and the distance from the nearest exemplar.

1. Introduction

Researchers highlight the environmental importance of the brush fire to form light-colored coniferous forest. However, the impact of the space structure (tree exposure, plant population) on the brush fire intensity needs to research.

2. Results and Discussion

Studying the flatland pinetum in National park "Shushenskiy bor", O.P.Kalenskaya [1] determines pleurocarpous moss wood type plantation after the severe fire demonstrates the dependence of scorch height on stem diameter and average distance among trees. Scorch height increases if stem diameter and distance among trees increase. Plant health considerably depends on an average distance in independent fire of moderate and high intensity comparing to fires of low and moderate intensity.

The research focuses on pinetum in the South of Central Siberia. The observations were performed in low-mountain suburban forest. The sample plots were provided on the sites where the fires of different intensity happened; the plantations withstanding to fire were in the control group. The autumn post-bush fire forest stand demonstrated post-fire plant litter in the crowded plantation.

We should mention all trees on the sample plots were mapped (coordinate areas). Every tree breast height diameter was measured, plant health was determined, scorch height and damages (fire scars, resinosis, deseases) are described. The plantation in the sample plots were subjected to the autumn bush fire. The fire intensity increased and ranged from moderate to high when the cutslope ratio and its altitude elevation increased. The fire intensity in the specified point was determined by scorch height. During the spring fires, group location of young regeneration provided their survival due to later snow pack melting, and consequently, less fuel drying near the groups. During autumn fires a considerable plant litter percentage of grouping trees was mentioned. The phenomenon can be explained by complete plant litter dry and greater amount of dried plant litter under the tree groups and young regeneration clusters. In addition, as a rule, we registered the increase in the debris amount of big-sized trees having bigger canopy. Maximum quantity of fuel results in increasing the burning power intensity and fire intensity, therefore, more damage. The observations highlight the impact of the factors on trees exposure to fire and tree size category [2, 3]. However, the observation needs mathematical proof.

Correlation analysis and variance analysis were performed to estimate complex and selective impact of space structure of forest stand and tree diameter on the fire intensity and tree condition. According to the plant health, the trees were grouped into three categories: vigorous trees, problematic trees, and dried up trees.

Index of competitiveness is a special coefficient to consider the 'social' factor (the plant population and exemplar size).

A tree location was determined by average distance within the social group (the central tree and its neighborhood)

The correlation analysis was performed to evaluate the closeness relationship among the factors above. Table 1 contains the results.

Table 1. Interrelation matrix among different indexes of pine tree exemplars.

Characters	Average distance, м	Competitiveness Index	Scorch height	Stem diameter	Plant health category
Average distance	1,0	-	-	-	-
Competitiveness Index	-0,028	1,0	-	-	-
Scorch height	0,018	-0,274	1,0	-	-
Stem diameter	0,332	-0,627	0,288	1,0	-
Plant health category	0,202	0,503	-0,026	-0,027	1,0

Remarkable interrelation was between plant health category and stem diameter (-0,827); plant health and index of competitiveness (0,503). These data confirms the impact of trees size and plant population on their post- fire condition. Index of competitiveness (-0,274) and stem diameter (0,288) demonstrate a low impact on selective fire intensity.

The regression analysis results in multivariate linear equation:

$$К= 2,773 - 0,068*d_{1.3} - 0,033*L_{ср}, \quad (1)$$

К – plant health category (1-2-3); $d_{1.3}$ – stem diameter, sm; $L_{ср}$ – average distance in the social group, m.

The equation obtains several features: adaptation factor – 0,628; capital error – 0,7 (within a category); Fisher's variance ratio – 119; they indicate an equation dependability.

We use the two-way analysis of variance to estimate a regressor effect degree on tree condition and selective fire intensity.

Fisher ratio test determines two constraints as indubitable: $К = f\ (CI, d_{1.3})$ и $H_н = f\ (L, d_{1.3})$, they correspond to $F_ф = 4,1 > F_{ст} = 1,6$ и $F_ф = 2,4 > F_{ст} = 1,5$ respectively.

66,4% means the impact percentage of competitiveness index and stem diameter to plant health category. 38,6% is scorch height under the impact of average distance and stem diameter.

Post fire trees condition dramatically depends on exemplar size and plant population. The tree fire damage (scorch height) depends on its size and the distance to the 'neighboring' exemplar. It could be explained by the considerable amount of plant litter of large-sized trees and groups of trees. Low plant population occurs due to unaccounted factors; the main is tree dependence on nano- and micro-relief elements [4, 5, 6].

References

[1] Kalenskaya O P 2006 Brush fire influence on the pinetum condition in flatland of National park "Shushenskiy bor" [Vliyanie nizovykh pozharov na sostoyanie sosnovykh nasazhdeniy v ravninnoy chasti natsional'nog parka «Shushenskiy bor»] cand. dis. (Krasnoyarsk) p 18

[2] Buryak L V 1999 Role of brush fire to form light coniferous forest in the South of Central Siberia [Rol' nizovykh pozharov v formirovanii svetlokhvoynykh nasazhdeniy yuga Sredney Sibiri] doc. diss. (Krasnoyarsk) p 26
[3] SHirnin Yu, A Zverev and I V Morzhanov A Yu 2012 Spatial distribution of trees in fire affected forest areas [Prostranstvennoe razmeshchenie derevev na porazhennyh pozharom lesnyh territoriyah] (Lesnoj vestnik) No 6 pp 176-180
[4] Buryak L V, Vays V V and Rossinin A A 2001 The impact of the brush fire intensity and spacious structure of forest crop on plant health in population [Vliyanie sily nizovogo pozhara i prostranstvennoy struktury drevostoya na sostoyanie derev'ev v nasazhdeniya] *Forest and chemical complexes - problems and solutions* (Krasnoyarsk, SibGTU) pp 39-42
[5] Ganina. N.V. Distribution of trees by diameter using the Weibull function [Raspredeleniye derevyev po diametru s pomoshchyu funktsii Veybulla] (Lesovedeniye) No 2 1984 pp 65-70
[6] Karmanova I V, Sudnitsyna T N and Ilina N A 1987 The spatial structure of complex pine forests: a monograph [Prostranstvennaya struktura slozhnykh sosnyakov: monografiya] (M.: Nauka) p 199

IOP Conf. Series: Materials Science and Engineering **822** (2020) 012019 doi:10.1088/1757-899X/822/1/012019

Simulation of forest ecosystems dynamic processes message

S V Ushanov, S L Shevelev, N N Kulakova, A A Vais and V N Nemitch

Reshetnev Siberian State University of Science and Technology, 31 Krasnoyarsky Rabochy Av., Krasnoyarsk, 660037, Russia

E-mail: taksator@sibstu.Ris.ru

Abstract. Dynamic characteristics of changes in the diameter of larch forests in the Lower Angara region are being analyzed. Larch stands of the herbal group of fox types is the object of the present research. This research is based on the data of field inventory 858 forest areas with larch predominance. The structure of the forest average diameter is approximated by the exponential function, statistical analysis of which has been carried out using a non-linear least-squares method and simulation methods by conducting 50000 statistical tests. Regularities of error distribution, parameters and calculation accuracy of approximating function have been defined. Estimations of parameter correlations approximating function have been suggested. Holetsky's method helped to receive two dimensional parameter generator. The defined regularities of the dynamics of average diameters allow to design economic activities in forest stands of larch herbal groups types in the region under study.

1. Introduction

Dynamics of forest stands is one of the main factors responsible for taking actions aimed at forestry development. Features of growth and formation of forest stand is determined by several factors, the main one is their growth conditions. Classification of stands productivity is valid within the territory with the same climate.

Productivity assessment of vegetation conditions for the forest stands is the basis for studying their dynamic patterns. In the domestic forestry for these purposes, the bonitet scale (usually M. M. Orlov's scale) and typological classifications (V. N. Sukachev, P. S. Pogrebnyak, etc.) are employed. Methods of dynamic simulation of plant communities have been considered in the works of J. K. Vanclay, P. J. Sands [1], A. Porte, H. H. Bartelink [2], Ch. Peng, X. Wen [3], V. V. Kuzmichev [4] etc. V.V. Kuzmichev [4] gives the following classification of methods to describe the dynamics of forest communities:

- simulation development model;
- eco-physiological simulation and application of stochastic models;
- tables of forest growth progress.

A distinctive feature of simulation is consideration of dynamics of separate elements of a complex biological system with a certain level of confidence without explaining the mechanism of their functioning.

Eco-physiological simulation pursues the achievement of a relatively complete understanding of trends in the dynamics of the object under study, by describing the mechanisms of its elements' functioning. At present eco-physiological models do not account for "all or even a large part of influencing the process of growth (forest) individual factors and their interactions" [4], therefore they cannot be regarded as a model which adequately captures the dynamic processes of forest stands and trees.

The tables of growth which present taxation regulations, despite repeatedly expressed doubts concerning their impartiality and adequacy of processes of forest vegetation dynamics, still remain the only effective tool in domestic forestry practices which allows the use of regularities of dynamic processes in forest stands for the purposes of practical forestry.

One of the major shortcomings of growth tables is application of simplified methods of curves alignment, characterizing the dynamics of inventory performance, without logical analysis of obtained results; another one is weakness of the used mathematical tools.

Regularities of the dynamics of inventory indices of larch stands in Central Siberia were investigated by B. N. Tikhomirov and A. I. Tishenkov [5], E. N. Falaleev and V. S. Polyakov [6], I. I. Krasikov and S. L. Shevelev [7], N. N. Kulakova, and S. L. Shevelev [8]. In their work N. N. Kulakov and S. L. Shevelev [8] note the relevance of growth tables in the study of quite an extensive herbal group of larch forests. The purpose of this study was to develop the method for establishing the patterns of inventory indices of forest stands on the example of average diameters, using mathematical tools of data processing and computer modeling in the construction of forest growth tables.

2. Materials and methods

The object of the study was stands of Siberian larch on the territory of Nizhneangarsk taiga area [9] where the tree stands with a predominance of this breed is about 24% of the total forested area.

Based on inventory materials, selection of 858 taxation plots applying typological approach was made. Forest stands are characterized by productivity, corresponding to bonitet class III. According to N. I. Tretyakov and I. V. Semechkin [10] recommendations, a series reflecting dynamics of average diameters of forest stands has been obtained; their variability (W, %) and accuracy of the test (P, %) (table 1).

Table 1. Statistical characteristics of dynamic series with average diameter.

age, number of years	10	30	50	70	90	110	130	150	170	190	210	230	250	270
diameter, cm	1.7	8.7	16.0	19.5	23.1	25.9	27.8	30.2	35.3	38.6	41.0	41.6	40.7	46.0
W, %	72.1	45.2	18.6	23.8	11.3	10.0	8.0	7.6	13.4	16.4	12.8	13.7	8.8	7.3
P, %	13.9	11.0	3.6	3.1	1.3	0.9	0.8	0.9	1.8	2.9	2.4	2.5	1.3	2.3

For mathematical description of the deterministic component or temporary trend of forest growth series different models have been used [11]. In the 30s of XX century L. von Bertalanffy (Von Bertalanffy, 1938) [12] formulated a scientific hypothesis, according to which the increase (growth rate) by the volume of the body is expressed by the difference between anabolic (synthesis of substances) and catabolic (decomposition of matter) growth rates. Anabolic growth rate is proportional to the surface area of the body, while catabolic - to the size of the biomass.

A preliminary analysis of the experimental data is presented in table 1. It allowed to describe age-related changes in the average diameter larch stands employing the Robertson-Ostwald (Robertson, Ostwald, 1908) [13, 14] model, namely its particular case - Bertalanffy model:

$$D(t) = Dmax \times (1 - \exp(-t/T)) , \tag{1}$$

where D (t) is the average diameter of larch stands, cm; t – the age of forest stands, years; Dmax – the maximum rating value of diameter, cm; T – time constant characterizing the rate of diameter change, years. Under t = T diameter equals 63%, and under t = 3T – 95% of the maximum value Dmax.

3. Results and Discussion

While differentiating (1) we evaluate the annual rate of average diameter change:

$$V_D(t) = Vmax \times \exp(-t/T)), \tag{2}$$

where $V_D(t)$ is the estimate of the average annual larch stands diameter increase cm/year; Vmax = Dmax/T is the estimate of the initial (maximum) velocity, cm/year.

Switching to generalized variables Do(t) = D(t)/Dmax, τ = t/t in expression (1) we obtain:

$$Do(\tau) = 1 - \exp(-\tau), \tag{3}$$

Evaluation of the rate of generalized variables Vo(τ) = VD(τ)/Vmax is:

$$Vo(\tau) = \exp(-\tau). \tag{4}$$

From (4) it follows that under t = T, the annual diameter growth rate reduces to 37%; under t = 3T – to 5% from initial (maximum) value.

Evaluation of equation coefficients (1) was performed with a nonlinear least-squares method [16], their standard errors and 95% confidence intervals of changes were defined in the simulation modeling by implementation of 50,000 statistical tests [16,17,18]. Statistical characteristics of model (1) are presented in table 2.

Table 2. Statistical characteristics of model (1).

Index	Symbol	Value
Freedom degrees number	f	12
Least-squares estimate of time constant, cm	Dmax	57.3 ± 3.9
Least-squares estimate of time constant, years	T	180 ± 21
Least-squares estimate of maximum velocity, cm/year	Vmax	0.32 ± 0.02
Standard error of the model, cm	S	1.5 ± 0.3
95% confidence variation rate, Dmax, cm		[51.2 66.5]
95% confidence variation rate T, years		[146 229]
95%-confidence variation rate S, cm		[0.9 2.1]
Determination coefficient of the model	R^2	0.988

Figure1 presents the empirical and theoretical (normal) distribution function of errors of model (1). The normality of error distribution was checked according to the Frozini consent criteria [19] for cases of complex hypotheses testing, where parameters of distribution are estimated based on empirical data. Statistics of consent criteria and critical values under 5% level of significance were evaluated based on 50000 statistical tests [17].

The analysis showed that the hypothesis of error distribution normality is not rejected at 5% significance level as under Frozini criteria and ω^2 (figure 2).

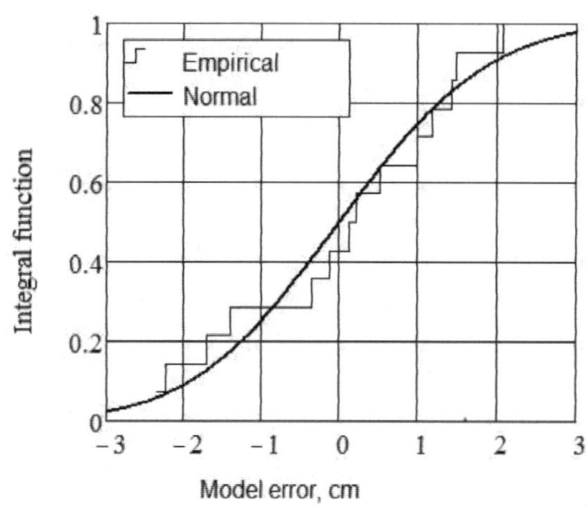

Figure 1. Cumulative error distribution function of the model (1).

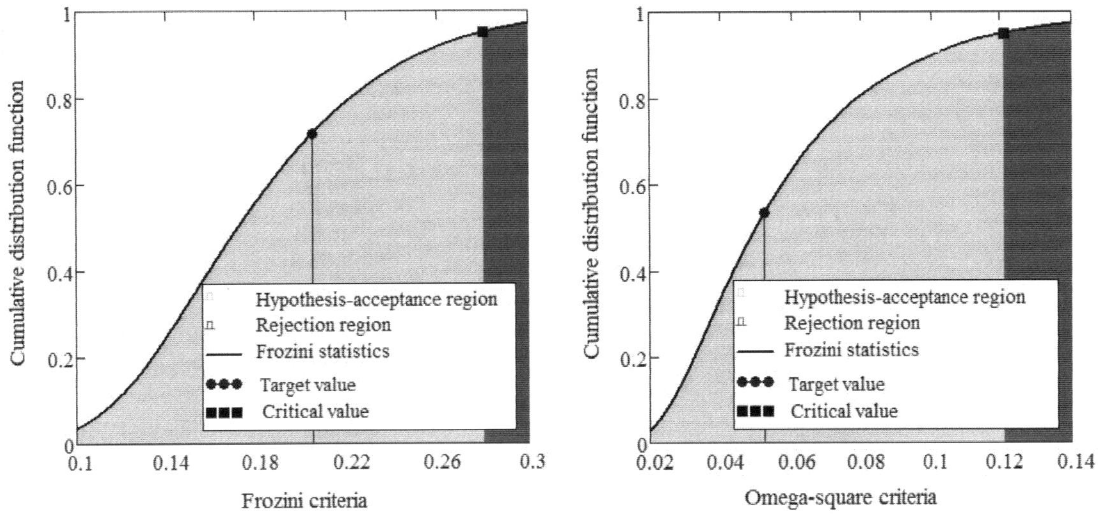

Figure 2. hypothesis test of error distribution normality under Frozini criteria and ω^2.

Since the hypothesis of error distribution normality is not rejected, estimates of model parameters distributions can be obtained by the simulation method conducting the following computer experiments: for each age group (table 1) random variables (errors) are superimposed on the results of diameter calculation according to the model (1) distributed under the law N(0, S^2). To obtain the values module parameters estimates are determined (1) [16].

Figure 3 - 6 present the results of distribution parameters functions estimation Dmax, T, Vmax and standard error of the model S based on the results of 50000 statistical tests.

The obtained results characterize statistical distribution of relevant variables and allow simulation modeling of larch stands' diameters.

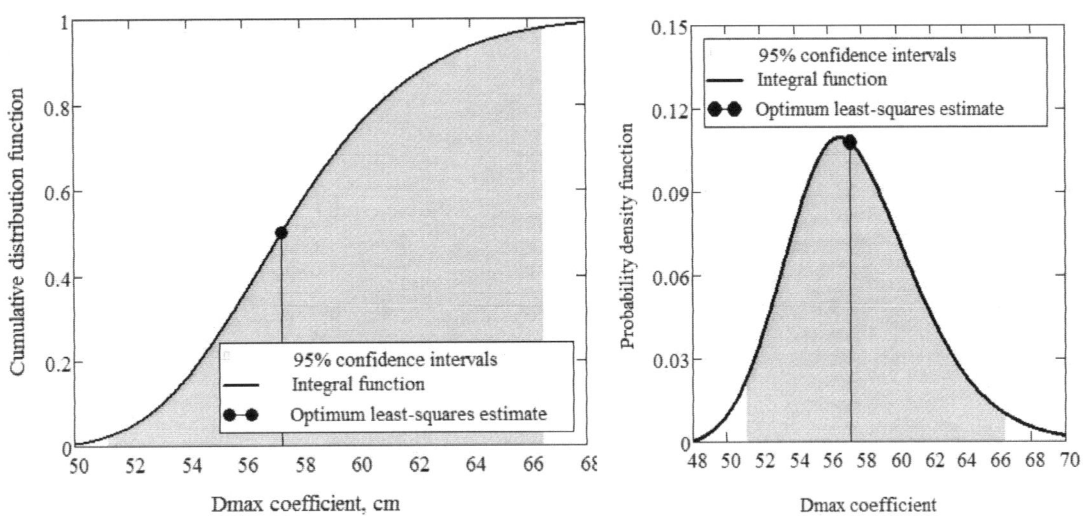

Figure 3 Estimates of coefficient distribution function Dmax with statistical tests method.

There is a close correlation between values of the maximum diameter Dmax and the time constant T (correlation coefficient RDT = 0.981). The correlation cloud of 95% confidence interval of these parameters variation is shown in figure 7a.

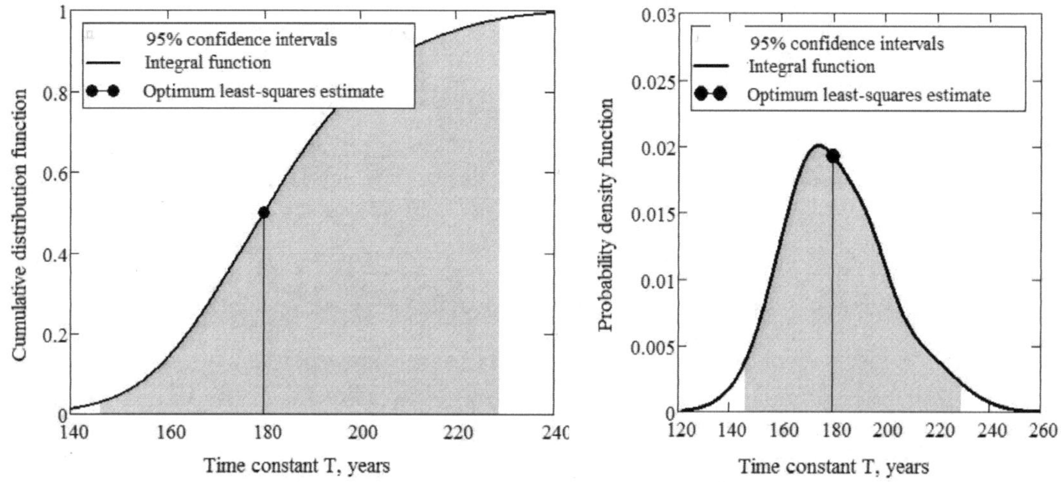

Figure 4. Estimates of coefficient distribution function T with statistical tests method.

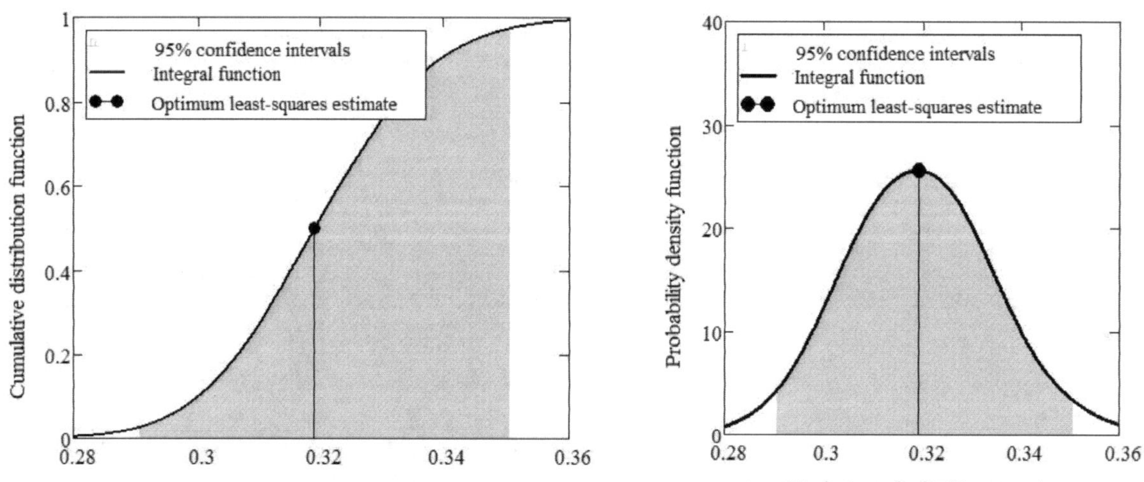

Figure 5. Estimates of coefficient distribution function Vmax with statistical tests method.

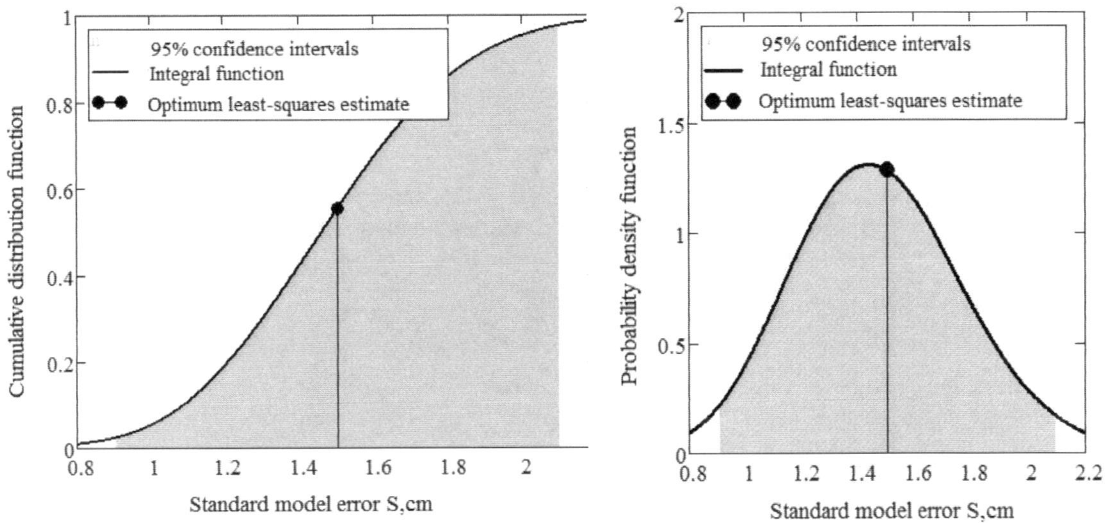

Figure 6. Estimates of the modal standard error distribution function S with statistical tests method.

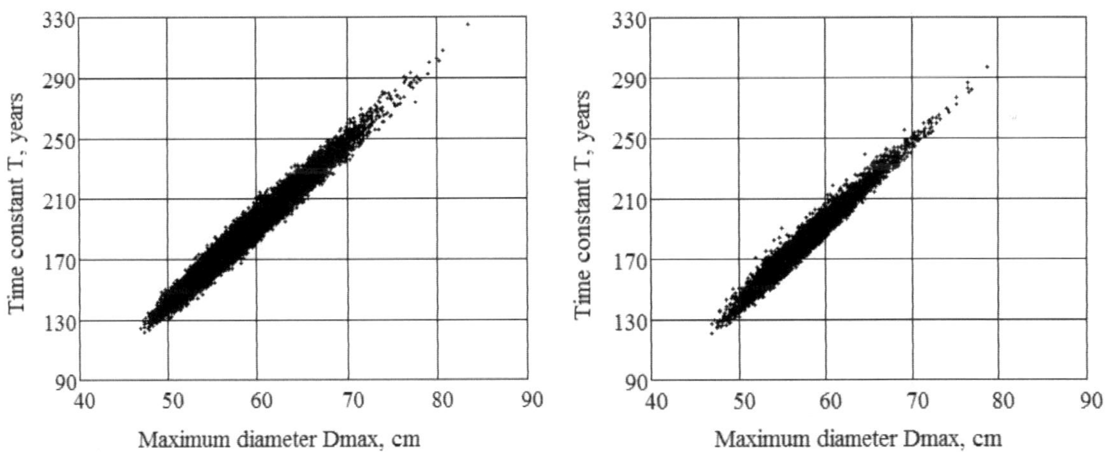

a) *b)*

Figure 7. Evaluation of 95%-confidence interval of coefficients variations Dmax and T: (a) method of statistical tests, (b) random values generator (5) – (7).

Applying the method of multidimensional random variables formation based on the Cholesky transformation generator of pseudo-random values to determine the Dmax and T has been obtained:

$$\begin{pmatrix} D_{\text{ГЕН}} \\ T_{\text{ГЕН}} \end{pmatrix} = \begin{pmatrix} M_{Dm} + S_{Dm} \times D_{\Gamma} \\ M_{T} \times S_{T} \times T_{\Gamma} \end{pmatrix}, \tag{5}$$

$$\begin{pmatrix} D_{\Gamma} \\ T_{\Gamma} \end{pmatrix} = \begin{pmatrix} D_{\varepsilon} \\ R_{DT} \times D_{\varepsilon} + \sqrt{1 - R_{DT}^{2}} \times T_{\varepsilon} \end{pmatrix}, \tag{6}$$

$$\begin{pmatrix} D_\varepsilon \\ T_\varepsilon \end{pmatrix} = \begin{pmatrix} (\varepsilon_{Dm} - M_{Dm}) / S_{Dm} \\ (\varepsilon_T - M_T) / S_T \end{pmatrix}, \tag{7}$$

where Dgen, Tgen – generated values of the maximum diameter and time constant; Dr, Tr – generated normalized values of the maximum diameter and time constant; Dε, Tε – normalized values of maximum diameter and a time constant corresponding to εDm, t; εDmax t – randomly selected values from the array of coefficients of T and Dmax obtained by the method of statistical tests (figure 3-4); M_{Dm} = 57.3, M_T = 180 – least-squares estimate of the maximum diameter and time constant (table. 2); S_{Dm} = 3.9, ST = 21 – standard estimates errors of the maximum diameter and time constant (table. 2); R_{DT} = 0.981.

Figure 7 b presents the results of 10000 pseudo-random values formation obtained with generator (5) to (7).

Experimental calculated values and 95% confidence variation interval of average larch stands diameters of the forest-type herbal group depending on the age of the stand is shown in figure 8. The cumulative distribution function of the average larch stands herbal groups of forest types diameters depending on the age of forest stands is shown in figure 9.

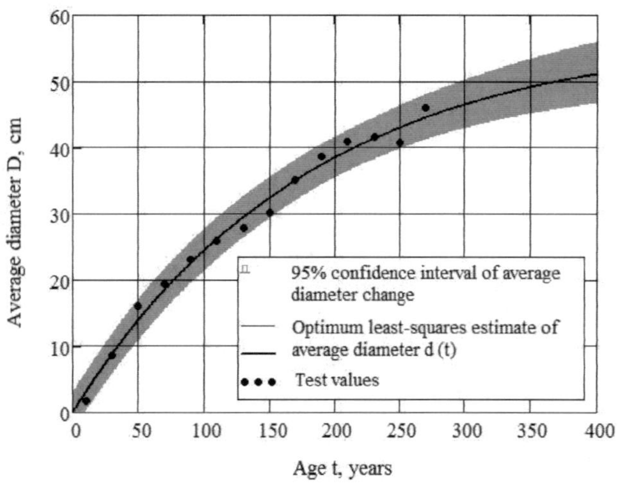

Figure 8. Dynamics of the average larch stands of the forest-type herbal group diameters

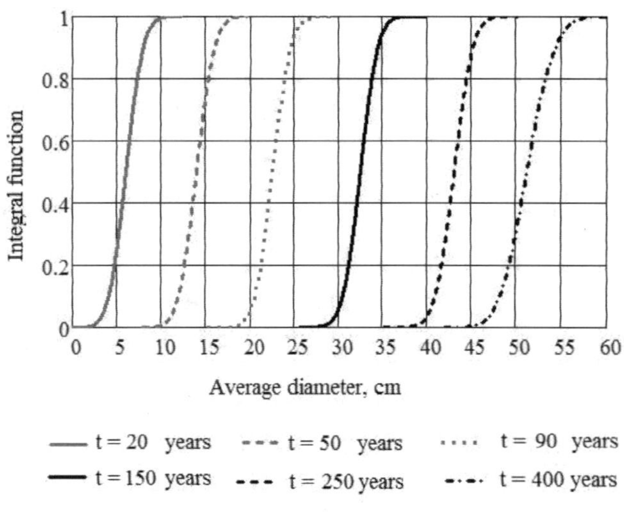

Figure 9. Integral function of the average larch stands of the forest-type herbal group

Conclusion

The proposed scheme estimates the dynamics of the average inventory indices, allowing to estimate the most probable values of the Robertson -Ostwald model coefficients, statistics of variations in these ratios, change statistics of projected inventory indices.

The use of the proposed scheme will improve the methods of constructing tables of growth and designing forest management activities.

References

[1] Tishenkov A N and Tihomirov B N 1929 Hod rosta sibirskoj listvennicy po issledovaniyam v Hakasskom okruge Sibirskogo kraya [The course of growth of Siberian larch according to research in the Khakass District of the Siberian Territory] *Trudy po lesnomu opytnomu delu* (Omsk: Sibirskij institut sel'skogo hozyajstva i lesovodstva) **2** pp 43

[2] Falaleev Eh N and Polyakov V S 1975 Hod rosta listvennichnyh drevostoev srednego i nizhnego techeniya r. Angary [The course of growth of larch stands of the middle and lower reaches of the r. Hangars] (Krasnoyarsk: Sibirskij tekhnologicheskij institut) **2** pp 13-17

[3] Krasikov I I and Shevelev S L 2010 Dinamika taksacionnyh pokazatelej modal'nyh listvennichnikov v respublike Tyva [Dynamics of taxation indicators of modal larch forests in the Republic of Tyva] *Hvojnye boreal'nye zony* (SibGTU, Krasnoyarsk) **27** No 3-4 pp 253-256

[4] Kulakova N N and Shevelev S L 2017 Dinamika taksacionnyh pokazatelej modal'nyh drevostoev listvennicy sibirskoj v Nizhnem Priangar'e [Dynamics of taxation indicators of modal tree stands of Siberian larch in the Lower Angara region] *Hvojnye boreal'nye zony.* (SibGTU, Krasnoyarsk) **35** No 1-2 pp 41-45

[5] Korotkov I A 1994 Forest vegetation zoning of Russia and the republics of the former USSR [Lesorastitel'noe rajonirovanie Rossii i respublik byvshego SSSR] Uglerod v ehkosisteme lesov i bolot Rossii (Krasnoyarsk) p 170

[6] Semechkin I V 1962 Opyt ispol'zovaniya dannyh glazomernoj taksacii dlya izucheniya dinamiki nasazhdenij [Opyt ispol'zovaniya dannyh glazomernoj taksacii dlya izucheniya dinamiki nasazhdenij] Trudy Instituta lesa i drevesinovedeniya (Krasnoyarsk) 8 pp 119-131

[7] Ushanov S V 2003 Primenenie mnogomernyh statisticheskih metodov pri prinyatii reshenij [Primenenie mnogomernyh statisticheskih metodov pri prinyatii reshenij] (Krasnoyarsk)

[8] Ushanov S V 2012 Parametricheskaya identifikaciya modelej [Parametric identification model] (Krasnoyarsk: Litera-Print,) p 199

[9] Ushanov S V 2015 Parametric identification model in MathCad [Parametricheskaya identifikaciya modelej v MathCad] *Vychislitel'nye ehksperimenty* (mnogomernye normal'no-raspredelennye velichiny i linejnye modeli) (Krasnoyarsk: Litera-Print) p 87

[10] Lemeshko B Yu, Lemeshko S B, Postovalov S N and Chimitova E V 2011 Statistical data analysis, modeling and investigation of probability laws [Statisticheskij analiz dannyh, modelirovanie i issledovanie veroyatnostnyh zakonomernostej] (Novosibirsk: NGTU, 2011) p 888

[11] Ushanov S V and Ogurcov D A 2018. Evaluation by the method of statistical tests of the statistics of the omega-square test for checking the normality of the distribution [Ocenka metodom statisticheskih ispytanij statistiki kriteriya «omega-kvadrat» proverki normal'nosti raspredeleniya] XIII Mezhdunarodnaya nauchno-prakticheskaya konf. «Sovremennye tekhnologii: aktual'nye voprosy, dostizheniya i innovacii» (Penza : MCNS «Nauka i Prosveshchenie») **1** pp 94-96

[12] Atroshchenko O A 2004 Simulation of forest growth and business processes [Modelirovanie rosta lesa i hozyajstvennyh processov] (Minsk: BGTU) p 216

[13] Bertalanffy L. 1938 von. A quantitative theory of organic growth: (Inquiries on growth laws. 2) [A quantitative theory of organic growth: (Inquiries on growth laws. 2)] *Human Biol.* Vol. 10 N 2 pp 181-213

[14] Kull' K and Kull' O 1989 Dynamic tree growth modeling [Dinamicheskoe modelirovanie rosta derev'ev] (Tallin: Valgus) pp 232

[15] Ostwald W 1908 *Uber die zeitlichen Eigenschaften der Entwickelungs-vorgänge /Vorträge und Aufsätze Über Entwickelungsmechanik der Organismen.* (Leipzig) pp 71
[16] Robertson T B 1908 On the normal rate of growth of an individual and its biochemical significance *Archiv für Entwickelungsmechanic der Organismen* 25(4) pp 581-614
[17] Vanclay J K and Sands P J 2009 Calibrating the self – thinning frontier *Forest Ecol. And Management* Vol 259 pp 870-872
[18] Porte A, Bartelink H H 2002 Modeling mixed forest growth: Areview of models for forest management *Ecol. Modeling* Vol. 150 pp 141-188
[19] Peng Ch and Wen X 2006 Forest simulation model *Computer applications in sustainable management: Including perspectives on collaboration and integration* Springer pp 101-125

Structural diagram of mass transfer in larch wood during convective drying

Sh G Zaripov[1], A P Chizhov[2] and Zh N Klevitova[3]

[1,2] Lesosibirsk Branch of Reshetnev Siberian State University of Science and Technologies, 29 Pobeda Str., 662543, Lesosibirsk, Russia
[3] JSC "Lesosibirsky LDK № 1", 16e Belinskogo str., 662543, Lesosibirsk, Russia

E-mail: [1]zaripov_sh@mail.ru, [2]ap_chizhov@mail.ru

Abstract. The article provides a substantiation of the structural diagram of the moisture removal mechanism, which is similar to baromembrane process,from larch sawnwood during convective drying. The concept proposed is based on such provisions as the presence of a semipermeable membranes system in larch wood and the chemical activity of larch wood, which leads to the formation of a gas-vapor mixture in volumes many times exceeding the volume of the sample itself. As a result, mass transfer of an aqueous solution of extractive substances from the center of the board to the surface is observed. The passage of an aqueous solution of extractives through a system of semi-permeable membranes leads to the separation of the solution into concentrate and permeate. The change in the physical state of permeate leads to the formation of the layer that greatly reduces the intensity of the moisture removal from the wood. Then the entire duration of drying should be considered as the sum of two clearly marked periods - intensive and slow. The aim of the work is to substantiate the structural diagram of mass transfer in larch wood during convective drying.

Introduction.

The modern theory of the convective drying of larch lumber is based on such provisions as the presence of permanent micro channels in larch wood, in which there is a redistribution of moisture [8], which eliminates the possibility of overpressure formation in the wood in the low-temperature process. Then, by using certain thermodynamic relations the micro mechanism of moisture transfer in the wood during drying was developed [5] and advanced in many other works. It is assumed that the flux density of moisture is proportional to the gradient of moisture content and coefficient of hydraulic conductivity [5].

It is possible to agree with this approach only partly, because the Board sawn from larch logs is some substance of plant origin. Larch wood contains a significant amount of extractives, which are able to dissolve in different types of solvents, including water. The influence of extractives on the drying process in the modern theory is not taken into consideration. The presence of the system of semi-permeable membranes, which provides vital activity of the tree also influences the water transfer in wood [6], [9], which is not adequately reflected in the modern theory of wood drying.

Chemical activity of larch wood, which manifests itself with increasing temperature [1], promotes the formation of overpressure in the center of the Board. The above material indicates the need in revision of the structural diagram of moisture removal from larch wood during drying. Then the aim of the paper is to justify a structural diagram of a mass transfer in larch wood during convective drying.

Experiments.

To reproduce dynamics of the process of the overpressure action in the centre of larch sawngoods a number of experiments were carried out in a specially designed drying installation. The results of the experiments showed (figure 1, 2) that the whole process of larch sawn goods drying should be considered as the sum of two clearly marked periods - intensive and slow. The intensive period is characterized by relatively low values of overpressure ($p_{excess.} \leq$ 20 kPa, figure1) and high specific drying speed (G =0.065 kg/(m^2·h)). Similar results were obtained in the study of the permeability of the membrane system of pine wood.

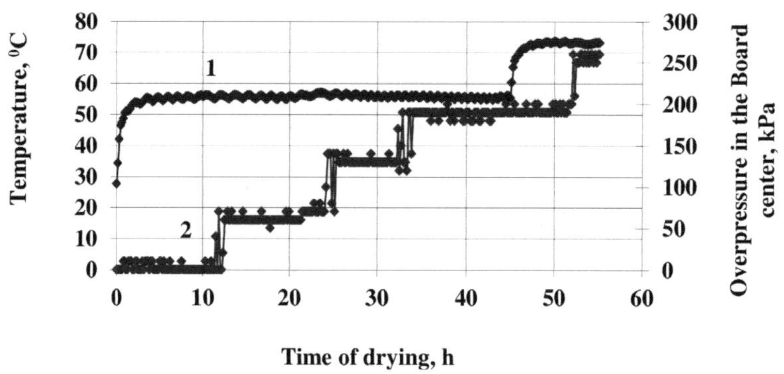

Figure 1. Overpressure p_c in the center of the Board at the temperature of drying agent in the initial period of drying t_c = 55^0 C; 1 – temperature of drying agent; 2 – pressure in the center of the Board.

The slow period is characterized by increased pressure in the center of the Board ($p_{excess.}$>> 20kPa, figure 1) with a relatively low index of specific drying rate G =0,011 kg/(m^2·h).

A similar pattern is observed for other components of an aqueous solution.

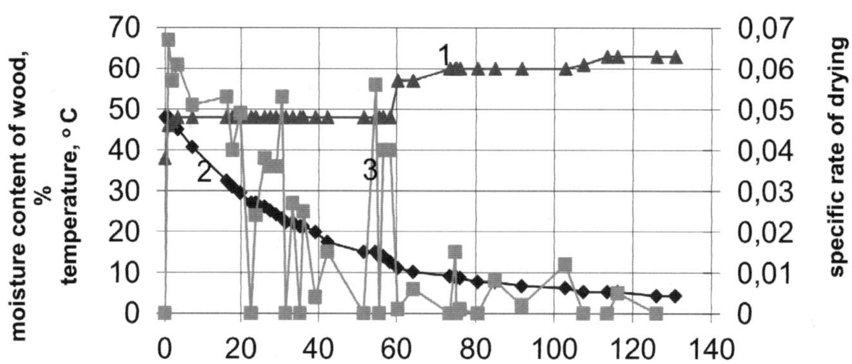

Figure 2. Kinetics of drying of larch lumber with 25x100 mm section: 1 – temperature; 2 – moisture content of wood; 3 – specific rate of drying.

Discussion.

At all variety of processes that occur in the larch wood during the drying process, the mass transfer of substances through non-porous membranes at the baromembrane process is of particular interest. The overpressure, which is pressurized into the cavity of the cell, equally creates the conditions of substances transfer both through cell walls and a membrane system of intercellular pores.

To analyze the conditions of transfer, the linear Darcy's law is used [7] as a product of power transfer in a particular area and the environmental resistance to the passage of the solution:

$$G = \frac{\Delta p}{x} \cdot \frac{\kappa}{\mu}, \quad (1)$$

where G - volume flow rate; κ - permeability of environment; μ - dynamic viscosity of liquid or gas; Δp - pressure drop across the length of the x environment.

The equation (1) shows that the shorter the path of this solution, the higher the volumetric flow. For a given parameter the membrane, which is located in an intercellular bordered pore, is preferred.

The main steps of the transfer of the aqueous solution of arabinogalactan in the larch wood during convective drying can be represented in a structural diagram of membrane separation (figure 3). To ascertain the type of separation is not possible, since the degree of blockage of the extracellular pores by extractive substances varies widely. The molecular weight of arabinogalactan also varies widely (3000 ... 94000 [2], [3]). So, one can equally speak of ultra-filtration and microfiltration [7].

Direct transfer of the molecules of elements in the water solution through the membrane can be represented as the result of successive periodic jumping of the diffusing molecules from one equilibrium position to another [4]. The possibility of this approach to the movement of molecules in non-porous medium is justified by the heat motion of the molecules in the polymer at temperatures above its glass transition temperature when micro cavities are continuously appearing and disappearing. Then the diffusion of elements in the aqueous solution is due to the movement of molecules from one cavity to another in the presence of a concentration gradient.

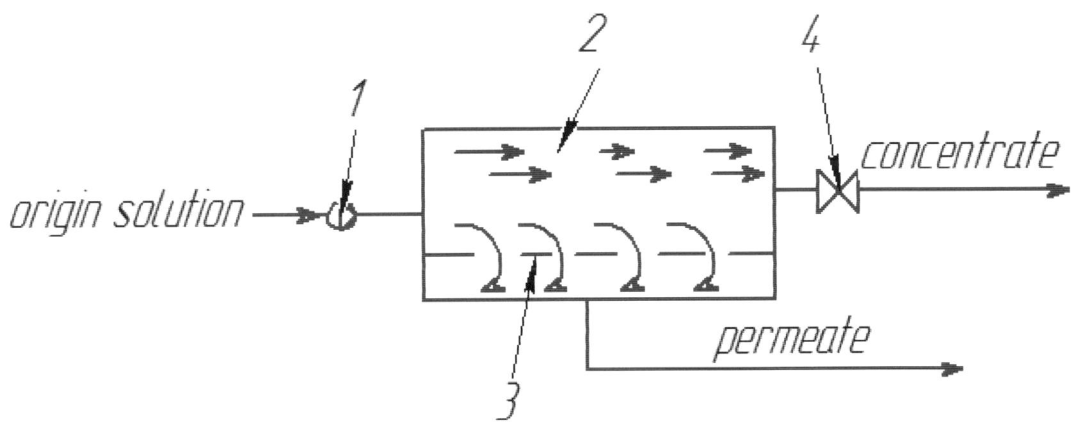

Figure 3. Structural diagram of mass transfer in larch wood during convective drying of sawngoods: 1 – pressure pump (overpressure); 2 – membrane apparatus (larch wood); 3 – semi-permeable (intercellular) membrane; 4 – throttle (wood temperature)

Consequently, the formation of microcavities of a certain size is set by the temperature level, and their filling (substitution) depends on the shearing force generated by overpressure. Then, such property of the membrane as selectivity depends on the temperature of wood, which relates to one of the core components of an aqueous solution of extractives - arabinogalactan. As a result, larger molecules of arabinogalactan remain in the wood to form a concentrate, and smaller macromolecules with moisture and vapor-gas mixture are carried through the system of intercellular membranes (figure 5, step 3) onto the Board's surface, forming the permeate.

After a certain period of time on the surface of larch Board a layer of water-soluble substances is formed. The intensity of the formation will be larger, the higher the temperature level is in the drying chamber [1]. Therefore, on the surface of larch boards regardless of the temperature of drying the layer of water-soluble extractives is formed, and it sets the intensity of moisture removal from larch wood during drying.

The second period of drying is characterized by degradation of intercellular membranes with regard to moisture up to 8 ... 10 times regardless of the Board thickness, and temperature level as well.This effect is explained by the formation of permeate on the external side of the system of membranes. The presence of permeate on the surface of the Board is well observed visually in the form of spots (figure 4).

Figure 4. Presence of extractives spots on the face of dried larch Board

Conclusions.

1. In the larch wood during drying a baromembrane process is being formed, i.e. there is a transfer of the aqueous solution elements from the center of the Board onto the surface.

2. The passage of an aqueous solution of extractives through a system of semi-permeable membranes leads to separation of molecules of arabinogalactan into the major fractions - large and small. A large fraction (concentrate) remains in the center of the Board, and a small one is moved onto the Board's surface, forming the permeate.

3. On the surface of the Board there is a division of the solution into fractions. Then, the gas-vapor mixture and water are removed, and extractives will accumulate (permeate).

4.The layer of permeate formed in the initial period of drying greatly hinders the removal of moisture from the drying lumber. The effect of restraining removal is manifested until the end of the drying process.

References

[1] Zaripov S G 2012 "Te Power Transfer" of Moisture in the Larch Wood during Convective *Actual problems of forestry complex: collection of scientific works* vol 34 pp 39-44
[2] Medvedeva A S, Aleksandrova G P and Tantsyrev A P 2002 Gel – Penetrating Chromatography of Arabinogalactan *Forest journal* **6** pp 108-113
[3] Ovodov Yu S 1998 Polysaccharides of Flower Plants: Structure and Physiological Activity *Bioorganic chemistry* vol 24 **7** pp 483-501
[4] Reitlinger S A 1974 Permeability of Polymeric Materials (Moscow: Chemistry) p 272
[5] Sergovsky P S and Rasev A I 1987 Hydrothermal Treatment and Preservation of Wood: textbook for higher schools – 4th ed., rev. and add. (Moscow: Forest industry) p 360
[6] Kharuk E V 1982 Silvicultural and Biological basis of the Permeability of Core and Ripe Wood of Conifers of Siberia [Text]: author. dis. ... Dr of agricultural – Leningrad p 36
[7] Hwang S T and Kammermeyer K 1981 Membrane Separation Processes (Moscow: Chemistry) p 464
[8] Chudinov B S, Tjurikov F T and Zuban P E 1965 Larch Wood and its Processin (Moscow: Forest industry) p 144
[9] Yakushkina N I and Bakhtenko E J 2004 Physiology of Plants (Moscow: Vlados) p 464

IOP Conf. Series: Materials Science and Engineering **822** (2020) 012021 doi:10.1088/1757-899X/822/1/012021

Low-temperature phase transition in bismuth ferrite films substituted by manganese

S S Aplesnin[1], A N Masyugin[1], U I Rybina[1], T N. Tarasenko[2] and K I Yanushkevich[3]

[1] Reshetnev Siberian State University of Science and Technology, 31 Krasnoyarsky Rabochy Av., Krasnoyarsk, 660037, Russia
[2] Donetsk Institute for Physics and Engineering named after A.A. Galkin, R. Luxembourg str., 72, Donetsk, 83114, Ukraine
[3] Scientific-Practical Materials Research Center NAS, Minsk 220072, Belarus

E-mail: apl@iph.krasn.ru

Abstract. Measurements of the capacitance and tangent of dielectric loss angle, impedance and resistance, magnetization in a magnetic field of 8.6 kOe $BiMn_xFe_{1-x}O_3$ films in the temperature range 80-600 K were carried out. The anomalies of the dielectric properties are explained in the Debye model and correlate with changes in the magnetic moment versus temperature.

1. Introduction

Materials with magnetoelectric properties are promising for technological applications in spintronics, sensor technology, information storage and recording devices [1-4]. Magnetically ordered crystals with antiferromagnetic ordering – multiferroic bismuth ferrite [5] were chosen as model objects of study. Today, bismuth ferrite $BiFeO_3$ or briefly (BFO) is one of the most popular compounds, based on which new magnetoelectric materials are created, it has high temperatures electrical (Tc = 820 C) and magnetic (T_N = 370 C) ordering [6]. Below 180 K, an increase in the reciprocal lattice vector of the film surface was found, which is absent in the bulk of the single crystal [7].

The main feature of this phase transition is a sharp change in volume without actually changing the symmetry. According to the authors of this work, the phase transition was observed due to the complexity of the phase diagram of the $BiFeO_3$ compound, which is very sensitive to even small changes, such as surface tension or deformation of the local field around vacancies and defects. The peculiarity of this compound $BiFeO_3$ is that it should not be considered as a homogeneous material. Structural, electrical, and magnetic properties of the surface layer differs from the bulk. The anomaly found at the 200K is attributed to the spin-glass state formation [8].

Increasing the ferroelectric and magnetic moment is achieved by substitution of iron into $BiFeO_3$ by manganese [9]. Studies on these compounds prove the presence of structural distortions. With an increase in the concentration of substitution, the residual magnetization is about 15 times higher than the magnetization for a pure sample of $BiFeO_3$ and is explained by the suppression of the cycloid spin structure and an increase in the angle of angularity of antiferromagnetically ordered spins [10].

The goal of the study is to establish the effect of the substitution of manganese ions on the low-temperature phase transition in $BiMn_xFe_{1-x}S$ films.

2. Results and Discussion

The bismuth ferrite solid solution films were formed by the burst-mode deposition of the preliminary synthesized solid solutions onto object glasses. The precursors used were powders with a grain size from 0.1 to 0.3 mm. The deposition was performed in a UVN-71P-2 vacuum facility. The pressure in a reaction chamber during deposition was 10^{-3} Pa. The tantalum evaporator temperature was kept at a level of ~2000°C. The substrates were placed at a distance of 10 cm from the evaporator. The substrate temperature ranged between 250–300°C. The films were 1240-nm-thick and had geometric sizes of 13 × 17 mm

The magnetic moment of the film was measured by the Faraday method in a magnetic field of 8.6 kOe [11, 12]. Figure 1 shows the temperature dependences of the magnetic moment for two compositions. As in bulk samples, the magnetic moment increases and reaches a maximum value at x = 0.05, then decreases by half. This is due to the destruction of the spin cycloid and the formation of antiferromagnetic order G type.

Below room temperature, the magnetization has an inflection point at T = 245 K and T = 218 K for concentrations x = 0.05 and 0.15. The temperature of the magnetic phase transition is in the high-temperature region of 560-580 K .

The substitution of iron for manganese forms electrically inhomogeneous states, including regions with local electric polarization. The electrically inhomogeneous states were established by impedance spectroscopy [13, 14]. We measured the active (real part Z') and reactive (imaginary part Z'') of the resistance using an AM-3028 component analyzer and determined full impedance $Z = Z' + iZ''$ of the films in the frequency range of ω = 0.1–1000 kHz at temperatures of 80–400 K.The imaginary part of the impedance (figure 2) measured at frequencies of 0.1–1000 kHz is described well by the function ImZ =1/ωC with the capacitance increasing upon heating.

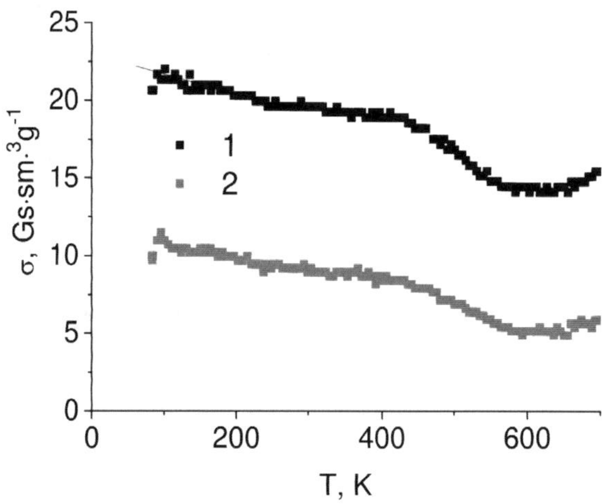

Figure 1. Magnetization thin films BiMnxFe1-xO3 for x=0.05 (1), 0.15 (2) versus temperature.

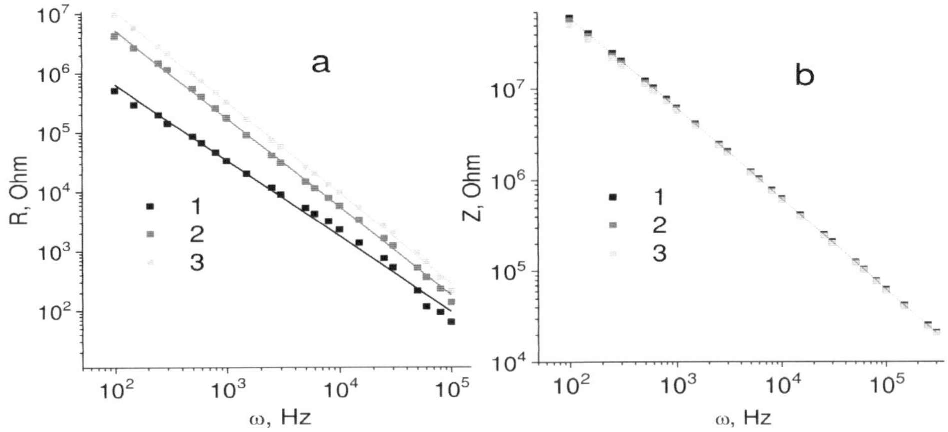

igure 2. The real part *Z'* (a) and full impedance *Z* (b) of the films $BiMn_xFe_{1-x}O_3$ for x=0.05 on the frequency at T=300 K(1), 340 K (2), 380 K (3).

The carrier transport mechanism can be established from the frequency and temperature dependences of the conductivity. The power dependence of the conductivity $\sigma(\omega) = C\omega^s$ is indicative of the hopping conductivity s=1; in particular, it is attributed to electron hoppings between electrically inhomogeneous states in the film, which are caused by spin polaron pinning over the localized states with phonons involved [15]. With increasing temperature, a smooth transition from a linear dependence associated with the Coulomb glass regime for interacting electrons to a quadratic dependence in the Fermi glass mode for noninteracting electrons is observed. Thus, at T = 380 K index s = 1.6. Below 220-240 K, Coulomb glass is realized. The capacity of the $BiMn_xFe_{1-x}O_3$ film for x = 0.05 undergoes small gaps at temperatures T = 214 K and T = 244 K with a wide maximum of the tangent of dielectric loss angle (figure 3).

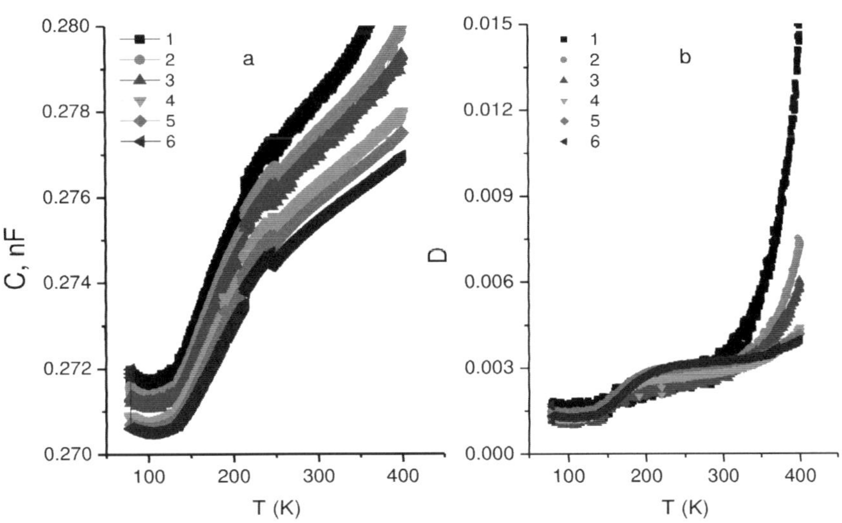

Figure 3. Capacity (a) and tangent of angle (b) dielectric loss of thin films $BiMn_xFe_{1-x}O_3$ for x=0.05 at frequencies ω = 1 kHz (1), 5 kHz (2), 10kHz (3), 50kHz (4), 100kHz (5) , 300kHz (6).

The independence of jump C(T, ω) temperature on frequency indicates the possibility of a first-order phase transition. Dielectric losses are associated with energy transfer to the magnetic and ferroelectric systems. The transition temperature interval shifts to high temperatures (278–316 K) with an increase in the concentration of substitution for manganese ions (figure 4). At T = 176 K, a dynamic transition at high frequencies was determined (figure 4b).

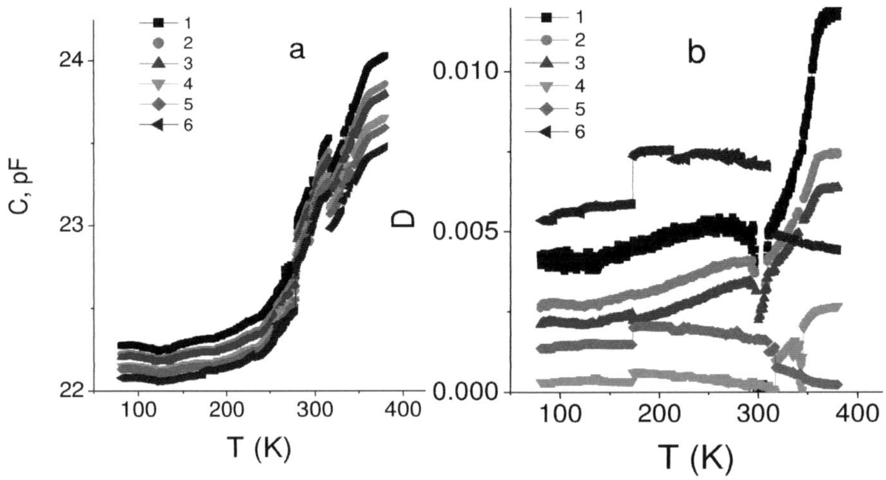

Figure 4. Capacity (a) and tangent of angle of dielectric loss (b) thin films $BiMn_xFe_{1-x}O_3$ for x = 0.15 at frequencies ω = 1 kHz (1), 5 kHz (2), 10kHz (3), 50kHz (4), 100kHz (5) , 300kHz (6).

3. Conclusion

In the low-temperature region, anomalies of magnetic and dielectric properties were found in thin films $BiMn_xFe_{1-x}O_3$. A jump in the capacity and the loss tangent for the composition with x = 0.2 is established. Perhaps these anomalies are associated with the pinning of lattice polarons and the formation of a Coulomb gap with the spectrum of impurity electronic excitations.

This study was supported by the Russian Foundation for Basic Research №18-52-00009, and the research program the state order № 3.5743.2017/6.7.

References

[1] Romanova O B, Sokolov V V, Har'kov A M, Pichugin A Y, Eremin E V, Balaev D A and Aplesnin S S 2011 *IEEE Transactions on Magnetics* **47** 4413
[2] Aplesnin S, Romanova O, Har'kov A, Balaev D, Gorev M, Vorotinov A and Pichugin A 2012 *Physica Status Solidi (B)* **249** 812
[3] Aplesnin S S, Kharkov A M, Eremin E V and Sokolov V V 2012 *Solid State Phenomena* Vol. **190** 105
[4] Aplesnin S S, Sitnikov M N, Romanova O B and Pichugin A Y 2016 *Physica Status Solidi (B)* **253** 1771
[5] Catalan G and Scott J F 2009 *Advanced Materials* **21** 2463.
[6] Aplesnin S S, Kretinin V V, Panasevich A M and Yanushkevich K I 2015 *Journal of Experimental and Theoretical Physics* **121** 422
[7] Jarrier R, Marti X, Herrero-Albillos J, Ferrer P, Haumont R, Gemeiner P and Dkhil B 2012 *Physical Review B* **85** 184104
[8] Teague J R, Gerson R and James W J 1970 *Solid State Communications* **8** 1073

[9] Lee J H, Vlahos E, Ramesh R, Xu X S, Ihlefeld J F, Schlom D G and Pan X Q 2010 *Applied Physics Letters* **96** 192901
[10] Zhao T, Scholl A, Zavaliche F, Lee K, Barry M, Doran A and Ramesh R 2006 *Nature Materials* **5** 823
[11] Aplesnin S S, Romanova O B and Yanushkevich K I 2015 *Physica Status Solidi (B)* **252** 1792
[12] Aplesnin S S, Demidenko O V, Panasevich A M, Sitnicov M N, Yanushkevich K I and Galyas A I 2016 *Bulletin of the Russian Academy of Sciences: Physics* **80** 1306
[13] Aplesnin S S, Romanova O B, Korolev V V, Sitnikov M N and Yanushkevich K I 2017 *Journal of Applied Physics* **121** 7
[14] Romanova O B, Aplesnin S S, Sitnikov M N, Kharkov A M, Masyugin A N and Yanushkevich K I 2019 *Solid State Communications* **287** 72
[15] Aplesnin S S, Romanova O B, Sitnikov M N, Kretinin V V, Galyas A I and Yanushkevich K I 2018 *Semiconductor Science and Technology* **33** 085006

IOP Conf. Series: Materials Science and Engineering **822** (2020) 012022 doi:10.1088/1757-899X/822/1/012022

The study of the light section canal anodic-abrasive polishing mechanism

V I Trifanov, M G Melkozerov, O A Sukhanova, I V Trifanov and D M Mednikov

Reshetnev Siberian State University of Science and Technology, 31 Krasnoyarsky Rabochy Av., Krasnoyarsk, 660037, Russia

E-mail: sibgau-uks@mail.ru

Abstract. The development of electronic technology, waveguide transmission lines of electromagnetic energy, as well as precision engineering require the development of polishing technologies. One of these technologies is the method of anodic abrasive polishing (AAP). The causal diagram of the influence of technological parameters on the output characteristics of AAP is presented, as well as the model of the process and the analysis of the mechanism of anodic abrasive polishing at the longitudinal vibration contact with the treated surface and the impact of the pulse current density used for electrochemical polishing of micro-irregularities with the help of a cathode integrated with an elastic abrasive tool

The development of laser technology, the transmission line of electromagnetic energy of the EHF-range and precision engineering place high demands on the surface treatment of metal parts (Ra 0.16-0.005 μm).

To reduce the surface roughness of the canals of light section tubes, an anodic-abrasive polishing (AAP) method can be applied [1,2]. In order to select rational modes of AAP, one should consider the influence of the parameters of the combined process of abrasive effect on the surface quality of the canals of light section, as well as the electrochemical parameters of anodic dissolution.

Figure 1 shows a diagram of the effect of the parameters of abrasive polishing combined with an electrochemical process on the quality of the layer surface and the output parameters of AAP.

In AAP the force of the abrasive effect of the bonded abrasive is used, and the force is directed tangentially to the surface being treated and makes an angle close to 90° with the normal. Under the abrasive effect on the micro-irregularity of the surface being processed, the movement of the elastic abrasive tool (EAT) can be carried out by the vibrocontact method. Research has established [3] that the amount of roughness of the treated surface Ra is largely influenced by the grit size of the abrasive and the working pressure P from the side of the EAT. The abrasive grains of the EAT slide over the surface being treated with a very low coefficient of friction due to the influence of the lubricating properties of the electrolyte and the small rake angle α=10-40°. At the same time, the influence of microgeometry and the energy of the surface layer being processed are also manifested [4].

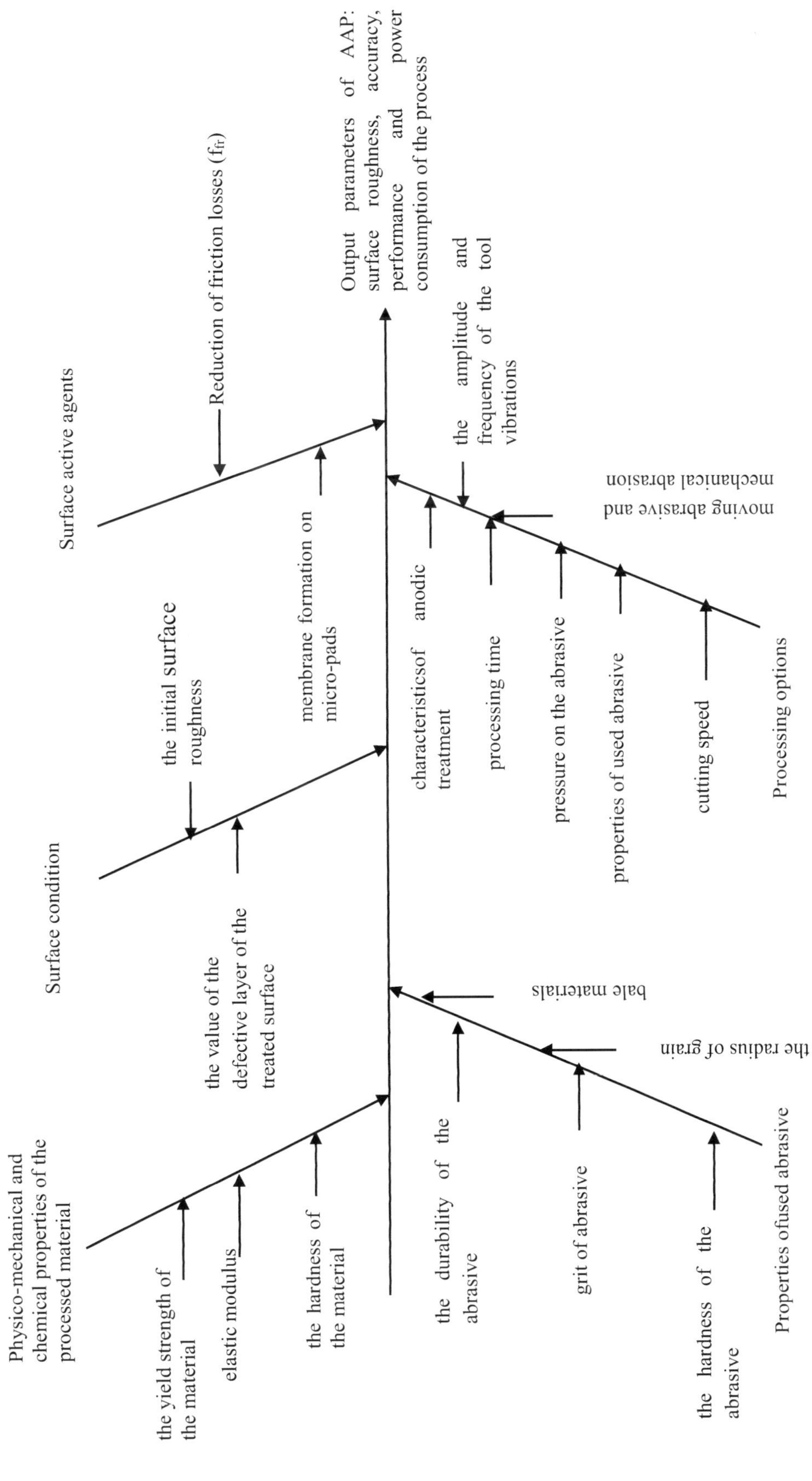

Figure 1. The causal diagram of the influence of technological parameters on the output characteristics of AAP.

The contact can be considered as the result of reinforcement of the engagement and deformation of the inter-penetrating roughness and asperities of two mating surfaces. With a relative longitudinal vibration displacement of the EAT with amplitude A, frequency ω, and pressing force P_H, elastoplastic states appear at the microcontacts of the roughness of the processed surface of the light section canal and an energy field appears. The energy field is caused not only by the energy impact of the interaction between two micro-irregularities, but also due to the concentration of electric charge on the activated microprotrusions of the treated surface when a voltage of an external electric field is applied during AAP. The potential for microcavities is reduced due to film formation.

Two types of effects: mechano-abrasive and electrochemical, can ensure the preservation of the initial accuracy and stable microprofile of the processed surface of the canal of light section due to selective alignment of the micro-irregularity.

Reducing the contact force of the abrasive grain with asperities due to liquid and oxide films, as well as aligning the peaks of the asperities can significantly reduce friction and microcutting forces with the abrasive grains of the treated surface, however, the impact of pulsed electric current and electrolyte pumping will allow localizing the process of levelling the asperities due to anodic dissolution within dimensional tolerance while maintaining the original accuracy of the canal of light section.

The scheme of the contact interaction of a single abrasive and the canal of the EAT with the asperities of the treated surface is shown in figure 2. With AAP, there is no impact interaction of the abrasive particles with the treated surface. In this regard, the destruction of abrasive particles is insignificant and the general nature of the contact interaction of abrasive particles with the treated surface is preserved.

In the case of AAP, a single grain contact circuit with the surface to be treated was selected, similar to the grinding grain pattern developed by E.N. Maslov [5]. We assume that microprotrusions M perform the role of independent cutting elements, and the top of the grain is approximated by a radius r_i. With an increase in the penetration depth h of the top of the abrasive grain into the metal with a cutting angle $\beta < 180^0$, a transition from plastic deformation to microcutting occurs. In this case, the normal force P and the friction force F_f will act on the abrasive grain. These forces are applied to the point A in the middle of the arc of grain contact with the processed micro-irregularity.

The force P_y, pressing the abrasive grain to the treated surface, as well as P_x, which produces microcutting, can be found by projecting the normal force P acting on the abrasive and friction force F_f on the direction of the x and y axes. After the transformations P_y and P_x can be represented as

$$P_y = P(sin\,\alpha + f\,cos\,\alpha)$$
$$P_x = P(cos\,\alpha - f\,sin\,\alpha) \tag{1}$$

Then the normal force P and the force of cutting P_x can be determined as:

$$P = \frac{P_p}{\sin\alpha + f\cos\alpha}$$
$$P_x = \frac{P_p(ctg\alpha - 1)}{1 + f\cos\alpha} \tag{2}$$

where α is the rake angle at the point of contact K of the abrasive grain with the micro-irregularity of the metal surface; f – the coefficient of sliding friction of the abrasive grain on the treated surface; P_p – the pressing force of the grain of abrasive.

The coefficient of sliding friction of the abrasive can be determined by taking into account the ratio proposed in the work of I.V. Kragelsky [6]:

$$f = c \cdot \sqrt{\frac{h_g}{r_i}} \tag{3}$$

where c is the coefficient taking into account the effect of contact vibration, film formation on the processed micro-irregularities and the lubricating properties of the electrolyte on friction;
r_i is the radius of rounding of the top of the grain; h_g is the depth of abrasive introduction.

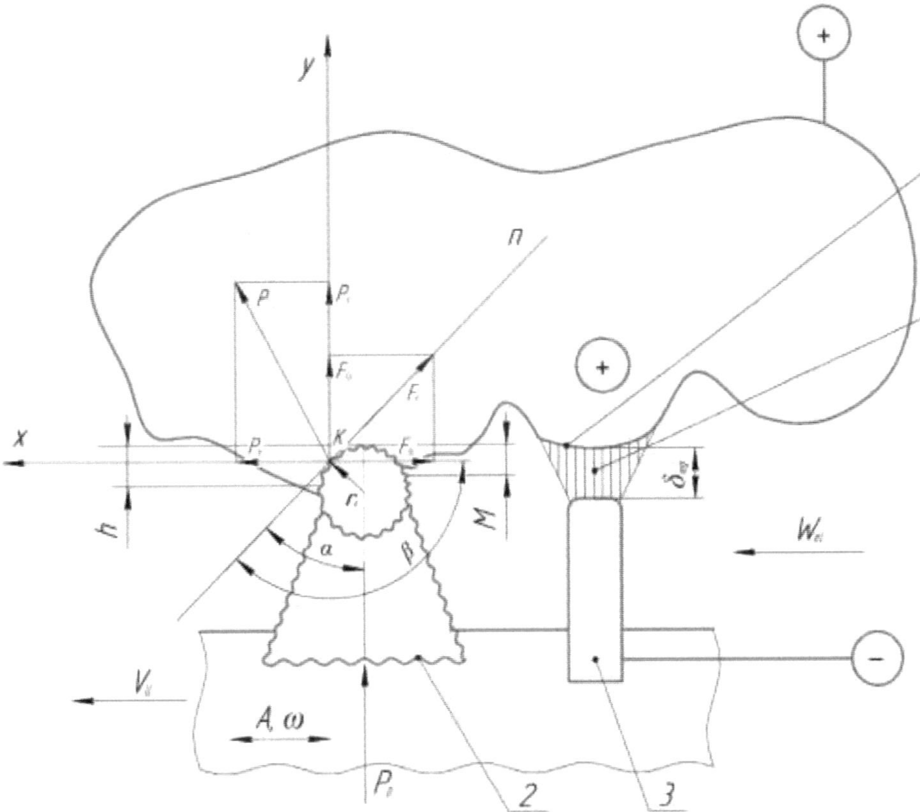

Figure 2. Scheme of the contact interaction of a single abrasive and the EAT cathode with the micro-irregularities of the treated surface: 1 is the micro-irregularity of the treated surface; 2 is an abrasive grain of the EAT; 3 is the cathode instrument; 4 is for electric field lines; α and β are rake angle and cutting angle at the point of contact K; r_i is the radius of the rounding of the top of the grain; $\delta_{МЭЗ}$ is the value of the interelectrode gap; h is the depth of penetration of the top of the abrasive grain; Pн is the pressing force of the abrasive, Fт is the friction force, P is the normal force acting on the abrasive grain; A, ω is the amplitude and frequency of vibration oscillations of the EAT; Vu is the linear velocity of movement of the EAT.

The repeated use of abrasive leads to rounding of sharp peaks and angles, as well as a change in the microrelief of the surfaces of abrasive grains, which leads to a change in the cutting ability of the aggregate of abrasive particles as a whole, therefore, r_i must be taken into account when calculating f_{mp} in APP.

The resulting cutting force Pp on the processed surface can be calculated by the formula:

$$P_p = P_a S_t f = P_a \cdot S_t \cdot c\sqrt{\frac{h_g}{r_i}}, H \tag{4}$$

where P_a is the external pressure on the abrasive particles; S_t is the area of EAT with an abrasive-bearing material in contact with the treated surface.

The power of abrasive cutting N_{cut} is estimated by the formula:

$$N_{cut} = M_{cut} \cdot \omega \tag{5}$$

where ω is the frequency of longitudinal vibrations of the EIA, 1/s; M_{cut} is the moment of cutting, N·m.

$$M_{cut} = P_{cut} \cdot B \cdot 0,5 = P_a \cdot 0,5B \cdot S_c \cdot c\sqrt{\frac{h_g}{r_i}} \tag{6}$$

where B is the thickness of the EAT, m.

It is determined that an increase in pressure P in the range of 0.2 ÷ 1.0 MPa raises Ra in 1.2 ÷ 1.4 times, which is probably due to an increase in thickness, cut off by a single grain of the layer and an increase in the work of cutting. The radius of rounding of the cutting edge of the abrasive grain has a certain role on Ra. It is established that the limiting values of the ratio of the thickness of the slice to the radius of rounding of the cutting edge of an abrasive grain is determined by the coefficient of friction f_{fr} of the abrasive grain with the processed surface [7]. The coefficient of friction can vary within $f_{fr} = 0,1 - 0,3$. Therefore, it is possible to reduce the surface roughness by reducing the friction coefficient of the abrasive grain with the processed material, which can be implemented in AAP, due to the fact that the abrasive method requires removing the anode films from the micro-irregularities.

Studies have shown that after vibrocontact polishing with lodgments with variable stiffness, deviations from the geometrical accuracy of the turbine blade feather profile were no more than 0.05 mm, and the charging intensity does not exceed 0.5% [3]. This is due to the fact that during the abrasive treatment a part of abrasive grains is destroyed, the particles roll freely and some of the grains can be fixed in the defects of the treated surface [8]. In the case of AAP, the process can be carried out with less effort on the EAT, mainly to remove the passivating films on ridges of micro-irregularities.

For finishing polishing an abrasive from 0,01-0,3 microns is mainly used (ultra micro powders are used). A multistage process of preliminary, finishing and fine finishing is also used. Polishing using fine-grained abrasive powder materials can be represented as a combination of processes: mechanical, adsorption, adhesive, wetting and oxidation of the surface layer [9].

The use of nano-dispersed tribochemical active abrasive materials changes the process of finishing polishing with the occurrence of chemical transformations and has a significant advantage over the known abrasive methods: it allows reducing the number of finishing operations with obtaining surface roughness with Ra 0.005 microns. From the point of view of solid mechanics, polishing can be attributed to non-slip friction processes. This is the basis of the AAP mechanism.

When an abrasive tool interacts with the treated surface, its single protrusion produces elastic and plastic deformation of the material [9]. After contact with the surface, elastic deformation begins, passing as it deepens into plastic. The depth of penetration of the grain depends on the kinetic energy of the movement in the direction normal to the surface [10]. The movement of grain in the direction parallel to the processed surface is accompanied by micro-cutting of the part.

The amount of material removed in the process of abrasive treatment consists of the volumes of chips that were shredded and destroyed as a result of polydeformation, which depends on the volume of plastically and elastically deformed material under the action of a single grain, as well as on the number of these impacts per unit of surface per unit of time. The achievable roughness is determined by the transverse and longitudinal sections of scratches applied by single grains to the surface. The model of interaction of a single grain with an elastoplastic half-space can be used to determine the normal P_N and tangential P_T of cutting forces.

$$P_N = \delta_{sh} \cdot r_i^2 \cdot f_T,\; P_T = \delta_{cд} \cdot r_i^2 \cdot f_N \tag{7}$$

where δ_{sh} is the shear yield strength of the material; r_i is the radius of rounding of the tops of the irregularities of the abrasive grain; f_T and f_N are dimensionless coefficients of forces characterizing the conditions of scratching of the material; P_N and P_T are tangential and normal cutting forces on a single grain [11].

The yield strength of various materials with elastoplastic properties varies differently in the scratch depth. For example, copper and stainless steel almost do not change their properties, however, after multilateral deformation of pipes of light section, the yield strength may change.

The interaction of the cutting microrelief of the EAT with the processed surface depends on many factors (the depth of introduction of the abrasive grains, the number of grains in contact, the radius of the grain tops).

Elastic abrasive tools allow can provide less surface roughness than rigid ones for the surface may be smoother when a larger number of elementary cutting profiles are involved in its formation. Elastic tools have significantly larger areas of contact with the treated surface than rigid ones, therefore their use is preferable for AAP.

For AAP, an elastic abrasive tool is used, in which a cathode is installed for anodic dissolution of micro-irregularities [2] that can reduce the cutting forces on the part of abrasive grains and process thin-walled canals of light section.

When vibrocontacting the EAT in the process of AAP, the roughness of the processed surface can be influenced by the amplitude and frequency of vibrations, grain size, density of distribution of grains over the abrasive surface, physical and electrochemical properties of the processed material, as well as the mechanism of anodic dissolution of micro-irregularities [12]. The influence of electrolyte is associated not only with the intensity of anodic dissolution of micro-irregularities and passivation of microdepressions, but also with the reduction of friction losses when exposed to abrasive particles when they are vibrated moving along the surface of the processed canal. Figure 2 shows the components of the forces acting on the surface of the part with an abrasive grain, as well as the principles of electrochemical polishing (ECP). The AAP process can provide the roughness of Ra 0.2–0.04 µm without deforming the treated surface with a low temperature effect with less effort on the EAT.

When AAP, the treatment process should be carried out with preservation of the initial accuracy of the light section canal. The AAP process can be affected by the physicomechanical and electrochemical properties of the processed material: microhardness, microstrength (tear resistance), modulus of elasticity, yield strength of the processed material, shear modulus, Poisson's ratio, material density, electrochemical equivalent, metal current output of the material processed by electrochemical method, the uniformity of anodic polishing associated with the electrochemical properties and structure of the processed material.

When calculating the technological parameters of abrasive processing with an EAT, it is necessary to apply a coefficient λ taking into account the influence of the physicochemical properties of the processed material [9], for example, when determining the total polishing time T_0 according to the formula:

$$T_0 = n_i \cdot \gamma \left(\lambda \cdot \frac{Ra_0}{Ra_n} \right)^{1/n_y} \tag{8}$$

where n_i is the number of impacts of the interaction of abrasive grains with microroughness per unit of time; λ is the coefficient taking into account the physicochemical properties of the processed material, Ra, Ra_n are initial and final roughness of the treated surface; γ is the coefficient taking into account the conditions of abrasive polishing:

$$\gamma = \frac{K\beta}{\alpha_r} \text{ мин}^{-1}, \tag{9}$$

where α_r is the coefficient taking into account changes in the radius of the abrasive after each treatment cycle; β is the coefficient of change of the roughness value after the *i*-th processing cycle; K is the coefficient depending on the geometric characteristics of the processed surface and the AAP scheme;

$$K = K_1 + K_2 + K_3 \tag{10}$$

where K_1 is the shape factor of the treated surface; K_2 is the coefficient taking into account the depth of the defective layer of the treated surface; K_3 is the coefficient depending on the technological parameters of the EAT of the adopted mechanism of the processing process, $K_2 \approx 3,73 \div 4$.

The number of impacts of interaction of abrasive grains with micro-irregularities during polishing is determined by the formula [9]:

$$n_i = 2,31 \cdot ln \frac{Rz_0}{Rz_n} \tag{11}$$

where Rz_0, Rz_n are the initial and final values of the roughness of the treated surface, determined by the established requirements.

Formula (12) shows that n depends on the initial and final roughness of the treated surface.

$$K_{ep} = \gamma \cdot \lambda \tag{12}$$

Where K_{ep} is the coefficient depending on the properties of elastic-plastic deformation of the processed material.

The life period T_{lp} of elastic abrasive tool is determined by the formula:

$$\mathrm{T}_{lp} = \frac{60 \cdot V_n}{Q} \tag{13}$$

where V_n is the volume of the steam space of the abrasive tool, mm^3; Q is efficiency in mm^3/min; V_n depends on the area of the treated canal surface (width a and length L), as well as on the diameter of the abrasive grain D and the technological density of the abrasive grains ρ_3 on the surface of the elastic tool.

The coefficient λ can be determined by the formula of one of the dependencies [9]:

$$\lambda = \left(\lambda_{M_h}, \lambda_R, \lambda_{EG}\right)^{1/3} \tag{14}$$

where E is the modulus of elasticity; G is the shear modulus; R is the microstrength; M_h is the microhardness.

The removal of material from a unit of area during anodic-abrasive polishing can be determined by the formula [10]:

$$Q = (Q_C + Q_{pl})T_0 \cdot n_i + Q_{AH} \tag{15}$$

where n_i is the number of impacts of the interaction of grains of abrasive with microroughness per unit of time, 1/s; Q_c is the volume of material displaced in the form of chips in one blow; Q_{pl} is the volume of material destroyed as a result of plastic deformation per one blow, mm^3; Q_{AH} is the additional volume of material removed by the anodic electrochemical dissolution of micro-irregularities in the combined AAP process.

$$Q_{AH} = \left(E_V \eta \cdot i\right) \mathrm{t}_p \tag{16}$$

where E_V is the volume electrochemical equivalent (mm³/A·min), η is the metal anode current yield in the treated electricity, i is the current density, A/cm², t_p is the time of anodic dissolution, min.

$$i = \frac{\left[U - (\varphi_c + \varphi_a)\right]\chi}{\delta_{ieg}}, \tag{17}$$

where U is the voltage at the electrodes, φ_c and φ_a are the electric potentials at the cathode and the anode, δ_{ieg} is the size of the inter-electric gap, χ is specific conductivity of electrolyte, cm·Ω^{-1} [12].

The removal of material Q_A per unit of time (m³/s) due to the impact of a vibrating abrasive-bearing elastic tool can be calculated by the formula:

$$Q_A = n_i \cdot K_z \cdot Q_z = A \cdot \omega \cdot WL \cdot K_z Q_z \tag{18}$$

where n_i is the number of impacts of abrasive granules in micro-irregularity, 1/s; Q_z is the amount of the metal removed when exposed to one abrasive grain, m³; K_z is a coefficient taking into account the number of impacts in micro-irregularity; A is the amplitude and frequency of oscillation of the EAT; W is the density of distribution of abrasive grains over the surface of an elastic tool, 1/m²; L is the length of the treated surface, m.

The amount of metal removed by a vibrating abrasive-bearing elastic instrument during processing T is calculated by the formula:

$$Q_a = Q \cdot T \tag{19}$$

The arithmetic average deviation of the profile of the established roughness of the treated surface after n_y cycles of abrasive processing can be determined:

$$Ra_{inst} = K_{ep}\left(\frac{h_{max}}{W \cdot A}\right)^{1/2}, \tag{20}$$

where h_{max} is the maximum depth of penetration of the grain of the elastic instrument into the treated surface, m; W is the density of distribution of abrasive grains over the treated surface, 1/m²; K_{ep} is a coefficient depending on the properties of elastoplastic deformation of the metal.

The maximum depth of introduction of the abrasive particle during AAP is proposed to be calculated by the formula:

$$h_{max} = \left(A\omega + V_t\right) 2r \cdot \sin\alpha \sqrt{\frac{m_t W}{3K_R \cdot c \cdot \delta_s \cdot A}} \tag{21}$$

where V_t is the speed of movement of the EAT; m_t is the mass of EAT; r is the radius of rounding of the abrasive particle; K_R is a coefficient taking into account the influence of the surface roughness of the part; δ_s is yield strength of the part material; c is a coefficient taking into account the influence of roughness on the surface of the part; α is therake angle.

The average depth of penetration of the grain of an elastic instrument h_a into the processed surface can be estimated by the formula:

$$\mathrm{h}_a = \left(\frac{2 \cdot N_c}{\pi \cdot r \cdot H_V} \right)^{1/2} \tag{22}$$

where N_c is a power of abrasive cutting of micro-irregularities; H_V is hardness of the treated surface by Vickers.

The roughness of the treated surface during an abrasive vibration contact with the surface can be determined by the formula:

$$Ra_i = Ra_{iw} - K_z \cdot \frac{Ra_r}{Ra_{iw}} \cdot \sin\beta \cdot \frac{\delta_{usa}}{\delta_{usm}} \cdot \left(\frac{2A \cdot \omega \cdot WL \cdot V_{cut} \cdot m_t}{\delta_{usa} \pi r} \right)^{1/2} \cdot T_c \tag{23}$$

where Ra_{iw}, Ra_r are the initial height of the irregularities of the workpiece and the elastic abrasive tool; K_z is a coefficient of abrasive grit; V_{cut}, β are speed and angle of cutting of micro-irregularity; δ_{usa}, δ_{usm} are ultimate strength of the abrasive particle and material; m_t is the mass of instrument; r is the radius of rounding of the abrasive; T_c is the processing cycle time.

Taking into account the formulas (17) and (24), we obtain an expression for calculating the roughness of the treated surface obtained as a result of the AAP during vibrocontact polishing:

$$Ra_n = Ra_{n-1} - (1 - n \cdot 10^{-1}) \left[\left(\frac{E_V x \eta (U - \Delta U_b) 10^{-2}}{\delta_{мээ}} \right) \tau_{ce} + K_z \frac{Ra_z}{Ra_0} \sin\beta \cdot \frac{\delta_{usa}}{\delta_{usm}} \cdot \left(\frac{2A\omega WL \cdot V_{cut} \cdot m_t}{\delta_{usa} \pi r} \right)^{1/2} \cdot T_c \right] \tag{24}$$

where $T_ц$ is the duration of one AAP cycle.

$$\Delta U = \varphi_c + \varphi_a \tag{25}$$

where φ_c and φ_a are the electric potentials at the cathode and the anode.

The duration of the pulse of electric current with pulsed anodic dissolution of asperities is determined by the formula [13]:

$$\tau_{pul} \leq \frac{\delta_{ieg} (\rho_e + \rho_g)}{2UK_g} \tag{26}$$

where ρ_e and ρ_g are the resistivity of the electrolyte and gas-liquid mixture; K_g is the coefficient connecting the volume of emitted gas with the thickness of the gas-liquid mixture. Anodic dissolution of micro-irregularities during AAP can be carried out in passivating electrolyte solutions of neutral salts ($NaNO_3$, Na_2SO_4, $NaNO_2$ and others) of the corresponding concentration (7 ÷ 15%).

The time between pulses of electric current [11] can be calculated by the formula:

$$\tau_{bp} \geq \frac{\pi \delta_{ieg}^{\ 2}}{4D_U} \tag{27}$$

where D_U is the diffusion coefficient of the ions of the treated metal, cm^2/s.

The cycle time of the electrochemical treatment τ_{ce} can be represented by the expression:

$$\tau_{ce} = \sum_{0}^{T_c} \tau_{pul}\ ;\ \tau_{ca} = \sum_{T_0}^{T_c} \left(\tau_{pul} + \tau_{bp} \right) = T_c \tag{28}$$

One cycle of anodic-abrasive polishing τ_{ca} is equal to the sum of the current time τ_{pul} and pauses τ_{bp} used for the electrochemical anodic dissolution of micro-irregularities of the treated surface. The electrical power spent on the removal of micro-irregularities by anodic dissolution with the pulse current while processing during the time T_c can be determined by the formula:

$$N_A = \int_0^{\tau_{ce}} i(t)U(t)dt \tag{29}$$

where $i(t)$ and $U(t)$ are current density pulse and voltage at the cathode and anode during the AAP cycle.

Pulsed modes of electrochemical processing increase the localization of the anodic dissolution of micro-irregularities, improve the surface quality as a result of discretization of the process of anodic dissolution and relaxation of the properties of the interelectrode medium. Reducing the temperature of the anode reduces the micro-irregularities height and the amount of allowance required to achieve a given roughness value compared to an uncooled electrode, and also eliminates etchings at the grain boundaries, the treated surface [12] with rationally selected pulse duration and amplitude density of electric current. Maintaining the active and passive state on the micro-irregularities and cavities is possible by abrasive exposure and removal of passivating films from the microprotrusions of the treated surface. The microlocal distribution of temperature and electrolyte properties, amplitude-time parameters of current density, active and passive state on microprotrusions and microcavities of the treated surface contribute to the alignment of the microrelief and improve the surface quality in the AAP process.

Conclusion

The mechanism of the combined process of anodic-abrasive polishing of light section canals is presented. It consists in electrochemical polishing of micro-irregularities with pulsed current during vibrocontact abrasive interaction of an elastic tool with the surface. The cause-effect relationship of the influence of technological parameters on the output characteristics of AAP has been established. A formula is presented for evaluating the surface roughness after AAP, taking into account the temporal parameters of electrochemical anodic polishing of micro-irregularities with pulsed current and their vibrocontact interaction with an elastic abrasive bearing tool.

References

[1] Ruban I V, Ruban V M and Gadzhiev Z A 2008Obrabotka detaley s zadannoy otrazhatel'noy sposobnost'yu [Processing of parts with a given reflectivity]*Traktory i sel'skokhozyaystvennye mashiny* No 5

[2] Trifanov I V, Oborina L I, Ryzhov D R, Sutyagin A V, Mal'ko L S and Trifanov V I Sposob anodno-abrazivnogo polirovaniya otverstiy [The method of anode-abrasive holes polishing] Patent 2588953 MPK H02N 11/00 (1995.01) reg. 7.07.2016

[3] Dimov Yu V 2007 Obrabotka detaley elastichnym instrumentom [Processing of parts with an elastic tool]. (Irkutsk: Izd-vo IrGTU) p 352

[4] Musokhranov M V, Kalmykov V V, Malyshev E N and Zenkin N V 2015 Energiya poverkhnostnogo sloya metallov kak instrument vozdeystviya na velichinu koeffitsienta treniya [The energy of the surface layer of metals as a tool to influence the value of the friction coefficient] *Fundamental'nye issledovaniya*No 2-2 pp 251-254

[5] Maslov EN 1974 Teoriya shlifovaniya materialov [Theory of grinding materials]. (M.: «Mashinostroenie») p 320

[6] Trenie, iznashivanie i smazka: Spravochnik.1978 [Friction, wear and lubrication: a Handbook] V 2 ed I V Kragel'skogo and V V Alisina(M.: Mashinostroenie) p 400

[7] Novikov F V, Shkurupiy V Gand Andilakhay V A 2014 Analiticheskoe opredelenie usloviy umen'sheniya sherokhovatosti poverkhnosti pri abrazivnom polirovanii [Analytical determination of conditions for reducing surface roughness with abrasive polishing] *Vestnik Priazovskogo gosudarstvennogo tekhnicheskogo universiteta. Seriya: Tekhnicheskie nauki*vol 28pp142-145

[8] Finishnoe polirovanie metallov nanodispersnymi tribokhimicheski aktivnymi oksidami, available at:http://pandia.ru,access30.01.2019

[9] Ruban I V 2015 Povyshenie effektivnosti finishnoy obrabotki metallicheskikh zerkal [Improving the efficiency of finishing metal mirrors: thesis]: dis. … kand. tekhn. nauk. Moskovskiy gosudarstvennyy otkrytyy universitet im. V.S. Chernomyrdina. (Moskva)pp 115-116

[10] Podashev D B 2014 Optimizatsiya finishnoy obrabotki detaley iz vysokoprochnykh alyuminievykh splavov elastichnym abrazivnym instrumentom [Optimization of finishing of high-strength aluminum alloys parts with an elastic abrasive tool: thesis]. dis. … kand. tekhn. nauk. Irkutskiy gosudarstvennyy tekhnicheskiy universitet (Irkutsk) p 154

[11] Dimov Yu V 2000 Obrabotka detaley svobodnym abrazivom [Processing of parts with free abrasive]. (Irkutsk: Izd-vo IrGTU)p293

[12] Moroz I I 1977Osnovy povysheniya tochnosti elektrokhimicheskogo formoobrazovaniya: monografiya [The basics of improving the accuracy of electrochemical shaping: a monograph] Shtiintsa (Kishinev)p 129

[13] Nadaraia Ts G, Babkina L A and ShestakovI Ya 2014 Elektrokhimicheskoe polirovanie v vodnykh rastvorakh neytral'nykh soley s vibratsiey elektroda: monografiya [Electrochemical polishing in aqueous solutions of neutral salts with electrode vibration: monograph] Sibirskiy gosudarstvennyy aerokosmicheskiy universitet(Krasnoyarsk)p 120

Disc friction to specify power balance of a turbopump unit of LPE

A A Zuev, V P Nazarov, D A Zhuikov, M I Tolstopyatov and M V Savelyeva

Department of Aircraft Engines, Reshetnev Siberian State University of Science and Technology, 31 Krasnoyarsky Rabochy Av., Krasnoyarsk, 660037, Russia

E-mail: dla2011@inbox.ru

Abstract. The research elaborates the main characteristics of the liquid propellant engines; it determines that some turbopump units have got the frequency of rotor spinning up to 100 000 and even 120 000 r.p.m. The authors analyse the current methodologies to specify losses connected with disc friction. The methodologies determining disc friction coefficient often obtain criterion-empirical character, and due to increasing a number of rotor spinning of turbopump unit of the liquid propellant engines, they reflect the specification reliability inaccurately. The research demonstrates equations to determine disk friction coefficients based on considering the theory of three-dimensional boundary layer to a rotational flow taking into account the turbulent process of velocity distribution. The authors give recommendations on determining an index of velocity distribution degree depending on Reynolds rotation criterion of a disc and plug flow.

1. Introduction

The major part of the existing methodologies to calculate flow parts of turbopump units (TPU) of the liquid propellant engines (LPE) is of criterion-empirical character; it is used for the nod boundary conditions and feeding units of the liquid propellant engines obtaining rotation frequency approximately 20 000 – 40 000 rotations per minute. The current TPU of the liquid propellant engines obtain comparatively high rotation frequency and it can be up to 120 000 rotations per minute. The fact does not consider the degree of the turbulent process of flow velocity distribution in the three-dimensional boundary layer of the elements of the flow range of the feeding units and it requires additional research and specifications of the used dependencies due to changing boundary conditions.

While designing current rocket engines and converting to environmental fuel components, there is a tendency to increase rotation frequency of TPU. That results in increasing in pressure produced by a pumping stage, growing a pump speed coefficient as upgrading the efficiency of the pump and the turbine, decreasing the dimensions such as reducing the axial and radial loads to the rotor and as a consequence it leads to TPU mass reduction in general.

Multiple discrepant values are provided by the numerous experimental and theoretical research conducted by scientists within a range of geometrical and performance parameters of flows in the rotor-stator gap [1-9]. Some methodologies to calculate resistant torque obtains various disadvantages first of all connected with applying empirical coefficients; they do not consider the availability of the radial component of the absolute velocity in the side cave.

One of the main tendencies to develop aerospace engineering is to improve the design quality based on using the current calculation methodologies and mathematical models reflecting the processes in both separate units and in the whole airborne vehicle in general.

While designing, the requirements to TPU configure due to the assignments performed by the power unit (PU), where TPU is an element constituent and with which it assembles as a single block. The requirements to the PU belong to TPU as well: to provide feeding fuel components with the flow requirement and pressure under high degree of reliability and efficiency in all operational modes of the engine; to provide minimal dimensions and mass; the construction simplicity and minimal cost.

2. Specifying disk friction coefficient

The power balance of a centrifugal-type pump are consider where the power output of the centrifugal-type pump is calculated as

$$N_{out} = N_{req} - N_{cont} - N_d - N_{leak} - N_{hydr},$$

where N_{req} - the power required of the centrifugal-type pump, N_{cont} - contact friction loss power, N_d - disc friction power of a rotor wheel, N_{leak} - lost power to the working fluid leakage through seal groups and N_{hydr} - hydrodynamic power loss.

Calculates a disk friction coefficient taking into account increasing rotation frequency of TPU of LPE. While realizing turbulent flow, originally, the power law of airfoil shape distribution of speed in the boundary layer [10] is applied

$$\frac{u}{U} = \left(\frac{y}{\delta}\right)^{\frac{1}{m}} \quad (1)$$

Depending on the shape of the velocity airfoil m, it is necessary to overdetermine the dependence of friction induced shear stress close to the inside wall in the boundary conditions of the turbulent boundary layer τ_0. Based on the expression for the friction stress [11, 3]

$$\frac{\tau_0}{\rho U^2} = 0,01256\left(\frac{U\delta^{**}}{\nu}\right)^{-0,25} \quad (2)$$

The considered flow in the side cave between rotor disk and TPU body of LPE belongs to rotational flow according to the "solid body" law $\frac{U}{R} = \omega = const$. The momentum losses thickness rotationally with a certain airfoil shape for the wall is determines as

$$\delta_{\alpha w}^{**} = 0,04535\left(\frac{4M^2 - 7L}{1+H}\right)^{0,4}\left(\frac{2}{J}+\frac{1}{L}\right)^{0.8}\left(\frac{\nu}{\omega_n}\right)^{0,2} R^{0,6} =$$
$$= 0,04535\left(\frac{18m^3 + 99m^2 - 171m}{m^3 + 18m^2 + 105m + 200}\right)^{\frac{2}{5}}\left(\frac{7m^3 + 76m^2 + 253m + 240}{18m^2 + 72m + 54}\right)^{\frac{4}{5}} \frac{1}{\mathrm{Re}_n^{\frac{1}{5}}} R^{\frac{4}{5}}, \quad (3)$$

where *M, L, H, J, K* – relative characteristic quantities of the dynamic three-dimensional boundary layer [10].

Denote the parameter

$$A = \left(\frac{18m^3 + 99m^2 - 171m}{m^3 + 18m^2 + 105m + 200}\right)^{\frac{1}{10}}\left(\frac{7m^3 + 76m^2 + 253m + 240}{18m^2 + 72m + 54}\right)^{\frac{1}{5}}.$$

The momentum losses thickness rotationally with a certain airfoil shape for a disc is

$$\delta_{\alpha d}^{**} = \left(\frac{5}{3} \frac{0{,}01256}{\sqrt{\frac{(1+H)J}{3LJ + 4L(K-2J)}}} \left(\frac{2}{J} + \frac{1}{L} \right) \right)^{0.8} \left(\frac{\nu}{\omega} \right)^{0{,}2} R^{0{,}6} = \tag{4}$$

$$= 4{,}492 \cdot 10^{-3} \left[\frac{(m+5)(7m^2+41m+48)}{\sqrt{\frac{(m+3)(m+5)(m+8)}{9m(2m^2+9m+5)}}(m+1)(m+3)} \right]^{\frac{4}{5}} \frac{1}{\mathrm{Re}_d^{\frac{1}{5}}} R^{\frac{4}{5}},$$

where $\omega = \omega_d - \omega_f$, ω_d – disk angular velocity, ω_f – flow angular velocity, and Reynolds criteria for disk $\mathrm{Re}_d = \frac{\omega R^2}{\nu}$.

Denote the parameter

$$B = \left[\frac{(m+5)(7m^2+41m+48)}{\sqrt{\frac{(m+3)(m+5)(m+8)}{9m(2m^2+9m+5)}}(m+1)(m+3)} \right]^{\frac{1}{5}}.$$

Assuming the expressions for the momentum thickness rotationally, the friction induced shear stress on the wall is determined as

$$\tau_{0\alpha}^{w} = \frac{0{,}02722 \rho \nu^2 \mathrm{Re}_n^{\frac{9}{5}}}{A R^2} \tag{5}$$

and on the disc it is

$$\tau_{0\alpha}^{d} = \frac{0{,}048515 \rho \nu^2 \mathrm{Re}_d^{\frac{9}{5}}}{B R^2} \tag{6}$$

The equation for disc friction loss coefficient for a rotor wheel is

$$C_{fr} = \frac{\tau R^2}{\rho \mathrm{Re}^2 \nu^2}. \tag{7}$$

Therefore, the friction coefficient equation for a rotational wall is

$$C_{fr\alpha}^{w} = \frac{0{,}02722}{A \mathrm{Re}_n^{\frac{1}{5}}}, \tag{8}$$

and for the friction coefficient for a disc rotationally, the equation is

$$C_{fr\alpha}^{d} = \frac{0{,}048515}{B \mathrm{Re}_d^{\frac{1}{5}}} \tag{9}$$

IOP Conf. Series: Materials Science and Engineering **822** (2020) 012023 doi:10.1088/1757-899X/822/1/012023

Should to take into account that when the disc losses is depend on the disc and wall frictions [5], therefore, the disc friction loss coefficient is determined as

$$C_M = C^w_{fr\alpha} + C^d_{fr\alpha}. \tag{10}$$

Also should to take into account the following: to determine disc friction loss coefficient and, thus, disc losses should be aware of distributing angular velocity of the plug flow ω_f depending on disc angular velocity ω_d, that requires to solve a differential equation [10]

$$\frac{d\omega_f}{dR} = -\frac{2\pi}{\rho \dot{V}_{leak}}\left(\tau^w_{0\alpha} + \tau^d_{0\alpha}\right) - \frac{2\omega_f}{R}, \tag{11}$$

or taking into account the equation for the shear stress of friction at the wall and the disc, the equation is

$$\frac{d\omega_f}{dR} = -\frac{2\pi}{\rho \dot{V}_{leak}}\left(\frac{0,02722\rho v^2 \mathrm{Re}_n^{\frac{9}{5}}}{AR^2} - \frac{0,048515\rho v^2 \mathrm{Re}_d^{\frac{9}{5}}}{BR^2}\right) - \frac{2\omega_f}{R}, \tag{12}$$

where $\dot{V}_{leak}$ - seal leakage.

A sophisticated assignment is to specify analytical dependence of angular velocity of the plug flow ω_f on the disc angular velocity ω_d. For practically important occasions as the first approximation for the non-leakage flow while determining disc friction coefficient should apply $\omega_f = 0,5\omega_d$, then expression is resulted as

$$C_M = \frac{1}{\left(0,5\mathrm{Re}_d\right)^{\frac{1}{5}}}\left(\frac{0,02722}{A} + \frac{0,048515}{B}\right). \tag{13}$$

Figure 1 presents the dependence of disc friction coefficient for the turbulent flow if Re>10^5.

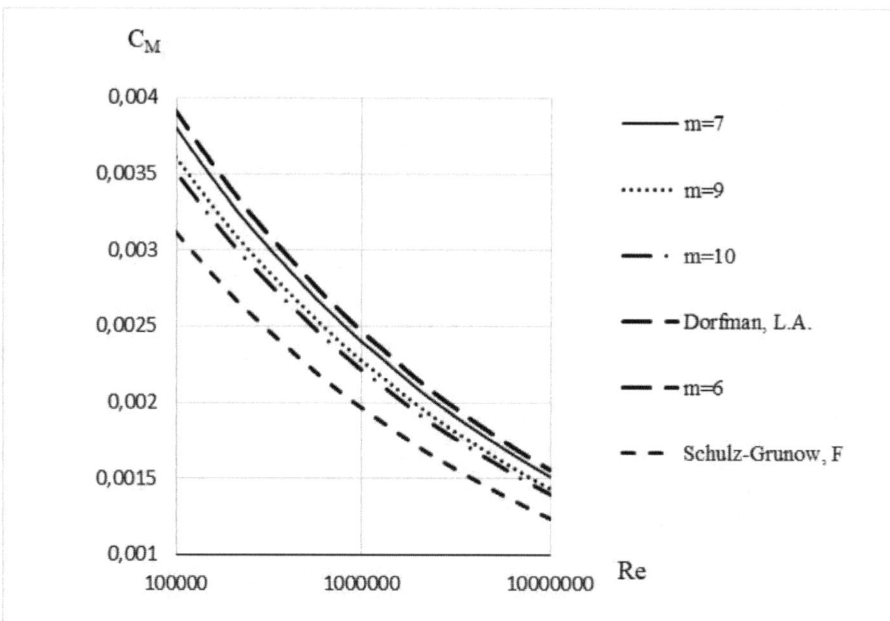

Figure 1. The disc friction coefficient of turbulent flow

Should be emphasized that if $m = 6$ function coincides with the dependence by L.A. Dorfman [4]. The dependence by F. Schulz-Grunow [2] is located lower than the obtained dependences. Should to take into account that the obtained equations demonstrate a good convergence with the results, obtained by other researchers and according to experimental data, for the different degrees of distributing the velocity traverse under the turbulence flow.

A separate assignment is to determine an airfoil shape m depending on the criterion Re of the plug flow and Re of the disc. When the disc friction and thermal efficiency in the rotary cavities are specified depending on the mode parameters (Reynolds criterion) of the earlier obtained airfoil shape degree of the function of distributing flow velocity, recommended to consider m=9÷11 at the disc and m=7÷9 at a wall. Reynolds criterion at the disc is determined with angular velocity of a pump wheel disc (4) or a turbine.

3. Conclusion

The obtained equations for the disc friction loss coefficients allow to determine the efficiency loss to the disc friction in the TPU of LPE centrifugal-type pumps. Comparing with the empirical dependencies obtained by other researchers, using degree m of the turbulent dynamic three-dimensional boundary layer significantly increase the range of the area and reliable determination of the disc friction.

References

[1] Von Karman Th 1921 Uber Laminare und Turbulente Reibung, Zeitschrift fiir Angewandte Mathematik und Mechanik (ZAMM) **1/4** pp 233-251

[2] Schulz-Grunow F 1935 Der Reibungswiderstand vortierender Scheilen in Gehausen, Zeitschrift fiir Angewandte Mathematik und Mechanik (ZAMM) **15/4** pp 191-204

[3] Schliechting G 1969 *Boundary layer theory* (Moscow: Nauka (Science))

[4] Dorfman L A 1963 *Hydrodynamic Resistance and the Heat Loss of Rotating Solids (Edinburgh: Oliver & Boyd)*

[5] Dorfman L A 1960 *Hydrodynamic Resistance and the Heat Loss of Rotating Bodies* (Moscow: Physmatgis)

[6] Jacques R Le, Quere P and Daube O 2002 Axisymmetric Numerical Simulations of Turbulent Flow in Rotor- Stator Enclosures *International Journal of Heat and Fluid Flow* **23/4** pp 381-397

[7] Owen J M, Haynes C M and Bayley F J 1974 Heat Transfer From an Air Cooled Rotating Disk *Proceedings of the Royal Society of London* **336** pp 453-473

[8] Kusyumov A N and Romanova E V 2011 Flow between Finite and Infinite Disks with Consideration of Surface Mobility *Russian Aeronautics* **54/2** pp 146-153

[9] Evgcn'cv S S, Petrosyan G G and Futin V A 2009 Calculation of Axial Gasodynamic Forces, Disk Friction Losses and Overflow in Semiopen Impellers of Centrifugal Compressors *Russian Aeronautics* **52/3** pp 319-326

[10] Kishkin A A, Zuev A A and Delkov A V 2016 Three-dimensional temperature boundary layer within the theory of convective heat transfer (*Siberian State Aerospace University, Krasnoyarsk*)

[11] Zhuikov D A, Kishkin A A and Zuev A A 2013 Calculation of axial force in case of flow in the slots of turbomachinery equipment *Scientific-Educational and Applied Journal University News. North-Caucasian Region. Technical Science Series* **170** pp 24-27

Magnetocapacity of manganese sulphides substituted by thulium ions

A M Kharkov, M N Sitnikov, V V Kretinin, O B Fisenko and G Yu Fillipson

Reshetnev Siberian State University of Science and Technology, 31 Krasnoyarsky Rabochy Av., Krasnoyarsk, 660037, Russia

E-mail: khark.anton@mail.ru

Abstract. The capacity and the dielectric loss tangent of a $Tm_xMn_{1-x}S$ (x=0.05, x=0.1) solid solution in the frequency range (1–300) kHz without field H=0 and in a magnetic field of H=8 kOe in the temperature range (80–500) K were measured. The increase of dielectric permittivity and the maximum of dielectric losses at high temperatures were found. The shift of the maximum of the imaginary part of the permittivity towards high temperatures with increasing concentration is found. The magnetocapacitance effect for two compositions and change the sign of magnetocapacity with increasing concentration were revealed. The magnetocapacity is explained in terms of the model with orbital electron ordering.

1. Introduction

Materials based on solid solutions, which show the relationship of magnetic and dielectric properties are interesting both from fundamental and practical standpoints. Of a particular attention are materials that demonstrate magnetoelectric properties at room and higher temperatures due to their practical application in microelectronics for data recording and storage [1].

Among such studied materials is $BiFeO_3$ bismuth ferrite [2]. The giant magnetocapacitance effect was observed in $LuFe_2O_4$ at room temperature and was explained by fluctuations of charges with different values of spins in Fe^{2+} и Fe^{3+} [3] ions as a result of splitting between two types of the charge order by an external magnetic field.

Orbital degeneracy in manganese sulfide MnS can occur as a result of the substitution of the divalent manganese ion with trivalent thulium ion at n-type doping. Thulium sulfide TmS is a metal and has the same crystal and magnetic structure as that of the MnS semiconductor. Orbital ordering is possible due to strong electron correlations in MnS.

Nonuniform electrical system with orbital degeneration can be obtained by the substitution of trivalent thulium ions for bivalent manganese ions. The manganese sulphide MnS is semiconductor and thulium sulphide TmS is degenerate semimetals, the electrical resistances of which differ by factors of 10^6–10^9 [4].

Thulium sulphide TmS and manganese sulphide MnS have an antiferromagnetic structure of the second-type ordering with the Neel temperatures T_N = 63 K and T_N = 137 K respectively. The spin-orbital and the Jahn–Teller interactions split t_{2g} – electronic states and induce the spin splitting of the spectrum of electron excitations. As a result, the dielectric properties can be controlled by electric and magnetic fields [5].

The aim of this work is to establish the effect of the magnetic field on the dielectric properties in the spin-disordered region in solid solutions $Tm_xMn_{1-x}S$.

2. Experimental results and discussion

The $Tm_xMn_{1-x}S$ solid solutions were synthesized by the solid-phase reaction method described in [6] from powders of the initial compounds in evacuated quartz ampoules in a single-zone resistance furnace. The phase composition and the crystal structure of the $Tm_xMn_{1-x}S$ samples were studied at 300 K using a DRON-3 diffractometer with CuK_α – radiation. The X-ray diffraction data showed that the synthesized compounds have a NaCl-type face-centered cubic (fcc) structure typical of manganese monosulphide. As the concentration of cation (*X*) increases, unit cell parameter *a* increases from a = 0.5440 nm for MnS to a = 0.5520 nm for $Tm_{0.1}Mn_{0.9}S$ [7, 8].

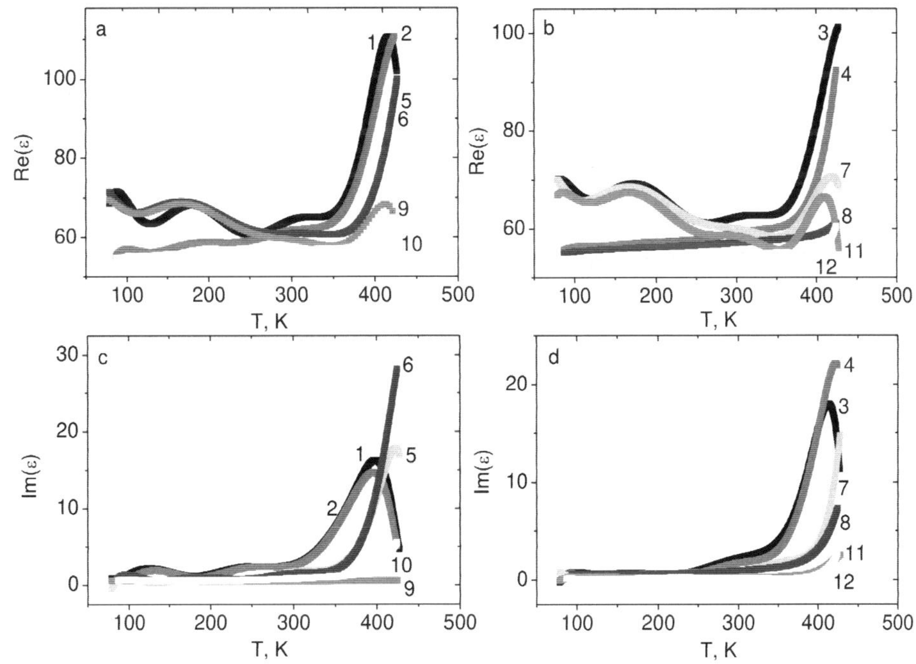

Figure 1. The real Re(ε) (*a*, *b*) and imaginary Im(ε) (*c*, *d*) components of the permittivity from temperature for the sample $Tm_{0.05}Mn_{0.95}S$ a without field H = 0 at frequencies of ω = 1 (1), 10 (5), 100 (9) kHz (a, c) and ω = 5 (3), 50 (7), 300 (11) kHz (*b*, *d*) and in a magnetic field of H = 8 kOe at frequencies of ω = 1 (2), 10 (6), 100 (10) kHz (a, c) and ω = 5 (4), 50 (8), 300 (12) kHz (*b*, *d*).

The capacity and the dielectric loss tangent (tg δ) were measured using an AM-3028 component analyzer in the temperature range 80–500 K in a zero magnetic field H=0 and in a magnetic field of H = 8 kOe. The magnetic field was applied parallel to the capacitor sheets [9, 10]. The magnetocapacitance effect $\delta\varepsilon$=(Reε(*H*)−Reε(0))/Reε(0) is determined as a result of the studies of the complex dielectric permittivity.

The spectral and temperature dependences of the dielectric constants can be used to the observation of the dipole electric moment and the determination of its characteristics, even when we are dealing with a local dipole moment in small clusters without a long-range order. The dielectric properties also give information of the charge transport and the processes of the charge ordering. The response of the dielectric properties on the action of a magnetic field gives a possibility of determining the main mechanisms that determine the correlation of the dielectric and the magnetic properties [11-13].

Figure 1 shows the temperature dependences of the real Re(ε) (a, b) and imaginary Im(ε) (c, d) parts of the dielectric permittivity of the $Tm_{0.05}Mn_{0.95}S$. sample. The real part of the dielectric permittivity is increased and the maximum of dielectric loss is shifted at T> 400 K to high temperatures with frequency increasing.

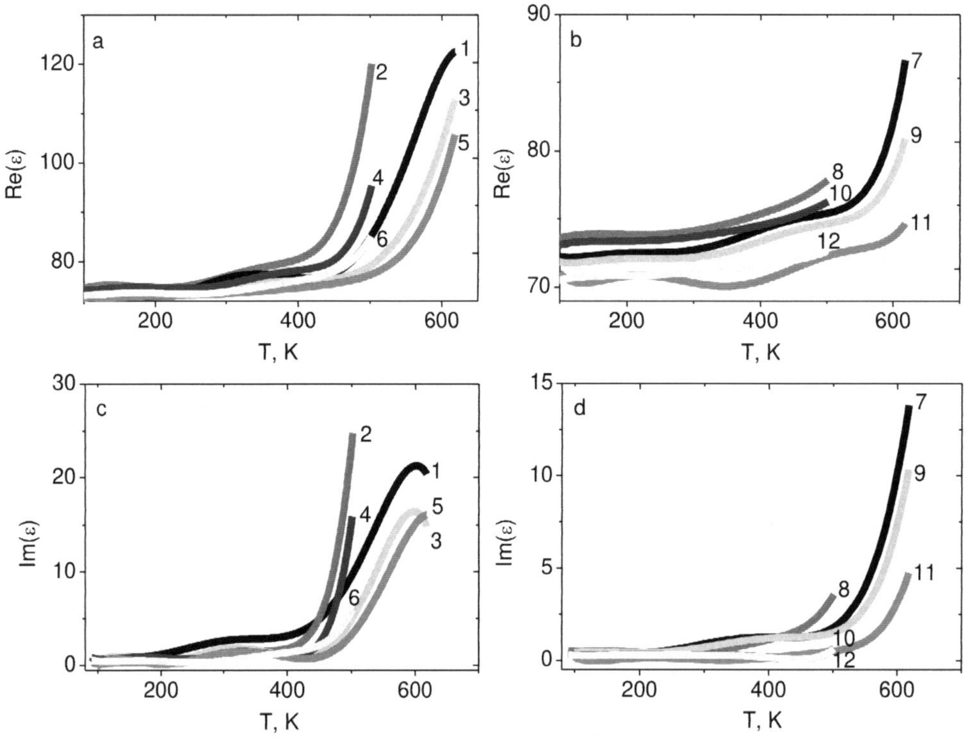

Figure 2. The real Re(ε) (*a*, *b*) and imaginary Im(ε) (*c*, *d*) components of the permittivity from temperature for the sample $Tm_{0.1}Mn_{0.9}S$ a without field H = 0 at frequencies of ω = 1 (1), 5 (3), 10 (5) kHz (a, c) and ω = 50 (7), 100 (9), 300 (11) kHz (*b*, *d*) and in a magnetic field of H = 8 kOe at frequencies of ω = 1 (2), 5 (4), 10 (6) kHz (a, c) and ω = 50 (8), 100 (10), 300 (12) kHz (*b*, *d*).

Magnetocapacity is shown in figure 3a and attend 18% for composition $Tm_{0.05}Mn_{0.95}S$. The magnetocapacity is decreased and reaches a minimum of about 20% at the temperature T=400 K for all frequencies. The decrease in the dielectric permittivity in the magnetic field is caused by the spin-orbit interaction and the formation of a dipole glass.

Figure 2 shows the temperature dependences of the real Re (ε) (a, b) and imaginary Im (ε) (c, d) parts of the dielectric permittivity of the sample $Tm_{0.1}Mn_{0.9}S$. The dispersion of inhomogeneous electronic states and local magnetic fields is growth at increasing of the concentration of thulium ions. Temperatures maximum of the imaginary part of the dielectric permittivity for x = 0.1 (figure 2) practically do not shift in a magnetic field, and the value of dielectric losses increases. The dielectric permittivity increases in the magnetic field (figure 2a), the magnetocapacity is positive for all temperature range (figure 3b).

The imaginary part Imε has an inflection point at T = 580 K without magnetic field at low frequencies for composition $Tm_{0.1}Mn_{0.9}S$ (figure 2c). At heating the imaginary part Imε decreases with increasing frequency (figure 2d) and the dielectric permittivity increases at the magnetic field.

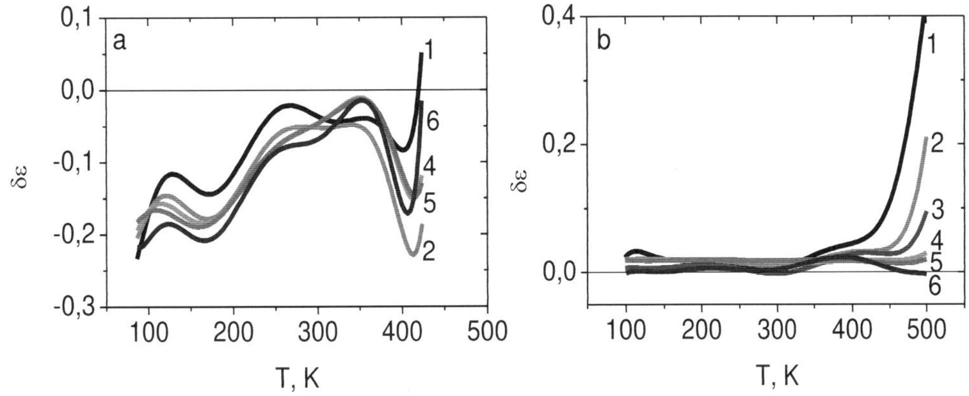

Figure 3. Dependence of magnetocapacity δε on temperature T for samples $Tm_xMn_{1-x}S$ with concentrations x = 0.05 (a), 0.1 (b) a without field H = 0 and in a magnetic field of H = 8 kOe at frequencies of ω = 1 (1), 5 (2), 10 (3), 50 (4), 100 (5), 300 (6) kHz.

The relative change in the dielectric permittivity $\delta\varepsilon=(\mathrm{Re}\varepsilon(H)-\mathrm{Re}\varepsilon(0))/\mathrm{Re}\varepsilon(0)$ at the magnetic field as a function of temperature for the sample $Tm_{0.1}Mn_{0.9}S$ is shown in figure 3b. Magnetocapacity $\delta\varepsilon$ for composition $Tm_{0.1}Mn_{0.9}S$ reaches a maximum of 40% at T = 500 K. This effect is explained by the pinning of conduction electrons at the Mn-Tm interface and the large electronic contribution to the dielectric permittivity.

3. Conclusion

Measurements of dielectric permittivity in solid solutions of $Tm_xMn_{1-x}S$ were carried out, a magnetocapacity effect was found at temperatures of 80–500 K at frequencies of 1–300 kHz. The experimental data are explained in terms of model with the orbital ordering. At temperatures lower than the Debye temperature, polarons are pinned at the interface with the formation of the orbital magnetic moment at a site and the dielectric permittivity anisotropy.

The orbital «glass» is created for thulium ion concentrations less than the percolation concentration and a long-range orbital order is induced at the interface at higher concentrations. The change in the orbital correlations of the magnetic angular moment at magnetic field lead to change in the dielectric permittivity anisotropy.

Acknowledgement
This study was supported by the Russian Foundation for Basic Research № 18-32-00079 mol_a.

References

[1] Zeches R J, Rossell M D, Zhang J X, Hatt A J, He Q and Yang C H 2009 *Science* **326** 977
[2] Yang J C, He Q, Suresha S J, Kuo C Y, Peng C Y, Haislmaier R C and Motyka M A 2012 *Phys. Rev. Lett.* **109** 247606
[3] Kambe T, Fukada Y, Kano J, Nagata T, Okazaki H, Yokoya T, Wakimoto S, Kakurai K and Ikeda N 2013 *Phys. Rev. Lett.* **110** 117602
[4] Udod L V, Sitnikov M N, Aplesnin S S and Molokeev M S 2014 *Solid State Phen.* **215** 503
[5] Aplesnin S S, Romanova O B, Galyas A I and Sokolov V V 2016 *Phys. Solid State* **58** 19
[6] Aplesnin S S, Kretinin V V, Ostapenko A A, Galyas A I and Yanushkevich K I 2014 *Solid State Phen.*, **215** 378
[7] Aplesnin S S and Sitnikov M N 2016 *Phys. Solid State* **58** 1148
[8] Aplesnin S S, Sitnikov M N, Romanova O B and Pichugin A Yu 2016 *Phys. Stat. Sol. B* **253** 1771

[9] Aplesnin S S, Kretinin V V, Panasevich A M and Yanushkevich K I 2017 *Phys. Solid State* **59** 667
[10] Romanova O B, Aplesnin S S, Kharkov A M, Masyugin A N and Yanushkevich K I 2017 *Phys. Solid State* **59** 1290
[11] Aplesnin S S, Udod L V and Sitnikov M N 2018 *Ceramics International* **44** 1614
[12] Aplesnin S S, Sitnikov M N and Zhivulko A M 2018 *Phys. Solid State* **60** 673
[13] Romanova O B, Aplesnin S S, Kharkov A M, Kretinin V V and Zhivulko A M 2018 *Phys. Solid State* **60** 1650

Magnetoelectric effect in bismuth - neodymium ferrite - garnet films

A N Masyugin, S S Aplesnin, Y Y Loginov and O N Bandurina

Reshetnev Siberian State University of Science and Technology, 31 Krasnoyarsky Rabochy Av., Krasnoyarsk, 660037, Russian Federation

E-mail: apl@iph.krasn.ru

Abstract. The prehistory of the dielectric properties of bismuth – neodymium ferrite – garnet films on a $Gd_3Ga_5O_{12}$ (111) substrate cooled in an electric and magnetic field were established. The dynamic characteristics of the electric polarization are studied at large times when the electric field is turned on and off. The influence of a magnetic field on the residual polarization was found. The dependence of the electric polarization on the electric field in a magnetic field is determined. Experimental data are explained in the Debye model.

1. Introduction

Multiferroics are characterized by a strong interrelation of the magnetic and electrical subsystems [1-6], which in single-phase ferromagnetic materials is realized due to the spin-lattice and electron-lattice interactions. Studies of single-phase materials are important for a deeper understanding of electromagnetic phenomena in solids and are of interest for the creation of a new generation of solid-state electronics devices.

Yttrium ferrite garnet has cubic symmetry with an inversion center [7] and at low temperatures (below 130 K) a structural transition with triclinic lattice distortion is detected [8, 9]. The electric polarization may be caused by deformation of the structure, which leads to breaking of the center inversion as a result of epitaxial film stress on the substrate or cationic substitution on dodecahedral nodes; surface electron states, magnetic domain structure. In films $(BiLu)_3(FeGa)_5O_{12}$ thickness of 10 μm, grown by liquid phase epitaxy on a substrate $Gd_3Ga_5O_{12}$ with a substrate orientation (210) is found the electric polarization of the domain walls, which can be switched by the external magnetic field [10]. Electric polarization is absent on films with substrate (111) orientation. These effects can be explained due to inhomogeneous magnetoelectric interaction and changes the magnetic anisotropy of the electric field [11]. The last factor can be neglected using external magnetic field an order of magnitude greater than the saturation field.

2 Magnetostriction

The purpose of the research is to establish the mechanism of the magnetoelectric interaction in a strong magnetic field in bismuth - neodymium ferrites - garnets films deposited on a garnet substrate. The $Nd_{0.5}Bi_{2.5}Fe_5O_{12}$ film (450 nm) was studied on a single-crystal substrate $Gd_3Ga_5O_{12}$ (GGG) in the (111) direction. The films were obtained by the method of epitaxial deposition [7].

The films have a maximum of magnetostriction and electrostriction at a temperature of 200 K. Expansion goes into compression of the film in a magnetic field at cooling . The experimental data are explained in terms of model of ferroelectric domains and the magnetoelectric interaction. The maximum of magnetostriction is associated with the formation of a dipole glass and domain pinning [12].

3. Capacitance

Cooling the films in a magnetic field H = 12 kOe and in an electric field E = 400 V/cm from room temperature to T = 80 K leads to a decrease in capacitance within one percent. After cooling and further heating to T = 300 K film capacity decreases by (3-4) % (figure 1).

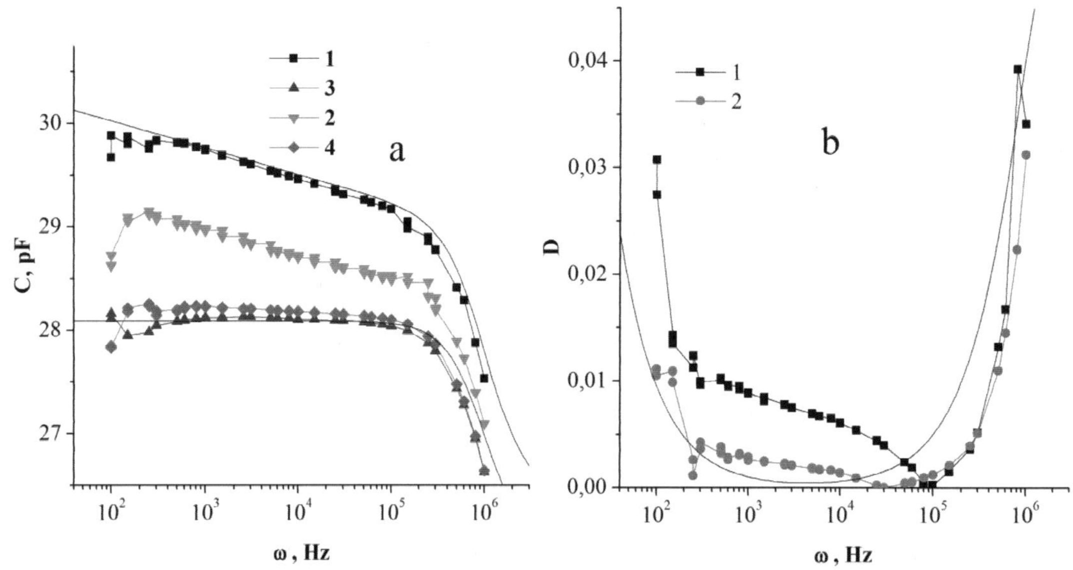

Figure 1. Frequency dependence of capacitance of $Nd_1Bi_2Fe_5O_{12}$(450nm) / $Nd_2Bi_1Fe_4Ga_1O_{12}$ (90nm) film on glass (a): 1 - initial state at T = 320 K, 2 - after cooling in a field E = 400 V / cm and heated to T = 320 K, 3- after cooling in a field E = 400 V / cm at T = 80 K, 4- after cooling in a magnetic field H = 12 kOe at T = 80 K; and dielectric loss tangent (b): 1 - initial state at T = 320 K, 2 - after cooling in a magnetic field H = 12 kOe at T = 80 K. Capacitance and dielectric losses calculated in the Debye model (1) (solid lines).

The dependence of the dielectric constant on frequency at low temperatures is satisfactorily described in the Debye model with a relaxation frequency of 5 MHz. The dielectric relaxation in the low frequency region ω <300 Hz, is caused by electron jumps in the film defects Im(ε)~σ/ω [13–15]. At temperatures above room temperature, the diffusion contribution of the domain boundaries is added to the dielectric susceptibility in the form:

$$\mathrm{Re}(\varepsilon)=\varepsilon_0+\chi_0/(1+(\omega\tau)^2)-v\lg(\omega),\ \mathrm{Im}(\varepsilon)=\chi_0\omega\tau/(1+(\omega\tau)^2)+\sigma/\omega \quad (1)$$

where τ is the relaxation time, σ is the conductivity, v is the relaxation rate, χ_0 is the dielectric susceptibility in a constant field.

4. Remanent polarization

The dynamic characteristics of the electric polarization at large times is determined by measuring the charge when a rectangular voltage pulse is switched on with amplitude E = 400 V / cm with frequency ω = 0.01, 0.003 Hz (figure 2). The residual polarization found after turning off the field disappears at a temperature T = 300 K (figure 3). The electric polarization depends on the direction of the magnetic field applied perpendicularly and parallel to the film. Coefficient magnetoelectric interaction is the second-rank tensor.

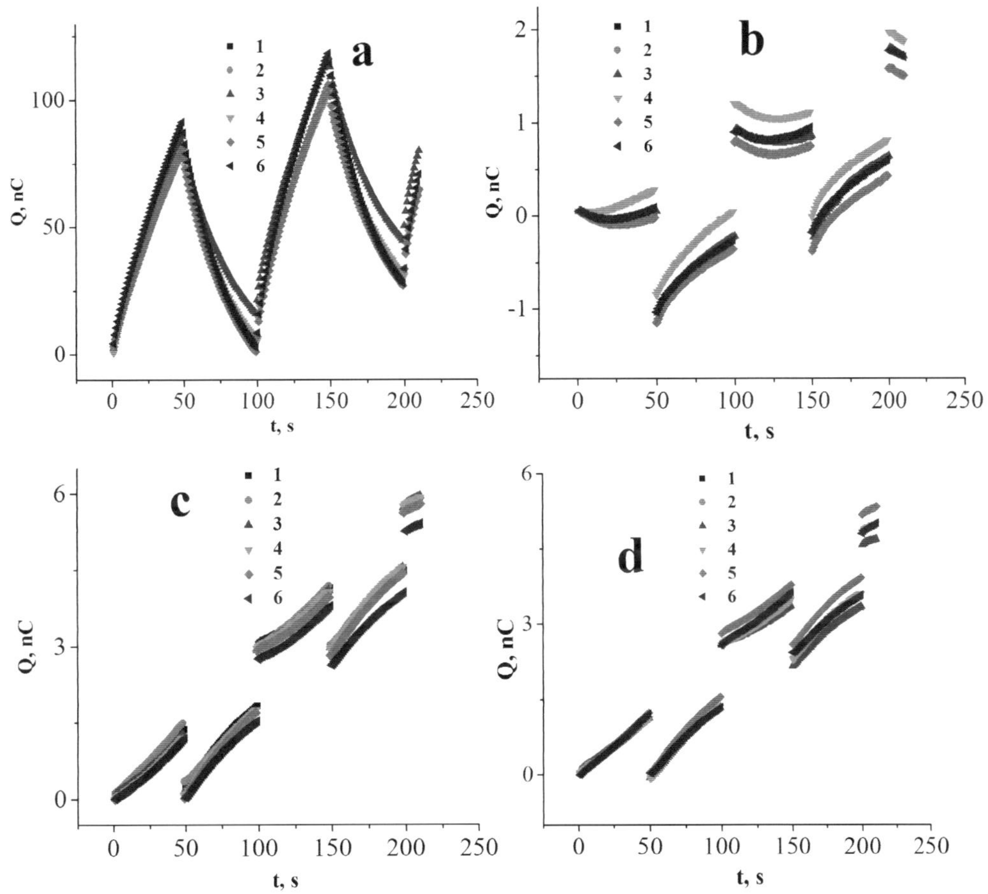

Figure 2. Dependences of the charge induced by periodic pulses of the electric field 400 V / cm from time, at 360 K (a), 300 K (b), 80 K (c), 200 K (d): 1, 6 - without a magnetic field; 2, 4, 5 - in a magnetic field of 12 kOe perpendicular to the film; 3 - in a magnetic field of 12 kOe along the film.

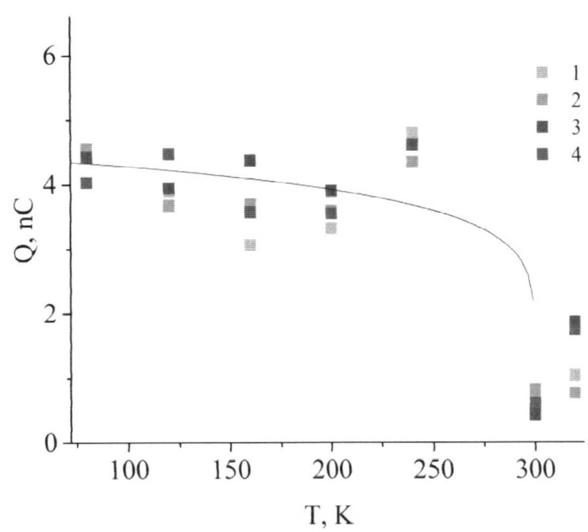

Figure 3. Residual charge after the impulse of the electric field of 400 V/cm: 1 - in a magnetic field of 12 kOe along the film; 2, 3 - in a magnetic field of 12 kOe perpendicular to the film; 4 - without magnetic field.

The electric polarization was found from the relation $P=\int jdt$ when measuring the current in an external quasi-periodic field with a frequency w = 0.01 Hz with different orientations of the magnetic field [16-18]. Electric polarization versus the external field is shown in figure 4. The shift of polarization is caused by negatively charged defects in the vicinity of the interface. Hysteresis is observed in the region of large electric fields and is associated with the formation of an electric field at the interface of the substrate – film. When heated, the relative polarization bias is halved from T = 80 K to T = 300 K. In the general case, the polarization is determined by the field of charged defects, the interface field (P_0) $P = P_0 + \chi\alpha E$. The dielectric susceptibility has a maximum at T = 280 K.

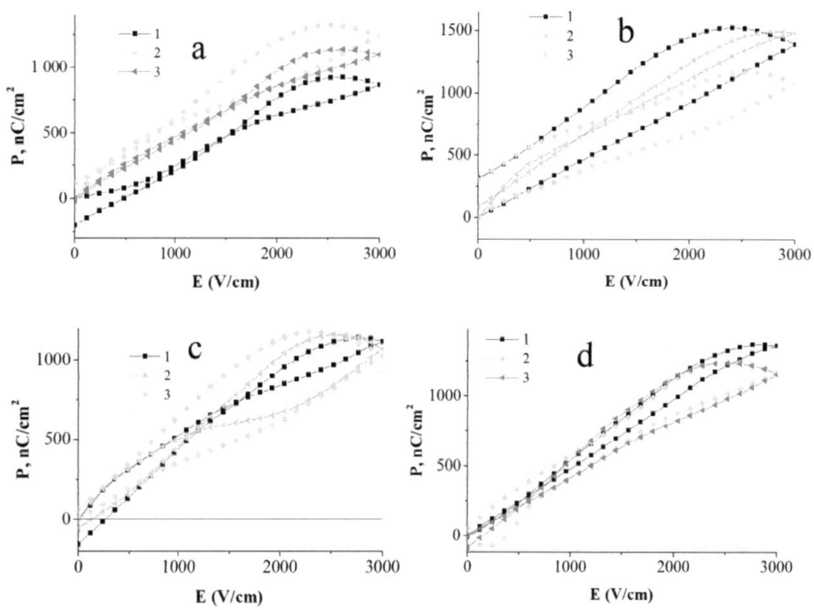

Figure 4. Electric polarization from an external electric field at 80K (a), 160K (b), 120K (c), 280K: 1 - without a magnetic field, 2 - in a magnetic field directed along the film, 3 - in a magnetic field directed perpendicular to the film.

5. Conclusion

A decrease in the capacity of the film after cooling from room temperature to T = 80 K in an electric field is found. The frequency dependence of the film capacity is described in the Debye model, the relaxation time is found. The critical temperature of the disappearance of the residual polarization and the anisotropy of the electric polarization in a magnetic field are found. The temperature of the maximum dielectric constant is established.

The reported study was funded by Russian Foundation for Basic Research, Government of Krasnoyarsk Territory, Krasnoyarsk Regional Fund of Science№ 18-42-240001 r_a, to the research project: «Inversion of the sign of the components of the magnetoelectric tensor on the temperature in films of bismuth garnet ferrite replaced by neodymium».

References

[1] Aplesnin S S, Ryabinkina L I, Romanova O B, Balaev D A, Demidenko O F, Yanushkevich K I and Miroshnichenko N S 2007 *Phys. Solid State* **49**1984

[2] Aplesnin S S, Romanova O B and Yanushkevich K I 2015 *Phys. Status Solidi B* **252** 1792

[3] Aplesnin S S, Romanova O B, Korolev V V, Sitnikov M N and Yanushkevich K I 2017 J. Appl. Phys. **121** 075701

[4] Udod L V, Aplesnin S S, Sitnikov M N and Molokeev M S 2014 *Physics of the Solid State* **56** 1315
[5] Aplesnin S S and Sitnikov M N 2014 *JETP Letters* **100** 95
[6] Yamakita J, Lou G, Nishikawa M, Kato T, Iwata S and Ishibashi T 2018 *Jpn. J. Appl. Phys.* **57** 09TC01
[7] Asada H, Kuwahara A, Sueyasu K, Ishibashi T, Liu Q, Lou G, Kishimoto K and Koyanagi T 2015 *Physics Procedia* **75** 932
[8] Kita E, Takano S, Tasaki A, Siratori K, Kohn K and Kimura S 1988 *J. Appl. Phys.* **64** 5659–61
[9] Pyatakov A P, Sechin D A, Sergeev A S, Nikolaev A V., Nikolaeva E P, Logginov A S and Zvezdin A K 2011 *EPL* **93** 17001
[10] Sergeev A S 2017 *J. Phys. Conf. Ser.* **929** 012085
[11] Aplesnin S S, Masyugin A N, Sitnicov M N, Rybina U I and Ishibashi T 2018 *J. Magn. Magn. Mater.* **464** 44–9
[12] Aplesnin S S and Sitnikov M N 2016 *Physics of the Solid State* **58** 1112-17
[13] Aplesnin S S, Kretinin V V, Panasevich A M and Yanushkevich K I 2017 *Physics of the Solid State.* **59**(4) 667-73
[14] Aplesnin S S, Udod L V, Sitnikov M N, Molokeev M S, Tarasova L S and Yanushkevich K I 2017 *Phys. Solid State* **59** 2268–73
[15] Shut V N, Syrtsov S R, Trublovsky V L, Poleyko A D, Kostomarov S V. and Mastyko L P 2009 *Ferroelectrics,* **386** 125-32
[16] Shut V N and Kashevich I F 2007 *Ferroelectrics* **450** 57-64
[17] Aplesnin S S, Udod L V., Sitnikov M N and Shestakov N P 2016 *Ceram. Int.* **42** 5177-83
[18] Aplesnin S S, Udod L V., Sitnikov M N, Kretinin V V., Molokeev M S and Mironova-Ulmane N 2018 *Mater. Res. Express* **5** 115202

IOP Conf. Series: Materials Science and Engineering **822** (2020) 012026 doi:10.1088/1757-899X/822/1/012026

Factor analysis of inelastic electron scattering cross section spectra of Si, Fe and iron silicides $FeSi_2$, Fe_5Si_3

A Yu Igumenov[1], A S Parshin[1], Yu L Mikhlin[2], T A Andryushchenko[1] and V O Kanzychakova[3]

[1] Reshetnev Siberian State University of Science and Technology, 31 Krasnoyarsky Rabochy Av., Krasnoyarsk, 660037, Russia
[2] Institute of Chemistry and Chemical Technologies of the Siberian Branch of the Russian Academy of Sciences, 660036, Krasnoyarsk, Russia
[3] OOO «RN-KrasnoyarskNIPIneft», 660022, Krasnoyarsk, Russia

E-mail: igumenovau@mail.ru

Abstract. The inelastic electron scattering cross-section spectra of Si, Fe and iron silicides $FeSi_2$, Fe_5Si_3 were calculated from the experimental reflected electron energy loss spectra as the product of the average inelastic mean free path and the differential cross section of the inelastic electron scattering. The inelastic electron scattering cross-section spectra were studied using factor analysis. This method allowed us to quantitatively separate the surface and volume contributions to the spectra, and determine the energy of the bulk plasmon more accurately than it is possible using traditional methods.

1. Introduction

The creation of nanoelectronics, nanophotonics and spintronics devices based on silicon and transition metals should be provided with a number of requirements for the structure synthesis methods, and for the analysis methods. Electron spectroscopy methods such as Auger electron spectroscopy (AES), X-ray photoelectron spectroscopy (XPS) and reflection electron energy loss spectroscopy (REELS) are widely used for the study of these structures. The research of Fe-Si system is perspective from fundamental and applied points of view [1]. However, the elemental analysis of iron silicides by XPS and REELS is difficult, because the energies of photoelectron and plasmon peaks for silicides of different compositions are close to each other [2].

The inelastic electron scattering cross section spectra (so-called $K\lambda$-spectra [3]) are the products of inelastic mean free path λ and inelastic scattering cross-section $K(E_0, T)$, where T is energy loss, calculated as the difference between the primary electron beam energy E_0 (zero loss) and the reflected electron energy E, $T = E_0 - E$. The advantage of inelastic electron scattering cross section spectra is that, unlike the REELS, the $K\lambda$-spectra exclude multiple excitations losses, and the intensities are determined in absolute units. These spectra are also more sensitive to changes in the of the primary electron energy and the emission angle [4–6]. Inelastic electron scattering cross section spectroscopy allows determining the elemental composition with much greater accuracy than the traditional REELS [6].

In this paper, we used factor analysis (FA) to study the inelastic electron scattering cross section spectra of Si, Fe and silicides $FeSi_2$, Fe_5Si_3. This method allowed us to solve the actual problem of decomposition spectra to different origin contributions, quantify it and determine the volume plasmon energies more accurately than the traditional methods. The study of electron energy loss processes by decomposition spectra to the different origin contributions is one of the most important problems of electron spectroscopy, which can be used to assess the effect of surface excitations on the REELS, XPS and AES [1-15]. Earlier, we studied Si, Fe, and $FeSi_2$, Fe_5Si_3 silicides in [16-21] using the author's method of spectra decomposition to loss peaks.

2. Experimental Part

We investigated commercial single crystal Si(100) substrates. The Fe samples were fabricated by high-vacuum thermal evaporation. Iron silicides $FeSi_2$, Fe_5Si_3 produced by melting a mixture of iron and silicon in atomic ratios of 1×2, 5×3 under high vacuum conditions using the vacuum deposition installation UVN-2M-1. The mixture was maintained at the melting temperature for 15 min, after which the annealed alloy was crushed and again melted for 15 min. The 1 mm thick washers were cut of bulk samples. Before the spectroscopic investigations the washers were polished.

The spectroscopic experiments were carried out using a photoelectron spectrometer SPECS (German production) equipped with the spherical energy analyzer PHOIBOS MCD9, double anode X-ray tube as an X-ray source, and a Microfocus EK-12-M electron gun (STAIB Instruments) for excitation of the REEL spectra. The probing and reflected electrons angles were respectively $\alpha_i = 59°$, $\alpha_d = 0°$. The surface contaminations, protecting layers and oxide layers were removed applying Ar^+ ion etching (accelerating voltage of 2.5 kV, ion current of 15 μA) with an ion source PU-IQE-12/38 (SPECS) in the analytical chamber of the spectrometer before the REELS measurements; the completeness of the procedure was controlled via the relevant photoelectron and Auger spectra. REEL spectra were acquired in the interval of 150 eV below the elastic peak with an energy step of 0.1 eV. Energy loss T was calculated as the difference between the primary electron beam energy E_0 (zero loss) and the reflected electron energy E, $T = E_0 - E$. The primary electron beam energies were 300, 600, 1200, 1900, 3000 eV, and the full width at half-maximum (FWHM) of the primary electron beam was less than 1 eV.

The inelastic electron scattering cross section spectra were calculated from the experimental reflection electron energy loss spectra with the software package QUASES^{TM} XS REELS (Quantitative Analysis of Surfaces by Electron Spectroscopy cross section determined by REELS) [22] based to the algorithm suggested in [23]. Inelastic electron scattering cross section spectra indicate the probability of electron energy loss T in a single scattering event. The values of the maxima of these spectra determine the probabilities of single energy loss for surface or volume excitations. Due to the absolute intensity values of the energy loss, the inelastic electron scattering cross section spectra, compared with the traditional reflected electron energy loss spectra, allow to compare and analyze the spectra of different materials not only by peak energies, but also by its intensities.

Factor analysis is a statistical method widely used in electron spectroscopy [10]. Factor analysis is using to study the spectra representing a linear combination of components with concentration varying during some process. This method is relevant for AES, where the simple decomposition of spectral lines into components is impossible, because in the Auger spectra, in addition to the chemical shift, the shape of the lines also changes [11–13]. In electronic Auger spectroscopy, FA is used for the processes of layer-by-layer etching of thin films [12], to study the processes of adsorption [13, 14].

In [15], the XPS method was used to study chemical changes induced in a natural pyrite crystal (FeS_2) under the influence of a He^+ beam. It is shown that the application of conventional analysis together with FA can be a powerful technique to identify elements, when changes in binding energy and broadening occur simultaneously. Through a combination of conventional analysis of the XPS spectra and FA, the full signal Fe 2p XPS was analyzed.

As remarked in [15], the important advantage that FA, in contrast with conventional analysis, uses the whole set of spectra, along the full ion bombardment process (while conventional fits each spectrum at a time), and it does not require neither preconditions nor predefined lineshapes to start with. FA does not require the previous knowledge of the shape and position of XPS yields, allowing us to obtain both these results just by using only the experimental yields. On the other hand, this is the reason of the noisy aspect and poorer definition of these lineshapes compared to conventional results.

The inelastic electron scattering cross section spectra obtained at different primary electron energies or different probing and reflected electron angles are data series. The relative contributions of surface and volume losses vary systematically depending on experimental conditions. In [5], factor analysis is used as a method for decomposing the inelastic electron scattering cross section spectra into two contributions (volume and surface) in the study of Si and SiO_2.

Factor analysis is performed as follows. The matrix with intensity values of energy loss at different energies of the primary electrons (columns) is prepared. At the first stage, the main factors (components) are selected, which together give a spectrum close to the original one. After, from the obtained matrix two columns with the highest intensities are distinguished - the main abstract factors (main components). Then the data is transformed by multiplying with the rotation matrix R (target rotation):

$$R = \begin{bmatrix} \cos\varphi & \sin\varphi \\ -\sin\varphi & \cos\varphi \end{bmatrix} \quad (1)$$

Target rotation transforms abstract factors into physically significant ones by achieving a predetermined criterion of peak shape. In [5] the Lorentz form of the peak corresponding to the excitation of a volume plasmon (so called volume-loss component or volume factor) was adopted, and by changing the angle φ, the shape of the component is reached close to the Lorentz peak, which has a maximum at the energy of the volume plasmon. Then additional rotation was achieved the most correct form of the surface-loss component (surface factor). Next, the source spectrum is reconstructed as a linear combination of two factors.

In this work, in contrast to [5], the shape of the peaks was chosen not close to the shape of the standard Lorentz peaks, but to the shape of the Tougaard peaks [24]:

$$\lambda K = \frac{BT}{\left(C - T^2\right)^2 + DT^2} \quad (2)$$

B, C, D are fitting parameters and have special values for different elements [21]. The peak intensity depends on parameter *B*, peak energy depends on parameter *C* but does not correspond to it explicitly, peak width and, indirectly, peak energy depend on parameter *D*. Tougaard functions (universal classes of inelastic electron scattering cross section) are used to describe inelastic electron scattering cross section spectra.

The Tougaard three-parameter function has been modified:

$$\lambda K = \frac{BT}{\left(\left(T_p\left(-T_p + \sqrt{4T_p^2 + D}\right)\right) - T^2\right)^2 + DT^2} \quad (3)$$

where the parameter T_p corresponds to the position of the peak maximum in absolute units.

3. Results and discussion

Figures 1-8 show the *Kλ*-spectra of Si, Fe and silicides $FeSi_2$, Fe_5Si_3 for E_0 = 300 eV and E_0 = 3000 eV. The main peak energy is 16.6 ± 0.3 eV for Si, 22.4 ± 0.4 eV for Fe, 20.6 ± 0.2 and 20.0 ± 0.7 eV for $FeSi_2$ and Fe_5Si_3 respectively. These values are close to the energies of the volume plasmons of these materials in the REELS spectra [25]. The energies are determined from spectra in a differential form, since this approach allows one to determine the energies more accurately. In the spectra of silicon at energy loss 10.8 ± 0.1 eV, a shoulder due to the presence of a surface plasmon is observed [25]. In the spectra of Fe peak at energy of 15.4 ± 0.4 eV corresponds to the surface plasmon, peak at energy of 6.1 ± 0.2 eV corresponds to the interband transition [25].

These spectra are approximated by a linear combination of surface electronic excitation (surface factor) and volume -loss components (volume factor). When real factors were obtained, the angle was selected in such a way that the volume factor was well approximated by the Tougaard peak, similarly for the surface factor, but its shape remained irregular, indicating the presence of several unresolved peaks of different origin.

Table 1 presents the energy positions of the factor maxima. The obtained values of the energies of the volume and surface factor are close to the experimental data for the energy of the volume and surface plasmon. The energies of the bulk plasmon for iron silicides in experimental REELS are often the same. As can be seen from table 1, the energies of the volume factor (corresponding to the volume plasmon) for $FeSi_2$ and Fe_5Si_3 have values that differ by 0.3 eV. This method gives a high accuracy in determining the energy of the bulk plasmon using the criterion of similarity of the volume factor shape and the Tougaard peak.

Table 1. Factor energies (eV).

Sample	Surface factor	Volume factor
Si	10.8	16.8
$FeSi_2$	14.9	20.9
Fe_5Si_3	13.7	21.2
Fe	6.8	23.2

The ratio of the volume factor energy loss to the surface factor energy loss is close to $\sqrt{2}$, which is well-known from the classical plasmon theory for free electron gas [26]. As shown in [18], for Fe in the low-energy region, the interband transition at energy loss 6.87 eV has the highest intensity, which corresponds to the surface factor energy obtained in this work.

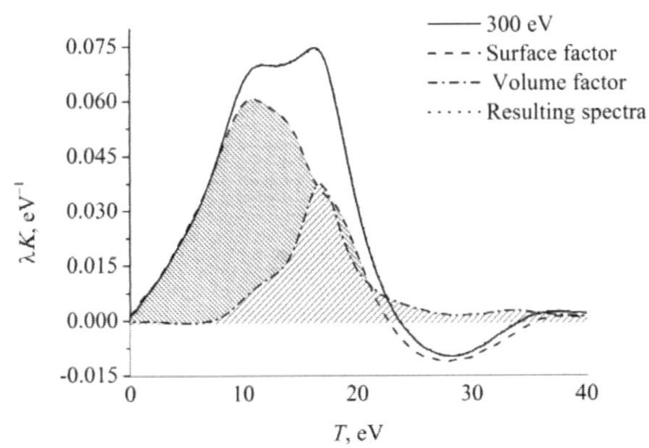

Figure 1. Si spectra at E_0 = 300 eV.

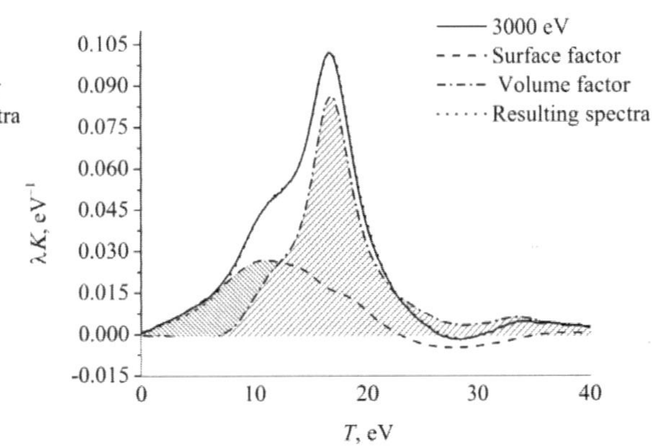

Figure 2. Si spectra at E_0 = 3000 eV.

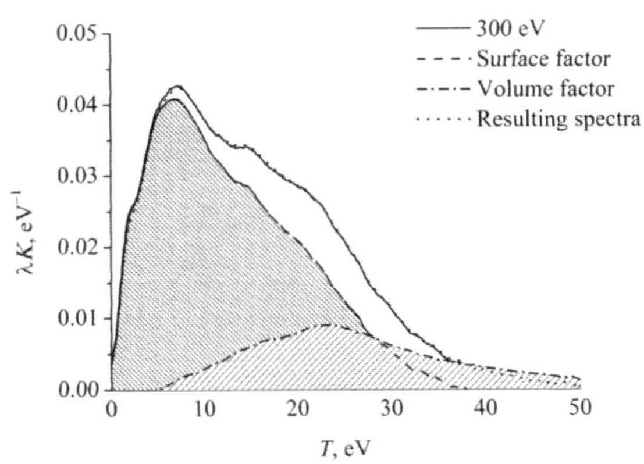

Figure 3. Fe spectra at E_0 = 300 eV.

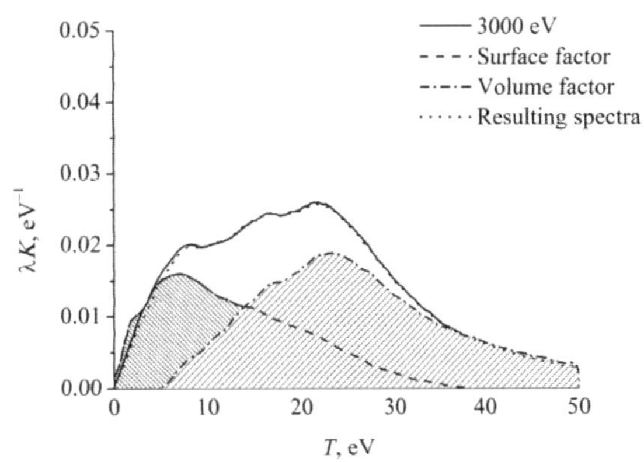

Figure 4. Fe spectra at E_0 = 3000 eV.

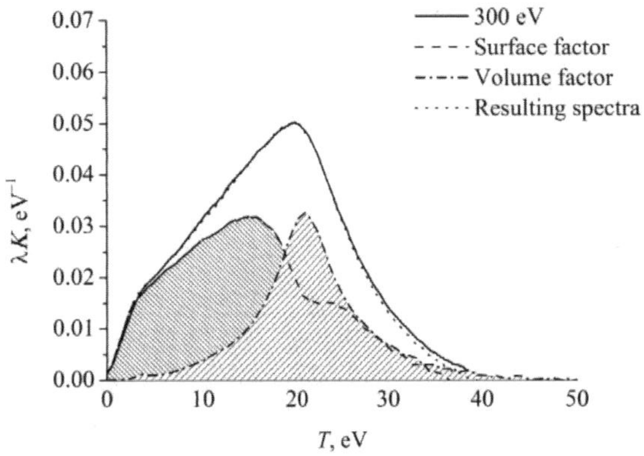

Figure 5. $FeSi_2$ spectra at E_0 = 300 eV.

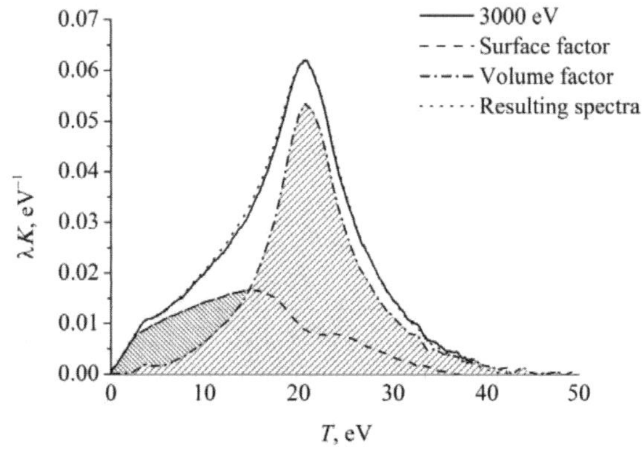

Figure 6. $FeSi_2$spectra at E_0 = 3000 eV.

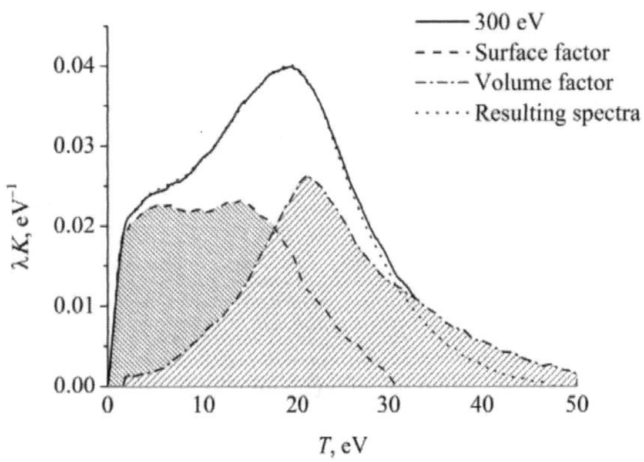

Figure 7. Fe_5Si_3 spectra at E_0 = 300 eV.

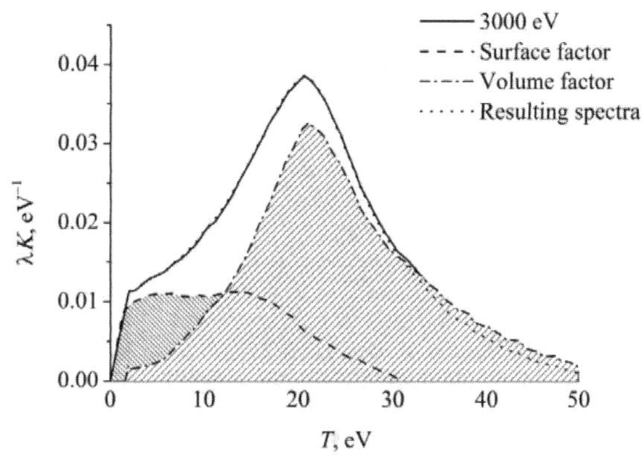

Figure 8. Fe_5Si_3 spectra at E_0 = 3000 eV.

The primary energy dependences of the factor amplitude are shown on the figures 7 and 8. Volume factor has gradually increasing primary energy dependence of amplitude which indicate its volume origin. Surface factor has gradually decreasing primary energy dependence of amplitude. Surface-loss component was unresolved in $K\lambda$-spectra for $FeSi_2$ and Fe_5Si_3 but FA allows determining its energy, intensity and origin.

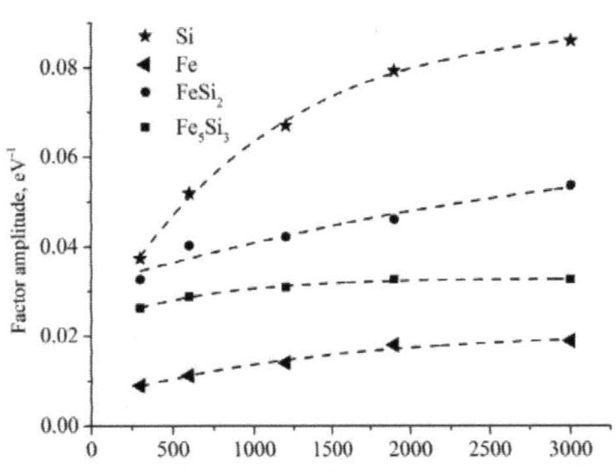

Figure 9. Primary electron energy dependence of volume factor amplitudes.

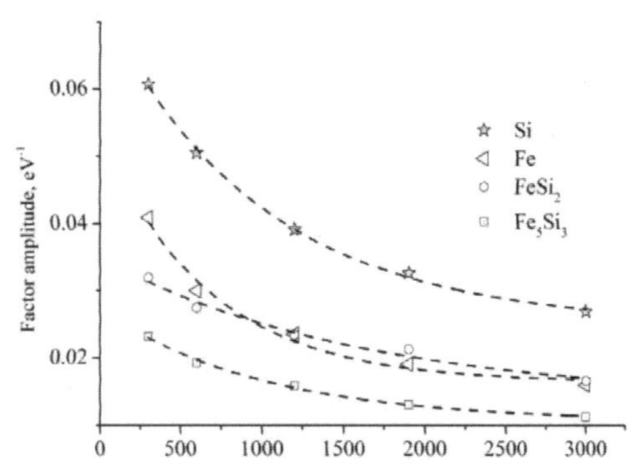

Figure 10. Primary electron energy dependence of surface factor amplitudes.

The factor analysis, along with the method of *Kλ*-spectra decomposition to loss components described by Tougaard functions [16-21], can be used to quantify changes in the contributions of various origin to the inelastic electron scattering cross section spectra.

4. Conclusions

The applicability of factor analysis to the separation of volume and surface contributions in the inelastic electron scattering cross section spectra of Si, Fe and silicides $FeSi_2$, Fe_5Si_3 is substantiated.

The separation and quantitative estimation of the surface and volume origin contributions in the inelastic electron scattering cross section spectra of Si, Fe and silicides $FeSi_2$, Fe_5Si_3 was carried out using factor analysis. The dependences of the amplitudes of the components on the energy of the primary electrons reflect its surface and volume origin.

References

[1] Galkin N G, Polyarnyi V O and Gouralnik A S 2004 *Thin Solid Films* **464-465** pp 199-203
[2] Ohtsu N, Oku M, Satoh K and Wagatsuma K 2013 *Appl. Surf. Sci.* **264** pp 219-224
[3] Tougaard S and Chorkendorff I 1987 *Physical Review B* **35** pp 6570-6577
[4] Jin H, Yoshikawa H, Iwai H, Tanuma S and Tougaard S 2009 *Journal of Surface Analysis* **15** p 321-324
[5] Jin H, Shinotsuka H, Yoshikawa H, Iwai H, Tanuma S and Tougaard S 2010 *Journal of applied physics* **107** pp 083709
[6] Jin H, Shinotsuka H, Yoshikawa H, Iwai H, Arai M, Tanuma S and Tougaard S 2013 *Surface and Interface Analysis* **45** pp 985–992
[7] Gergely G 2002 *Prog. Surf. Sci.* **71** pp 31-88
[8] Orosz G T, Gergely G, Gurban S, Menyhard M, Toth J, Varga D and Tougaard S 2003 *Vacuum* **71** pp 147–152
[9] Tougaard S 1988 *Surface And Interface Analysis* **11** pp 453-472
[10] Andryushechkin B.V., El'cov K.N., Klimov A.N. 2003 Eksperimental'naya tekhnika, metody i metodika raboty s galogenami na poverhnosti metallov (Moscow, Nauka)
[11] Malinowski E R and Howery D G 1980 *Factor Analysis in Chemistry (*New York, Wiley)
[12] Gaarenstroom S W 1982 *Journal of Vacuum Science and Technology* **20** pp 458–461
[13] Steffen J and Hofmann S 1988 *Surface Science* **202** pp. L607–L611
[14] Hofmann S and Steffen J 1989 *Surface And Interface Analysis* **14** pp 59–65
[15] Ruano S, Gustavo D and Pomiro F 2018 *Surface Science* **667** pp 138-147
[16] Parshin A S, Igumenov A Yu, Mikhlin Yu L, Pchelyakov O P and Zhigalov V S 2016 *Izvestiya Vysshikh Uchebnykh Zavedenii, Fizika* **59 10** pp 82-86
[17] Parshin A S, Igumenov A Yu, Mikhlin Yu L, Pchelyakov O P and Zhigalov V S 2017 *IOP Conference Series: Materials Science and Engineering* **255** pp 012019 1-7
[18] Parshin A S, Igumenov A Yu, Mikhlin Yu L , Pchelyakov O P and Zhigalov V S 2016 *Physics of the Solid State* **58** pp 908-914
[19] Parshin A S, Igumenov A Yu, Mikhlin Yu L, Pchelyakov O P and Zhigalov V S 2016 Technical Physic **61 (9)** pp 1418-1422
[20] Parshin A S, Igumenov A Yu, Mikhlin Yu L, Pchelyakov O P, Nikiforov A I and Timofeev V A 2015 *Semiconductors* **49 4** pp 423-427
[21] Parshin A S, Igumenov A Yu, Mikhlin Yu L, Pchelyakov O P, Nikiforov A I and Timofeev V A 2014 *Vestn. Sib. Gos. Aerokosm. Univ.* **4 (56)** pp 230-235
[22] Tougaard S QUASES – Software packages to characterize surface nano-structures by analysis of electron spectra, available at: http://www.quases.com/
[23] Tougaard S and Chorkendorff I 1987 *Physical Review B* **35** pp 6570-6577
[24] Tougaard S 1997 *Surface And Interface Analysis* **25** pp 137-154
[25] Lifshits V G and Lunyakov Yu V 2004 Electron Energy Loss Spectra of Surface Phases on Silicon (Vladivostok: Dal'nauka)
[26] Raether H 1988 Surface Plasmons on Smooth and Rough Surfaces and on Gratings (Berlin, Springer-Verlag)

Analysis and forecasting strategies of attacks on game resources for distributed cybersecurity training games

K V Safonov, V V Zolotarev and A M Derben

Reshetnev Siberian State University of Science and Technology, 31 Krasnoyarsky Rabochy Av., Krasnoyarsk, 660037, Russia

E-mail: amida.2@yandex.ru

Abstract. Game tasks are vulnerable to methods of obtaining benefits based on the results of attacks on game resources. In such conditions, it is critical to determine, monitor and evaluate the possibility of such attacks. It is also advisable to select methods to prevent them in advance, constructing the gaming environment accordingly.

The paper describes various types of attacks on the gaming environment, its vulnerable components. Some possibilities for building gaming environment were shown. To improve the architecture of the gaming environment, identify possible channels of key information leakage, game theory methods were chosen. The analysis was performed and the possibility of predicting the attack on game resources for a certain type of gaming environment was shown.

For a distributed multiplayer security awareness game based on a real case, the estimates for the selected types of attacks are shown, the attacker's selection criteria for the attack are calculated, and countermeasures are proposed. The article presents the results of two computational experiments with minor differences in the network infrastructure and the stored amount of game key information.

The results of the modelling of the main types of threats completely coincided with the obtained empirical data. The results shown can be used to evaluate various types of information security strategies in case of limited resources. The application of the approach is possible in various gamified media, including additional restrictions.

1. Introduction

There is a serious problem of countering various cyber attacks on game resources aor certain types of game tasks used in information security for training. These tasks include quest game tasks [1], serious games on critical management spheres, as early-warning systems [2], digital card games [3], games of the Capture the Flag format [4], as well as, to some extent, various types of gamified MOOC resources and Web-resources of universities [5, 6].

The main problem of the game task in this case is the presence of a key, answer or hint, integrated into the task or located on a separate game server. Also it has a problem in security dimension of bring-your-own-device (BYOD) conception, as the future of one modern challenge of serious games [7]. To obtain a key from a task by attacking game resources will mean a violation of the logic of the game or an unfair advantage if we talk about quest type games; for other types of games - the solution of tasks that would otherwise require the development of certain skills, and disturbance of the game balance.

Consider specific examples of game problems and their technical features. In the article [8], the authors give an example of a quest-type game based on an attack of social engineering aimed at imitating a company office. The logic of the game involves the consistent receipt of information elements and their comparison to form the correct conclusion (answer). Attack on game resources in this case is possible both by distributed cyber attacks, using viruses and phishing [9], and by violating the actions specified by the scenario. In addition, the vulnerabilities of the software can be used for bypassing the "game" protective countermeasures and gain an unfair advantage.

In the [4, 10] articles, the authors consider examples of the game based on the technical features of the protection measures, setting a specific key as an answer. The types of attacks here are also different, starting from the mentioned above advanced persistent threat (APT) and ending with identity theft (command identifier). Moreover, it is possible to substitute or steal a key value, intercepting traffic.

Figure 1 provides some insight into the types of attacks on game resources.

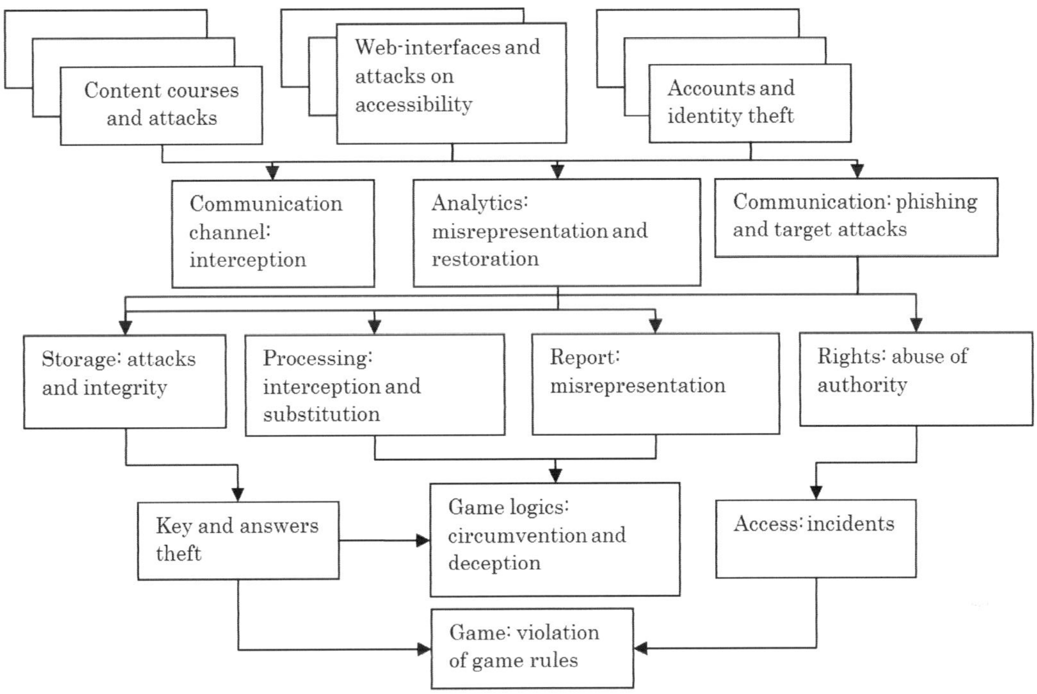

Figure 1. Several types of attacks on game resources.

There is a certain set of defensive measures that can counteract several types of attacks.

2. Model of attack on game resources

Distributed cyberattack was chosen as the attack model adopted for analyzing and predicting attacks on game resources. This type of attack involves the consistent violation of several game resources protection boundaries using phishing or other types of social engineering, identity theft, viral technologies and Trojans.

The game, considered as an attack target, has the structure shown in figure 2.

Game logics is built in the way that forecasting of right scenario of actions for an attacker can be more valuable than directly key theft or attack on a task generator.

Technically, the essence of the game considered is as follows: a random set of key sequences is generated, located in the documentation and tasks of the gaming environment. The managing structure of the gaming environment, represented by a set of modules, distributes tasks based on the scenario and adjusts them to reflect the current game analytics. The gaming environment can be either substantially automated or fully controlled from individual accounts. The same goes for the gaming environment. Game reports and gamified environment interfaces are available to participating teams.

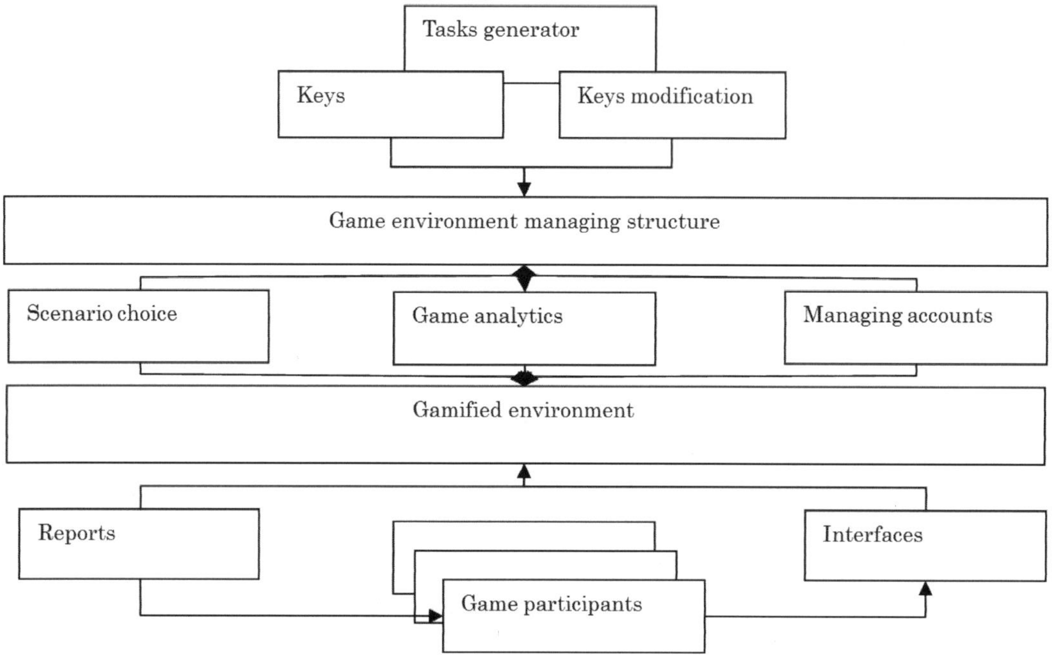

Figure 2. Game structure.

Distributed cyberattacks (advanced persistent threat - APT) [11] until the end of 2015 - the beginning of 2016 did not represent significant risks for information systems [12], since only 2-3 criminal groups acted on the "black market" of viruses and Trojans distributors. The situation has changed dramatically over the past year [11], [13]: several dozen criminal groups are already offering their services on the black market, methods have changed, new "tools" have emerged and the old ones have improved significantly. The traditional system of ensuring the information protection involves the preparation of a passport of vulnerabilities. At the same time, the list of current vulnerabilities can reach several hundred units [14]. Closing them means to attract limited financial and organizational resources, using various methods, such as expert assessments of various types, based on the requirements and risk analysis [15], game theory [16] and others.

Taking into account game logics and its technical basis, as well as the chosen type of attack and its features (secrecy, detection complexity), it has been suggested that the defense mechanisms do not adapt to the actions of the attacker by performing predetermined actions. Thus, the game "man against nature" is considered, where the role of "man" is assigned to the attacker (A), and by "nature" (B) is meant the state of the protected gamified environment. It is specified that player A is partially aware of the valuable intangible assets of object B, understands in general the infrastructure of the gamified environment, maximizes the benefits from using the information received, does not attack randomly.

3. Experimental evaluation

The security awareness game was chosen as an experiment to counter attacks against the resources of educational institutions. Several scenarios were selected for possible actions of the game participants. The main content of the game was the following:

1) The participants of the game can act both individually and as a team. In the first case, it is considered that the team consists of one player.

2) The game interface is a server to which it is possible to send certain information for an attack (for example, a phishing email).

3) Report of the game is data containing information about the benefit gained by the team from the attack, points earned in which the keys obtained are converted.

4) The keys were hashes of articles and personal data of "employees". It was assumed In the legend of the game that the main value are unpublished articles.

5) Keys had different value. Points were determined as follows: for one record of personal data - 0.1 points; for one article type 1 (on natural sciences) - 350 points; for one article of type II (in the humanities) - 150 points.

6) The game defines the periods of each key relevance. If the key is received when it is irrelevant, points are not awarded for it. The period of relevance is chosen with a certain periodicity, unknown in advance to the players.

Consequently, there are 8 scenarios in the game that a participant can play. these scenarios can be as follows: Player A can get into the interval when useful material has already been published and there is no actual content (scenario 1), there is actual content only from natural science (scenario 2), there is actual the content is only a humanitarian profile (scenario 3), there are both types of content (scenario 4).

There are also 4 scenarios for personal data case:

- Under scenario 1, player A does not have access to personal data storage.
- With scenario 2, he gets access to a smaller part of personal data storage (about 10%).
- With scenario 3 - to most of personal data (about 90%).
- Scenario 4 assumes access to 100% personal data storage.

4. Types of attacks and benefit assessment

Let us figure out the types of actions that an attacker can choose: attack on external resources (A1); attack on a server that provides access to the Internet (proxy server) (A2); attack on the messaging server (communication or mail server) (A3); APT attack (A4); phishing (A5). A ratio 5%, 3%, 1%, 0.5% of the total information cost of the natural science and humanities articles was adopted respectively to scenario 1, 2, 3, 4.

Obviously, the attacks will correspond to the action scenario in the game, if we exclude the attacks A1 and A2; in this case, the attacker's goal is to gain an unfair advantage by getting access to key information without following the game's logic and overcoming game defenses.

The initial data is represented in tables below (tables 1-2)

Table 1. Initial data for experiment 1*.

A1	0	0	0	0
A2	0	1568	1813	0
A3	245	66850	10050	77145
A4	0	10050	66850	76900
A5	3845	2330	769	3845

Table 2. Initial data for experiment 2*.

A1	0	1400	0	0
A2	0	2450	2597	0
A3	13800	117950	2597	134347
A4	0	13800	117950	131750
A5	6588	3953	1318	659

*Here and hereafter made for the benefit for the attacker the number of points he can receive for a successful investment targeted attacks on resources. The ability to score points in parallel in a regular way is not taken into account.

The difference in the experimental data is the result of the following factors: the number of hashes articles (in first experiment - 67 humanitarian profile, 191 natural science, in the second - 92 and 337 respectively); the number and structure of PD processed (experiment 1):

- PD 13.5% (2500) entries for scenario 1

- 86.5% (16000) for scenario 2,

(experiment 2):

- 5.6% (1500) for scenario 1,
- 94.4% (25000) for scenario 2.

In addition, the generated differences in the network infrastructure were considered to account the random factor. External servers are allocated to a separate subnet that does not have access to the local network of object B In experiment 1. In experiment 2, external servers also have a local network interface through which it is possible to access three articles of the natural science profile; in addition, the mail server in Experiment 2 acts as a domain controller, which allows Attacker A, to install any malware on all computers in the domain.

5. Analysis and prediction of attack scenario

Initial data were assessed using the following criteria: Maximax, Bayes, Laplace, Wald, Savage, Hurwitz, Hodge-Lehmann. According to the results, the optimal strategy was chosen (table 3).

Table 3. Experiment results.

	Criteria						
Experiment	Maximax	Bayes	Laplace	Wald	Savage	Hurwitz	Hodge-Lehmann
1	A3	A3	A3	A5	A3	A3	A3
2	A3	A3	A3	A3	A4	A3	A3

As can be seen (table 3), the A3 attack is optimal for player A, the A5 attack for Experiment 1 and the A4 attack for Experiment 2 are in the second place. Therefore, it is necessary to protect management accounts from "identity theft" attacks to prevent attacks on game resources in practice, (two-factor authentication, control of connection logs, biometrics).

In addition, the key problems are phishing messages sent on management and game accounts, tracking of keywords and frequency parameters, outgoing IP is required. The use of targeted attacks on individual accounts and game resources is possible from the attacker point of view as can be seen from the modeling results.

6. Future research and conclusion

Taking into account the dynamics of gaming environments development, further research may focus on the development of attack models that involve role-playing interaction in gaming environments; the possibility of increasing or decreasing authorities. It is also interesting to investigate a variant of the gamified environment, built with account of group interaction in solving individual problems.

However, the final goal is to create a series of models that show the development of attacks on game resources of various types of gamified environments.

The results of the research allow ranking the threat model according to the following criteria: "most probable" predicted by most methods, "probable" – the prediction of which appeared at least once, and all others – threats with a "low probability". The methodology proposed in the paper provides a mechanism for the allocation of resources for the organization of information protection in multi-user distributed gaming environments.

This method can be successfully implemented in managing information security risks, as well as in distributing the responsibilities of security personnel according to various schemes and models.

Acknowledgement

This work was supported by the Russian Foundation for Basic Research project No 19-013-00711 a.

References

[1] Kaivola T, Salomaki T and Taina J 2012 In quest for better understanding of student learning experiences *Procedia - Social and Behavioral Sciences* **46** 8-12

[2] Roblek V, Kresal F, Pejic B and Meško M 2018 Early warning systems as a paradigm for understanding organizational behaviour in extreme environment *5th International Symposium: Cocreating responsible futures in the digital age*
[3] Kordaki M and Gousiou A 2017 Digital card games in education: A ten year systematic review *Computers & Education* **109** 122–161
[4] Trickel E, Disperati F and Gustafson E 2017 Shell We Play A Game? CTF-as-a-service for Security Education *USENIX Workshop on Advances in Security Education (ASE)*
[5] Klemke R, Eradze M and Antonaci A 2018 The Flipped MOOC: Using Gamification and Learning Analytics in MOOC Design – A Conceptual Approach *Education Sciences* **8(1)** 25
[6] Tikhomirov V, Dneprovskaya N and Yankovskaya E 2015 Development of University's Web-Services *Smart Education and Smart e-Learning, Smart Innovation, Systems and Technologies* **41** 265-271
[7] Fernández-Manjón B, Moreno Ger P, Martinez-Ortiz I and Freire M 2015 Challenges of serious games *EAI Endorsed Transactions on Game-Based Learning* **2** 150611
[8] Zolotarev V, Maro E and Kulikova S 2018 New approach to activity evaluation for social network based student collaboration *12th IEEE International Conference on Application of Information and Communication Technologies (AICT 2018)* 374-380
[9] Chiew K L, Yong K S and Tan C L 2018 A survey of phishing attacks: Their types, vectors and technical approaches *Expert Systems with Applications* **106** 1–20
[10] Pham C, Tang D, Chinen K and Beuran R 2016 CyRIS: a cyber range instantiation system for facilitating security training *International Symposium on Information and Communication Technology* 251-258
[11] Kaspersky Laboratory: Report on targeted attacks and complex cyber threats for the second quarter of 2017, available at: https://www.kaspersky.ru/about/press-releases/2017_how-targeted-attacks-developed-in-q2-of-2017
[12] InfoWatch: Global survey of confidential information leaks in 2016 , available at: https://www. infowatch.ru/sites/default/files/report/analytics/russ/InfoWatch_Global_Report_2016_year.pdf?rel=1
[13] Vault 7: Projects, Wikileaks, available at: https://wikileaks.org/vault7/
[14] Cisco 2017 Midyear Cybersecurity Report, MCR, available at: https://www.cisco.com/c/ru_ru/about/press/press-releases/2017/07-21.html
[15] Zolotareva G, Zolotarev V and Filko S 2019 New approach to risk controlling in information security *IOP Conference Series – Materials Science and Engineering* in press
[16] Rass S, König S, and Schauer S 2017 Defending Against Advanced Persistent Threats Using Game-Theory *PLOS ONE* **12(1)** e0168675

Features of large-scale thin foils fabrication for transmission electron microscopy by focused ion beam

M N Volochaev[1], M S Shcheglova[2], Yu Yu Balashov[1] and Yu Yu Loginov[2]

[1] Kirensky Institute of Physics, Federal Research Center KSC Siberian Branch Russian Academy of Sciences, 660036, Krasnoyarsk, Russia
[2] Reshetnev Siberian State University of Science and Technology, 31 Krasnoyarsky Rabochy Av., Krasnoyarsk, 660037, Russia

E-mail: volochaev@iph.krasn.ru

Abstract. Focused ion beam (FIB) sample preparation for transmission electron microscopy is a well established technique. However there are some problems related with high-quality large-scale (over 10 um) sample preparation of thin foils. In this work we demonstrate the successful preparation of such samples on example of ferrite steel sample and present the refined lift-out preparation technique, which was applied here. All sample preparation has been carried out in conventional single-beam FIB Hitachi FB-2100 with additional low-kV Ar^+ ion polishing.

1. Introduction

Transmission electron microscopy (TEM) is one of the main direct methods for materials structure studying down to the atomic level. For successful study TEM specimens, depending on the objective of the study, should be from a few atomic layers to 100 nm thick (high resolution TEM - from a few atomic layers to 20 nm, crystal lattice defects and diffraction analysis - from 20 to 60 nm, elemental analysis - from 60 to 100 nm) and minimum preparation artefacts.

Nowadays, focused ion beam (FIB) microscopy is a well established tool for TEM specimens preparation due to ability to remove material in a highly localized area. Sample preparation capabilities of FIBs can be further extended by introducing micro-manipulators and gas- injection systems (GIS) into the microscope chamber. The first allows in-situ manipulations such as lift-outs, rotations and transfers of microscopic sample parts during preparation, while the second enables both site-specific material deposition and material-specific preferential milling by introducing reactive gases in the vicinity of the electron or ion probe during operation. Fully-equipped dual-beam FIBs (DB-FIBs) offer great flexibility in TEM sample preparation and many preparation techniques have been developed and reviewed over the years [1-5]. There are several, often unique, advantages of FIB preparation. The ability to cut and extract small quantities of any hardness material (including such materials as sapphire and diamond) and mount these pieces at fairly exposed positions on dedicated support grids generally helps minimize detrimental bulk-material effects such as charging, magnetic fields or surface migration of contaminants. Finally, FIB sample preparation is generally considered a fairly fast technique, capable of producing a complete sample from a bulk specimen within hours. However, it is sometimes useful to keep in mind that the pursuit of speed often comes at the cost of reduced sample quality [6].

The disadvantages of this technique include amorphization of samples (damage layer thickness is up to 30 nm) [7], the introduction of defects and the locality of the method (TEM area is up to 2-3 μm). Another disadvantage of TEM sample preparation by FIB is the inability to prepare a thin foil (lamella) with a size larger than 5 um and thinner than 100 nm. Since, when trying to obtain a larger sample, lamella is bent due to internal mechanical stresses, as well as due to the impact of an ion beam with high energies, and the sample becomes unsuitable for further research.

Modern DB-FIBs allow low-kV Ga+ ion polishing, which removes the damage layer and preparation defects and obtain 20 nm thick lamella. However, not every science center has such facilities. Therefore, when preparing samples using a conventional FIB (including single-beam FIB), additional polishing with low-energy Ar + ions is used to remove the re-precipitated and destroyed layers, but lamella larger than 5 um will bend for the reasons described above.

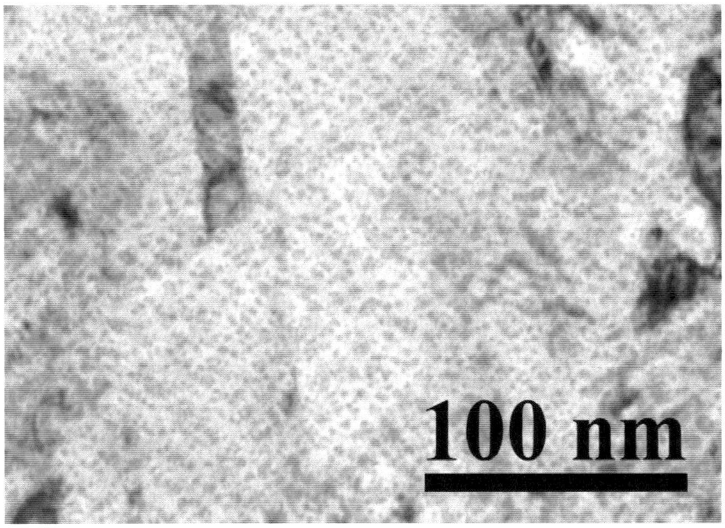

Figure 1. Re-deposition (nanoparticles) of the support grid material onto sample (austenite steel) surface.

Also, the re-deposition (figure 1) of the support grid material on the sample occurs when Ar^+ ion polishing, if the classic method of lamella mounting on support grid [6] is used. The re-deposition caused by different etching speed and angle of support grid and lamella.

2. Experimental and Results

FIB sample preparation was carried out on conventional single-beam FIB system Hitachi FB-2100 at 40 kV Ga^+ ion source. Final 2 kV Ar^+ ion polishing was carried out on Linda Technoorg ion mill system. TEM studies were carried out on conventional TEM Hitachi HT 7700 at 100 kV (W-sourse).

The proposed solution of described above problems is modification of the lamella support grid (figure 2 a). Immediately note that molybdenum support grid is better to use, because molybdenum sputtering ratio is much less than that of copper. Firstly is mechanically thinning the lamella support grid to ~ 5 μm. Secondly is high-intensity ion beam etching groove (figure 2 b) in which the sample is mounted. Mechanical thinning of lamella support grid avoids the excess consumption of gallium when etching the groove. In this position, the sample is firmly fixed on both sides with a deposition mask (figure 2 c) and is not deformed with further thinning. At the final stage of thinning lamella is tilted at a small angle (1-1.5 degrees) to obtain a uniform sample thickness throughout the depth.

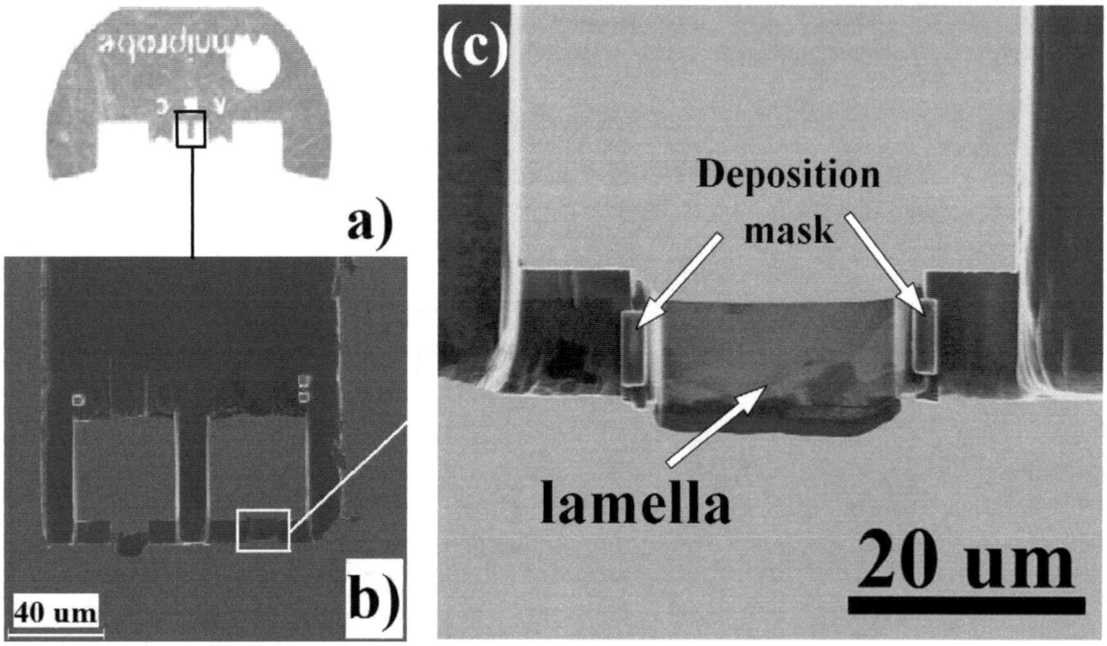

Figure 2. Modification of the lamella support grid: a) overview of support grid; b) enlarged fragment; c) the lamella mounting position.

After final Ga^+ ion milling lamella thick is consist ~ 150 nm. The sample of this quality is unsuitable for TEM studies (figure 3 a). Additional low-kV Ar^+ ion polishing at small angle (5-7 °) allows to thin the sample to the desired thickness (figure 3 b) also remove amorphous layer and Ga^+ ion beam artefacts.

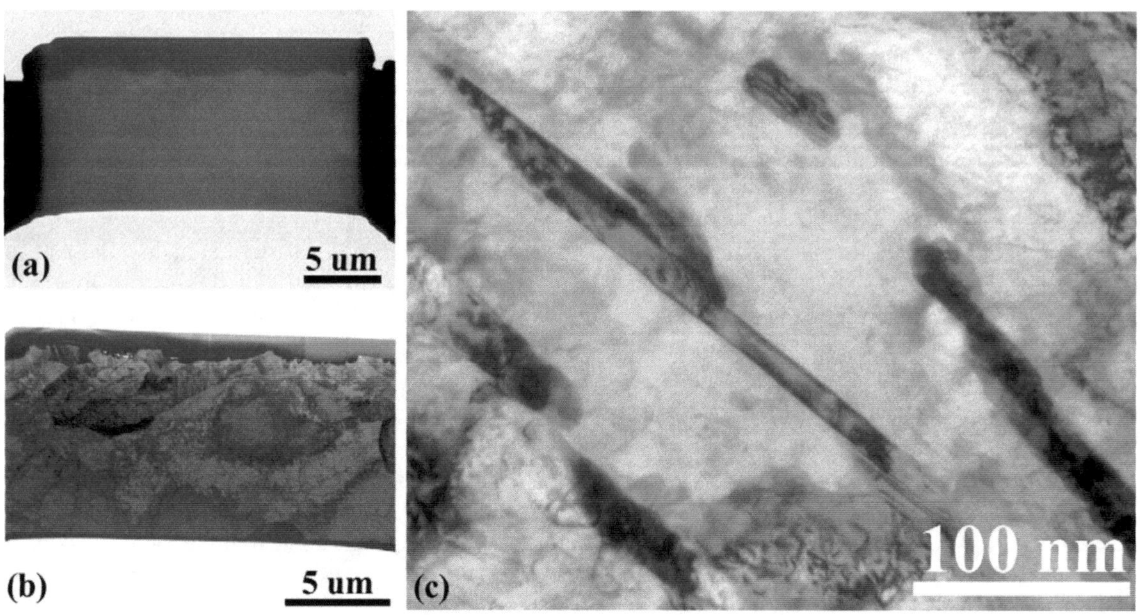

Figure 3. TEM studies of ferrite steel lamella: (a) Low-magnification TEM image after FIB preparation, (b) Low-magnification TEM image after 2 kV Ar^+ ion polishing, (c) Medium magnification TEM image of austenite grain fragment with carbide phase inclusions.

Medium magnification TEM image (figure 3 c) clearly demonstrate high-quality sample preparation without amorphization, Ga^+ ion beam artefacts and re-deposition.

3. Conclusion

In this work, we demonstrate that conventional FIB system together with low-kV Ar^+ ion polishing allows to obtained high-quality large-scale (~ 20x10 um and more) samples for TEM examinations. It is possible due to special samples support grid modification.

FIB and TEM studies were carried out at the Center for Collective Use of the Krasnoyarsk Scientific Center of the Siberian Branch of the Russian Academy of Sciences.

References

[1] Giannuzzi L A and Stevie F A 1999 *Micron* **30** 197
[2] Langford R M, Huang Y Z, Lozano-Perez S, Titchmarsh J M and Petford-Long A K 2001 *J. Vac. Sci. Technol.* **19** 755
[3] Sugiyama M and Sigesato G 2004 *J. Electron Microsc.* **53** 527
[4] Langford R M 2006 *Microsc. Res. Tech.* **69** 538
[5] Floresca H C, Jeon J, Wang J G and Kim M J 2009 *Microsc. Microanal.* **15** 558
[6] Schaffer M, Schaffer B and Ramasse Q 2012 *Ultramicrosc.* **114** 62
[7] Kato N I 2004 *J. Electron Microsc.* **53** 451

IOP Conf. Series: Materials Science and Engineering **822** (2020) 012029 doi:10.1088/1757-899X/822/1/012029

Conditions for residents attracting of the priority territories social and economic development of closed administrative and territorial unions of the Rosatom State Corporation

N T Avramchikova, M B Puchkin, M N Chuvashova

Reshetnev Siberian State University of Science and Technology,
Av. Krasnoyarsky Rabochy 31, Krasnoyarsk 660031, Russia

E-mail: dersaturn06@gmail.com

Abstract. The paper considers a special legal regime for conducting business and other activities in the field of taxation in order to attract residents of territories of priority socio-economic development (TPSED) of closed administrative-territorial unions (CATU) of the Rosatom State Corporation. This contributes to the accelerated development of the economy and improvement of the quality of life. The authors describe a brief feature of TPSED in terms of the interests of their potential resident and identify promising technical areas for attracting innovative small and medium companies as residents of TPSED of Rosatom State Corporation, which includes using the approaches of the circular economy in the implementation of innovative projects. The conditions that must be corresponded by representatives of small and medium businesses to obtain the status of a TPSED resident of Rosatom State Corporation. The existing criteria for the selection of residents in the TPSED of Rosatom State Corporation were considered. In addition, proposals to strengthen the motivation to attract representatives of small businesses and individual entrepreneurs as residents of TPSED are given.

According to a number of leading world scientists, global changes in almost all life spheres, due to the Fourth Industrial Revolution. Within the framework of this revolution, a cardinal transformation of the economy and the social sphere will take place all over the world. This will sharply raise the issues of ensuring the competitiveness and quality of economic growth not only in the region, but also in its separate territories.

Currently, the main factor of economic growth is scientific and technical progress and innovations. The basis of the strategy of economic growth in many developed countries is the creation of breakthrough technologies, as well as the finding and rapid development within the country of points of growth of the innovation economy. In the framework of these trends, the Russian Federation political leadership has set the ambitious tasks to transmit the country on the innovative development way, digitalization of the economy and the priority socio-economic development of separate regions and territories. A significant role in solving this task is assigned to the mechanism for realizing the existing potential for economic growth in the regions through the formation of territory social and economic development (TPSED). The TPSED created in the closed administrative-territorial unions (CATU) of the Rosatom State Corporation deserve special attention in the Russian Federation. In modern time, Rosatom State Corporation occupies a leading position in the global markets of the "peaceful atom", possessing in the regions of its presence a huge technological, industrial, scientific, innovative and personnel potential for further development [1].

At the same time, CATF have highly qualified personnel, production facilities, free land plots, engineering and municipal infrastructure, developed social sphere and are able to attract residents from high-tech spheres and industries. The change of the paradigm of spatial development at the present stage is due to the release in the economic space of the region of growth point of ensuring the accelerated development of the local territory [2].

According to the Federal Law of December 29, 2014 No. 473 "On the territories of priority socio-economic development in the Russian Federation", it is determined that the TPSED is a part of the territory of the subject of the Russian Federation. That includes a closed administrative-territorial union (CATU), within which a special legal regime is established business and other activities in the taxation field.

TPSED is created to attract investments in the regions of the Russian Federation. This create for accelerating economic development and improving the quality of standard of living. The goals of creating TPSED are to "level off" the level of socio-economic development of the Russian regions by creating conditions for their advanced growth. During the creation of the TPSED in the regions of Russia, including the closed administrative-territorial unions (CATU), the following tasks are solved [3, 4]:

- Formation in the territories necessary for priority of infrastructure development;
- Creating a favorable investment climate;
- Implementation of public-private partnerships;
- Creation of import-substituting industries with export potential;
- Creation of new high-tech jobs;
- Human development on the territory;
- Improving the quality of living standards of citizens.

Due to the use of tax privileges and simplified administrative and customs regimes by business entities, TPSED becomes a regional points of not only industrial, technological, innovative, but also socio-economic development. In many ways, this is happening not only because of the provision of preferential tax, customs and administrative regimes created by the TPSED by the state to residents, but also by creating the necessary infrastructure: energy; communal; engineering; social; innovation; investment; transport; logistic; managerial; informational [5].

At the same time, analyzing the interests of TPSED residents from the point of tax preferences view, the greatest interest is represented by the concessions on profits tax and corporate property tax.

For the first five years, TPSED residents are either fully exempted or pay these taxes at minimum rates. These tax preferences are beneficial for TPSED residents (significantly reduce the payback period of investment projects) and for the regions (improve the investment environment).

However, tax breaks granted to TPSED residents cannot fully compensate the regions for their distance from sales markets, poor availability of credit resources, and poor infrastructure [6].

In addition, scientists note the problem in the form of the existing personnel "hunger", as well as inconsistencies of the existing personnel with the requirements of TPSED residents. The issue of training for promising innovative areas of activity of TPSED residents has not yet been resolved either legally or methodologically or organizationally [7].

For example, the labor market analysis of single-industry towns (including the Rosatom State Corporation) of the Russian Federation with the status of TPSED showed a number of specific problems, such as female unemployment, hidden unemployment, informal employment.

However, the job placement for the stated vacancies of TPSED residents does not coincide with the free and declared applicants in the local labor markets for education, qualifications, skills and work experience. That is, temporarily unemployed citizens do not meet the requirements for newly created workplaces that seriously complicates the search for the right specialists and the implementation of projects by TPSED residents. As a result, when choosing a site for opening a new production, TPSED residents are taken into account as system-forming and system-forming factors influencing the satisfaction of their interests and needs [8, 9].

Thus, the interests of potential residents of TPSED are in the field of tax benefits, preferential administrative and customs regimes, infrastructure and personnel support for their activities.

Due to tax privileges and simplified administrative and customs regimes, TPSED residents become pioneers of industrial, technological, innovative development in the regions of the Russian Federation [10]. Creation of a favorable investment climate for small and medium businesses contributes to the organization in TPSED not only the production of the 4th technological mode, represented by such industries as microelectronics, computer science, biotechnology, genetic engineering, new types of energy, new materials, space exploration, satellite communications and etc. In addition, it includes more high-tech industries such as robotics, biotechnology, genetic engineering, nanotechnology, artificial intelligence and neural networks, global information networks, integrated high-speed transport and flexible automated production, hydrogen and renewable energy, etc.

One of their effective mechanisms for the implementation of promising technical areas of small and medium businesses as TPSED residents can be innovative clusters created in the closed city of Rosatom State Corporation (for example, the city of Zheleznogorsk, Krasnoyarsk Territory). The main goal of such clusters of innovative technologies is the development of an innovative territory that enhances the importance of the city-forming enterprises of the Closed State Corporation Rosatom, as well as the creation around them of a belt of high-tech innovative enterprises to TPSED residents using high technologies in the production of civilian products.

The general procedure for attracting residents to TPSED created provides for the preparation and sending by a potential resident to an authorized body (Government of the Russian Federation) for consideration of the application and documents proposed for it. In general, restrictions are imposed on potential residents of TPSED in the form of a minimum amount of investment and the creation of new jobs in the framework of new projects. In general, companies and individual entrepreneurs who meet the following conditions become residents of TPSED:

- carrying out the types of activity established for a specific TPSED;
- planning investments in the project on the territory of TPSED in the amount exceeding the minimum established size;
- planning to create a new project in the territory of TPSED over the minimum set number of new jobs.

The Government of the Russian Federation focuses on the coincidence of interests of business and state and claims some residents of the territories of priority of socio-economic development. An infrastructure provision of sites for the location of future production is of key importance for potential residents of TPSED. Therefore, financial support for the deployment of infrastructure facilities for the territory of advanced socio-economic development is provided by the federal budget, the budget of the constituent entity of the Russian Federation and local budgets, as well as extra budgetary sources of funding, for example, for TPSED in the closed city of Rosatom.

Regulations that define the conditions for attracting residents to the TPSED of Rosatom State Corporation set specific, different from general requirements and restrictions for the selection of residents. Only Russian companies and individual entrepreneurs registered and operating directly in TPSED can be residents of Rosatom State Corporation TPSED. In addition, residents of TPSED of Rosatom State Corporation should not have branches and representative offices outside of TPSED. Capital investments of TPSED residents of Rosatom State Corporation is at least 5 million rubles [11].

For obtaining the TPSED status of resident of the State Atomic Energy Corporation ("Rosatom") needs to:

- to carry out state registration of a legal entity or an individual entrepreneur on the territory of the TPSED;
- to provide the government agency (management company Open Joint Stock Company "Atom") with a business plan in the form approved by the Order of

the Ministry of Economic Development of the Russian Federation dated December 19, 2016. No. 813;

- to conclude an agreement with the management company "Atom" on the conduct of activities on the territory of TPSED and the inclusion in the register of residents.

The government agency of TPSED selects residents according to the following criteria:

1. Presence of a free land on the territory of the closed city of Rosatom State Corporation for the implementation of the proposed activity of a potential resident;

2. The ability to satisfaction of a potential resident in terms of engineering, utilities and transport infrastructure, as well as the availability of the necessary energy capacity;

3. Opportunities for issuing admission of employees of a potential TPSED resident to the closed city of Rosatom.

Thus, critically important criterion for potential residents of TPSED in the Closed State Unity of Rosatom State Corporation is only the type of activity of the General Economic Activity Classifier (this system is used to record activities in Russian statistic). The minimum capital investment is 5 million rubles, this isn`t essential for the innovative projects and the minimum number of new jobs created is absent.

In turn, the list of permitted types of an economic activity for TPSED residents in the Closed State Unity of Rosatom State Corporation is quite wide and allows us to attract a wide range of innovative and high-tech companies as residents. For residents of TPSED of the closed city Zheleznogorsk, Krasnoyarsk Territory, the Government of the Russian Federation has established the following list of economic activities in accordance with the General Economic Activity Classifier:

1. Research and development (R&D);
2. Development of computer software;
3. Consulting services in the field of computer software;
4. Manufacture of computers, electronic and optical products;
5. Activities in the field of telecommunications;
6. Manufacture of electrical equipment;
7. Manufacture of machinery and equipment not included in other groups;
8. Manufacture of vehicles and equipment;
9. Production of finished metal products, except machinery and equipment;
10. Production of chemicals and products;
11. Production of other non-metallic mineral products;
12. Furniture production.

All permitted economic activities have a high import substituting and export potential. In the future, the world economy will have another phase transition within the framework of the "Fourth Industrial Revolution", which implies a total secondary processing of all resources, as well as significant energy savings, which are also called the circular economies, the acute problem of the total recycling of all resources. The problem of the total recycling of all resources, a significant energy savings, including in the Closed State Unity of Rosatom State Corporation, will become acute.

According to many experts, circular economy is a common key to solving economic problems and environmental problems. The main risks of a circular economy are its complexity, adaptability, complexity, and the cost of implementation in the first stage. It is not only necessary to change individual elements of the economic model. A complete structural and ideological restructuring is required, including the consciousness of both consumers and producers of goods. This direction is of particular relevance for the regions of presence of the city-forming enterprises of the state corporation Rosatom.

Taking into account the existing criteria for the selection of small and medium businesses, the potential residents of TPSED of Rosatom State Corporation can be considered [12]:

- parts of R & D of the departments of the city-forming enterprises of the closed nuclear power industry of the nuclear industry that are separated into separate legal entities (with the separation of the property complex);

- innovative and high-tech small and medium enterprises of the TPSED region and neighboring territories, working in the areas of specialization "TPSED" (taking into account industry specifics) and using the potential of the region (taking into account regional specifics);
- «Technological gazelles» (Russian high-tech companies in competitive industries with annual growth rates that are many times higher than the average in the economy).

It should be noted that the existing procedure and conditions for attracting residents to the TPSED of Rosatom State Corporation do not fully take into account the regional characteristics and specifics of the nuclear industry.

In modern time, the weak investment activity of small business representatives and individual entrepreneurs to become a resident of the TPSED of Rosatom State Corporation was remarked. The authors proposed to increase the motivation of their entrepreneurial activity:

- lowering the threshold of the volume of capital investments;
- expansion of the list of activities in accordance with the allocation of priority areas with the highest export potential.
- using of the digital information technologies, as well as the principles and approaches of the circular economy in the implementation of innovative projects [13].

Research centers of federal and supporting universities, the Russian Academy of Sciences, as well as development institutions operating in the region, such as business incubators, business accelerators, technology and industrial parks, etc., should contribute to the active involvement of high-tech regional companies in the TPSED of ROSATOM.

The closed administrative-territorial unions (CATU) of Rosatom State Corporation has sufficient industrial, technological and human potential and it will become “points” of growth of the innovation sector of the economy in the regions of the Russian Federation in opinion some Russian scientists and economists. The existing criteria for the selection of residents in TPSED do not fully take into account regional peculiarities and the specifics of the nuclear industry. Activation of attracting new residents to the TPSED of Rosatom State Corporation will significantly expand the innovative sector of the economy in CATF, improve the business environment in CATU and improve the population welfare.

References

[1] Rosatom State Corporation, available at: http://www.rosatom.ru)

[2] Avramchikova N T 2006 Historical aspects of the evolution of the regional policy of Russia and the change of the paradigm of spatial development at the present stage *Vestnik SibSAU* **6(13)** pp 148-152

[3] Association of Closed Administrative-Territorial Unions of the Russian Federation, available at: http://zato.tv/association

[4] Federal Law 29.12.2014 No 473 2014 *On the territories of advanced socio-economic development in the Russian Federation (federal regulatory act)*

[5] Potapova A N and Zhirova V N 2018 On the territories of advanced socio-economic development in the Russian Federation *Collection of articles of the IV International Scientific and Practical Conference Modern socio-economic processes: problems, patterns, prospects. (Publisher: Science and Education)* pp 321–323

[6] Karataeva G E 2018 Comparative analysis of tax preferences of territories of advanced socio-economic development in Russia *Innovative development of the economy* **2(44)** pp 112-120

[7] Sushkova T V 2018 Territories of advanced socio-economic development: questions of theory and practice *Materials of the II All-Russian Scientific and Practical Conference with international participation (Publisher: Cognition)* pp 151–152

[8] Galiullina G F 2018 The choice of a potential resident: the pros and cons of the territories of advanced socio-economic development *the II Russian scientific-practical conference with international participation* pp 34–39

[9] Puchkin M B 2018 Global and regional challenges of the IV industrial revolution and their impact on the development of the TOSED nuclear industry *Vestnik of Tomsk State University: Economy* **2** pp 341–353
[10] Avramchikova N T, Volkov D O and Zakharova L N 2017 Methods and forms of state support of innovative activities in region (on the example of Krasnoyarsk region) *Siberian Journal of Science and Technology* **18(2)** pp 442–451
[11] Puchkin M B 2017 Classification of territories of advanced socio-economic development in the regions of the Russian Federation *Management of socio-economic systems* **3** pp 39–51
[12] Puchkin M B Russian technological "gazelles" as residents of the territories of advanced socio-economic development of the nuclear industry *The IV scientific-practical conference Actual problems of aviation and cosmonautics (Krasnoyarsk: SibGU)* pp 212–216
[13] Gerasimenko D and Nikolaev I The Circular Economy in Russia in the Context of the UN Sustainable Development Goals and the Year of Ecology available at: https://www.ictsd.org/bridges-news/bridges/news/

Symmetries and conservation laws in the theory of plasticity

S I Senashov[1], O V Gomonova[2], I L Savostyanova[1] and O N Cherepanova[3]

[1] Department of Economic Information Systems, Reshetnev Siberian State University of Science and Technology, 31 Krasnoyarsky Rabochy Av., Krasnoyarsk, 660037, Russia
[2] Department of Higher Mathematics, Reshetnev Siberian State University of Science and Technology, 31 Krasnoyarsky Rabochy Av., Krasnoyarsk, 660037, Russia
[3] Department of Mathematical Analysis and Differential Equations, Siberian Federal University, 79 Svobodny Av., Krasnoyarsk, 660041, Russia

E-mail: sen@sibsau.ru

Abstract. This article is a review of the main results that were obtained by the authors while studying equations of plasticity using symmetries and conservation laws from the year 1975 to the present day.

1. Introduction
L. V. Ovsyannikov, on the basis of S. Lie's works, has developed the methodology of researching and solving of differential equations of mechanics using point symmetries [1]. For the first time these methods were effectively applied to differential equations of the theory of plasticity by B. D. Annin [2]. He found a group of point symmetries admitted by the equations of ideal plasticity in axisymmetric case, and provided a new exact solution of these equations. B. D. Annin gave a suggestion to investigate the group properties of spatial equations of the ideal plasticity theory to one of the authors of the present article, and to build new solutions for these equations. S. I. Senashov and his disciples keep working till now in this direction. As a result, groups of point symmetries of the main plasticity equations were found [2]. These groups were used to build the new classes of exact solutions which have mechanical applications. Also, conservation laws for the equations of plasticity in case of plane strain were found [3]. These laws were used to solve the main boundary value problems for the equations of plasticity in case of plane strain [4]. Nowadays, it is shown that the conservation laws can also be used to solve elastoplastic problems [5, 6]. All these results are described in the present article.

2. Main equations of the theory of plasticity
Assume $x_1 = x,\ x_2 = y,\ x_3 = z$ is an orthogonal Cartesian coordinate system. The main equations of the theory of plasticity in stationary case under Mises yield criterion are written as

$$\partial_j S_{ij} = \partial_i P,\ \ S_{ij} = \lambda e_{ij} = \lambda\left(\partial_i u_j + \partial_j u_j\right),\ \ S_{ij}S_{ij} = 2k^2,\ \ i,j = 1,2,3, \tag{1}$$

where e_{ij}, S_{ij} are components of velocity tensor and stress deviator, P is hydrostatic pressure, u_1, u_2, u_3 are components of velocity vector, k is a constant (pure shear yield point), λ is nonnegative function. Summation is intended over the repeated indices.

A group of point symmetries has been found for equations (1). It is generated by the following operators [2]:

$$X_i = \partial_i,\ Y_i = \partial_{u_i},\ T_1 = x_2\partial_{u_3} - x_3\partial_{u_2},\ N = x_i\partial_{x_i},\ M = u_i\partial_{u_i},$$
$$Z_1 = x_2\partial_{x_3} - x_3\partial_{x_2} + u_2\partial_{u_3} - u_3\partial_{u_2},\ S = \partial_P. \tag{2}$$

Four more operators T_2, T_3, Z_2, Z_3 come out from T_1, Z_1 by cyclic permutation of indices. System of equations (1) in spatial case allows only classical conservation laws: mass conservation law, momentum conservation law and angular momentum conservation law.

3. Some classes of exact solutions

The presence of symmetries (2) permits to build new classes of exact solutions. In this way, the new classes of helical solutions were built [2, 3]:

$$u_i = u_i(r, z+\alpha\theta),\ P = P(r, z+\alpha\theta), \tag{3}$$

where α is a constant, r, θ, z is cylindrical coordinate system; $i = 1, 2, 3$. These solutions can be used to analyse plastic flows in helical canals, and they also describe plastic flow of rods at complex loading. In particular, the following new class of solutions was found:

$$u_1 = ax_1,\ u_2 = ax_2,\ u_3 = -2ax_3 + f(x_1, x_2), \tag{4}$$

where a is an arbitrary constant, f is an arbitrary solution of the minimal surfaces equation [2]. These solutions find applications in the analysis of plastic flows of material compressed by rigid plates.

4. Homotopy of the Prandtl solution

Consider equations of the theory of plasticity describing plane strain under Saint-Venant – Mises yield criterion. These equations are statically determinate, that is why it is sufficient to consider only equations relating components of stress deviator and hydrostatic pressure.

$$\partial_1 S_{11} + \partial_2 S_{12} = \partial_1 P,\quad \partial_1 S_{12} + \partial_2 S_{22} = \partial_2 P,\quad (S_{11} - S_{22})^2 + 4S_{12}^2 = 4k^2. \tag{5}$$

Equations (5), in contrast to the system (1), possess an infinite group of point symmetries, an infinite group of higher symmetries and an infinite series of conservation laws [4]. These symmetries were able to use for construction of new solutions of the equations (5), and the conservation laws were applied to solve the main boundary value problems for these equations [4].

The existence of the infinite group of point symmetries permits to «deform» continuously one of the exact solutions of equations (5) into any other. During the process of deformation, an infinite series of exact solutions is obtained. This process is convenient to observe on the slip lines of the system (5). In this manner one can «deform» the Prandtl solution describing compression of a plastic layer with rigid plates

$$\cos\left(\frac{S_{11} - S_{22}}{S_{12}}\right) = y,\ P = -x + \sqrt{1 - y^2} \tag{6}$$

into the Nadai solution that describes stress state around a circular hole (see figures 1–3).

IOP Conf. Series: Materials Science and Engineering **822** (2020) 012030 doi:10.1088/1757-899X/822/1/012030

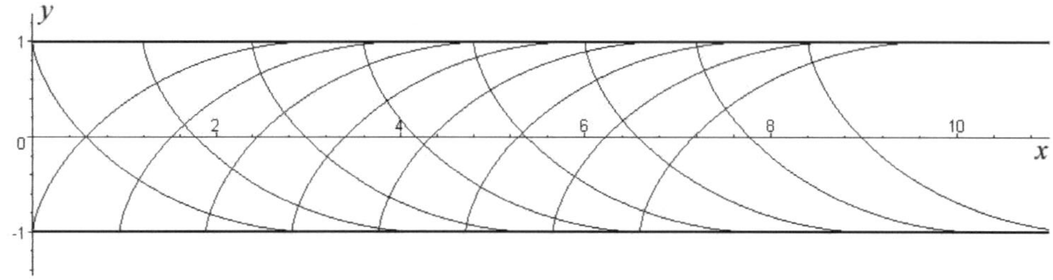

Figure 1. Slip lines of the Prandtl solution.

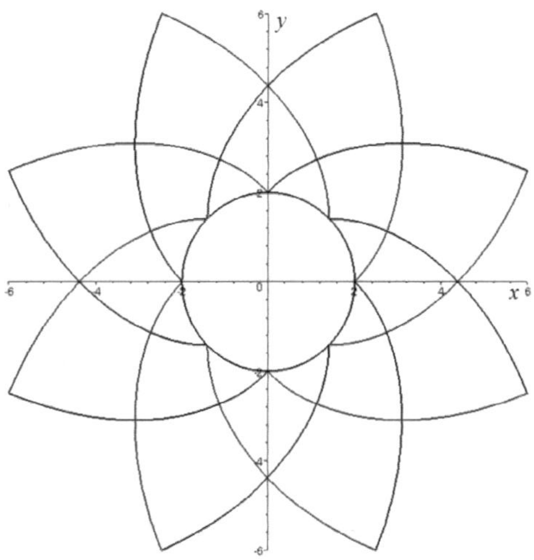

Figure 2. Slip lines of the Nadai solution.

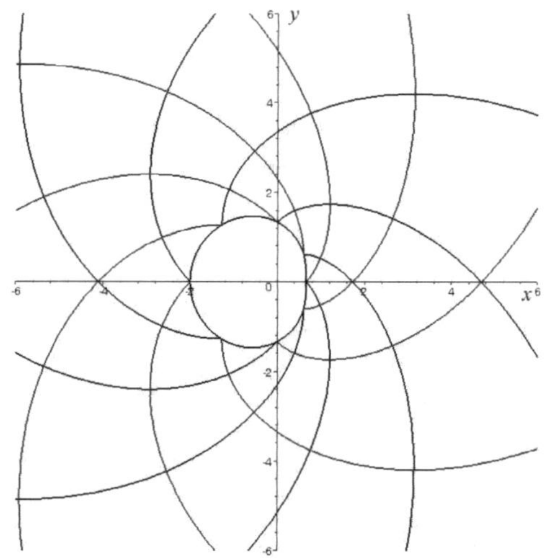

Figure 3. Slip lines of one of intermediate solutions. The hole has the form of Pascal's limacon.

Picture 1 shows the slip lines of the Prandtl solution, Picture 2 displays the slip lines of the Nadai solution, and Picture 3 illustrates the slip lines of one of intermediate solutions [5–11]. Presence of the infinite series of conservation laws has allowed solving the main boundary value problems for the equations (5). Herewith, the solutions of these problems are given by explicit analytic formulas [5].

5. Application of the conservation laws to elastoplastic problems solving

Let describe, in concluding, some new results that are related to solving of elastoplastic problems and to building of an unknown elastoplastic boundary. Such boundaries were able to build for rods under torsion, and for deformed plates with holes.

5.1. Elastoplastic boundary in a rectilinear rod under torsion

Let consider the elastoplastic torsion of a straight rod with the cross-section limited by a convex contour Γ. At a rather large value of the torsion torque, an elastic zone F and plastic zone P arise in the rod (Figure 4).

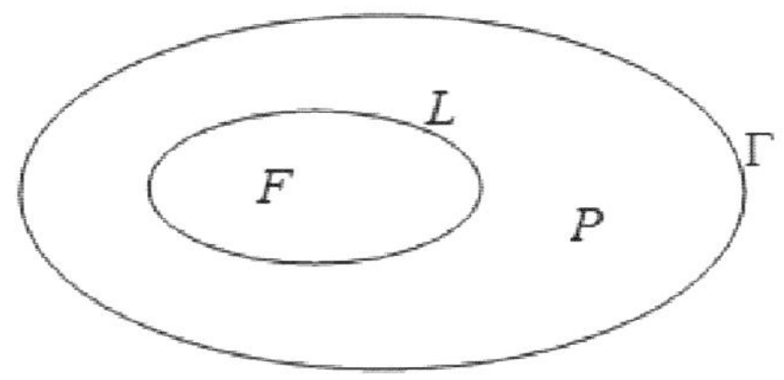

Figure 4. Cross-section of a rod under torsion.

It is known that the plastic zone arises on the outer contour Γ; suppose that the plastic zone fully involves the outer contour. Assume L is the boundary line between the elastic and the plastic zones.

In the elastic zone, stress components satisfy the following equations:

$$\partial_1 S_{13} + \partial_2 S_{23} = 0,\ \ \partial_2 S_{13} - \partial_1 S_{23} = -2\theta, \tag{7}$$

in the plastic zone, these components satisfy the equations

$$\partial_1 S_{13} + \partial_2 S_{23} = 0,\ \ S_{13}^2 + S_{23}^2 = k^2, \tag{8}$$

and on the outer contour Γ the stress components satisfy the boundary condition

$$S_{13} l + S_{23} m = 0 \tag{9}$$

and the yield criterion $S_{13}^2 + S_{23}^2 = k^2$. Here θ is a torsion angle, l, m are components of the outer normal vector to the contour Γ.

From Relations (8) – (9) one can determine tensor components S_{13} and S_{23} on the contour Γ.

The following relations are derived from the conservation laws obtained in [12]:

$$2\pi S_{13}(x_0, y_0) = -\oint_{\Gamma} A_1 dy - B_1 dx, \tag{10}$$

and

$$2\pi S_{23}(x_0, y_0) = -\oint_{\Gamma} A_2 dy - B_2 dx, \tag{11}$$

where A_i, B_i $(i = 1, 2)$ are the conserved current components.

Relations (10) – (11) permit to find the values S_{13} and S_{23} at all the inner points of the rod. Further, the yield criterion (8) is checked. The points, at which the expression in the right side part of the yield criterion (8) is less than k, belong to the elastic zone, and the rest of the points belong to the plastic one. These calculations allow restoring the sought border L with any required accuracy. Examples of building of elastoplastic boundaries for various types of rolled sections are considered in [12]. The case of multiply connected cross-section for isotropic medium is given in [13].

5.2. Elastoplastic boundary in a cantilever of uniform cross-section

Consider a cantilever of a uniform cross-section limited by contour Γ. The cantilever is subjected to the concentrated force P on its free ending; the force is parallel to axis Ox.

Under the action of a rather large value of the force P, the elastic and plastic zones appear in the cantilever. Let l is the length of this cantilever, and let the plastic zone fully involves the contour Γ.

With these assumptions, the stress components satisfy the following system of equations in the elastic zone:

$$\partial_1 S_{13} + \partial_2 S_{23} = -\frac{Px}{l},\ \ \partial_2 S_{13} - \partial_1 S_{23} = -\frac{Py}{l}, \tag{12}$$

and on the lateral surface, i.e. on the contour Γ, the conditions (8) and (9) are held. It means that the lateral surface is free from stresses and is in plastic state.

Formulas (10)–(11) that permit to find the values of stress components in arbitrary point (x_0, y_0) of the cross-section were obtained in the previous paragraph. Inserting the values A_i, B_i into these formulas, one can find the values of the stress in the sought point (x_0, y_0). The only remaining thing is to check whether this point is located in the plastic or elastic zone. If the value of the left side part of the yield criterion (8) is less than *k*, then the point belongs to the elastic zone, otherwise the point is in the plastic zone.

5.3. Elastoplastic boundary in a rectangular plate weakened by circular holes
Consider now a rectangular plate of the length $a = 10$ cm and the width $b = 2$ cm weakened by two circular holes with radii $r_1 = r_2 = 2.5\,\text{mm}$ (figure 5) [13].

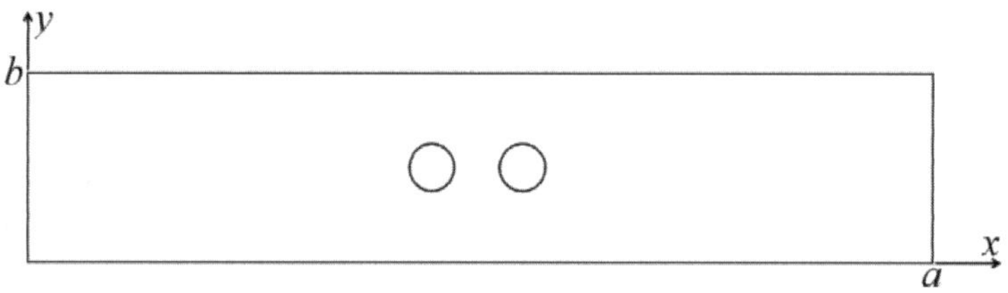

Figure 5. Rectangular plate weakened by two circular holes.

One have to find an elastoplastic boundary for the domain closed by Γ: $\Gamma = \Gamma_1 \cup \Gamma_2 \cup \Gamma_3$, where Γ_1 is a rectangular, Γ_2 and Γ_3 are circles. Contour Γ is in the plastic state.

Yield criterion for the plane stress state on the boundary of the plate is written in the following form:

$$\sigma_x^2 + \sigma_y^2 - \sigma_x\sigma_y + 3\tau^2 = 3k^2, \quad (13)$$

where σ_x, σ_y, τ are components of the stress tensor for the plane stress state, *k* is a constant of plasticity.

Boundary conditions for the plate have the form

$$\sigma_x n_1 + \tau n_2 = X,\ \sigma_y n_2 + \tau n_1 = Y, \quad (14)$$

where $\vec{n} = (n_1, n_2)$ is the unit normal vector to the contour Γ_1; *X*, *Y* are components of the vector of external forces.

The plate is subjected to a tensile load in the axis *x* direction and to a compressive one in direction of the axis *y*.

Construction of the elastoplastic boundary for the considered plate contains three main stages.

- Laplace equation solving

This stage lies in solving of Laplace equation $\Delta F = 0$ with boundary conditions $F|_\Gamma = \sigma_x + \sigma_y$ (here σ_x, σ_y are functions from (13)–(14), $F = \sigma_x + \sigma_y$ is a harmonic function) and in finding the values of the function *F* in inner points of the domain bounded by Γ.

- Application of conservation laws to find σ_x, σ_y, τ

The purpose of the stage is in finding of the functions σ_x, σ_y, τ in each point (x_0, y_0) of the considered domain by formulas derived using conservation laws method [13]:

$$\frac{\partial \sigma_x}{\partial x}+\frac{\partial \tau}{\partial y}=0,\ \frac{\partial \tau}{\partial x}+\frac{\partial F}{\partial y}-\frac{\partial \sigma_x}{\partial y}=0,\ \sigma_y=F-\sigma_x. \quad (15)$$

- Yield criterion criterion verification

On this stage of the elastoplastic boundary construction, the yield criterion for each inner point of the domain bounded by the contour Γ is verified. If the value of the left side of the yield criterion (13) is less than $3k^2$ for the stresses at the inner point, then this point is in the elastic zone; all the rest points, with stresses which do not satisfy this condition, are in the zone of plasticity of the examined domain.

The following figure shows the distributions of the points in elastic and plastic zones calculated for the plate:

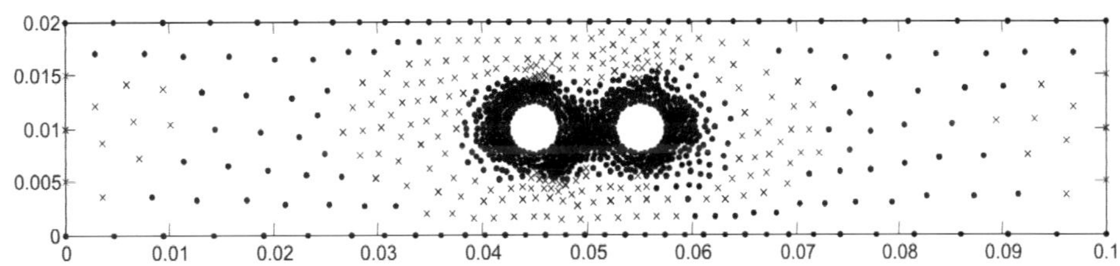

Figure 6. Distribution of elastic (crosses) and plastic (points) zones in the plate.

Note that detailed information can be found in [13].

6. Conclusion

Investigation of the plasticity equations using the symmetries and the conservation laws reveals availability of application of these methods to studying and solving the other equations of solid mechanics.

References

[1] Ovsyannikov L V 1982 *Group Analysis of Differential Equations* (New York: Academic Press)
[2] Annin B D, Bytev V O and Senashov S I 1985 *Group Analysis of Elasticity and Plasticity Equations* (Novosibirsk: Siberian Branch of the USSR Academy of Sciences) (In Russian)
[3] Senashov S I 2004 Plastic Flows of a Mises Medium with Helical Symmetry *J. Appl. Math. Mech.* **68 (1)** pp 133–37
[4] Senashov S I and Vinogradov A M 1988 Symmetries and Conservation Laws of 2-Dimensional Ideal Plasticity *Proc. of Edinb. Math. Soc* **31** pp 415–39
[5] Kiryakov P P, Senashov S I and Yakhno A N 2001 *Application of Symmetries and Conservation Laws to Differential Equations* (Novosibirsk: Siberian Branch of the Russian Academy of Sciences) (in Russian)
[6] Senashov S I and Yakhno A N 2007 Reproduction of Solutions for Bidimensional Ideal Plasticity *Int. J. NonLin Mech.* **42** pp 500–03
[7] Senashov S I and Yakchno A N 2009 Deformation of Characteristic Curves of the Plane Ideal Plasticity Equations by Point Symmetries *Nonlinear Anal.* **71** pp 1274–84
[8] Senashov S I and Yakchno A N 2012 Conservation Laws, Hodograph Transformation and Boundary Value Problems of Plane Plasticity *SIGMA* **8**, 071, 16 p
[9] Senashov S I and Yakchno A N 2013 Some Symmetry Group Aspects of a Perfect Plane Plasticity System *J. Phys. A: Math. Theor.* ***46*** No 35 p 355202
[10] Senashov S I and Yakchno A N 2013 Conservation Laws of Three-Dimensional Perfect Plasticity Equations under von Mises Yield Criterion *Abstract and Applied Analysis* **Volume 2013**, Article ID 702132, 8 p

[11] Senashov S I, Cherepanova O N and Kondrin A V 2014 Elasto-plastic Bending of a Beam. *J. Siberian Federal Univ., Math & Physics* **7 (2)** pp 203–08

[12] Senashov S I, Cherepanova O N and Kondrin A V 2015 On Elastoplastic Torsion of a Rod with Multiply Connected Cross-Section *J. Siberian Federal Univ., Math. & Physics* **7(1)** pp 343–51

[13] Senashov S I and Gomonova O V 2019 Construction of Elastoplastic Boundary in Problem of Tension of a Plate Weakened by Holes *Int. J. NonLin Mech.* **108** pp 7–9

IOP Conf. Series: Materials Science and Engineering **822** (2020) 012031 doi:10.1088/1757-899X/822/1/012031

Hook's law as Lie group

S I Senashov and I L Savostyanova

Department of Economic Information Systems, Reshetnev Siberian State University of Science and Technology, 31 Krasnoyarsky Rabochy Av., Krasnoyarsk, 660037, Russia

E-mail: sen@sibsau.ru

Abstract. In this work classification of anisotropic linear elastic materials with residual deformations assuming that the material constants depend on the residual deformations is offered. Each generalised Hook's law is assigned some Lie point group, and this group in its turn is assigned Lie algebra.

1. Introduction

One of the basic tasks of mathematics is classification. An example of such a task very well solved by means of theory of groups is classification of crystals.

A series of authors noticed analogies between the theory of elasticity and groups of continuous transformations (Lie groups). First to be mentioned is A. Love [1]. He showed that linear homogeneous deformations of an elastic body generate a group of continuous transformations. And also one the authors of the monography shared similar views. Some thoughts on this matter are included in Works [2, 3].

The very idea of classification of elastic materials on the basis of Hook's law is not new. It goes back to classification of crystals. It is believed that the elastic material inherits symmetry properties of its component crystals. That is why elastic materials are classified according to syngonies. Let us remind that Nikola Stenton (1669) devided all the crystals into seven syngonies (similar angled): triclinic, monoclinic, rhombic, trigonal, hexagonal, tetragonal and cubic. Then, using the notion of crystal lattice Auguste Brave (1848) divided crystals into 14 classes. In 1867 Aksel Vilgelmovich Gadolin introduced 32 crystal classes and in 1890-1894 Evgraf Stepanovich Fyodorov completed the classification by introducing 230 crystal groups. It is necessary to highlight that all these ingenious findings were made long before experimental discovery of the crystal lattice with the help of X-rays. This classification is still up-to-date and is widely used in various spheres of science.

Elastic materials that are described by linear Hook's law can also be successfully classified according to the number of constants included in Hook's law in accordance with the crystal symmetry (syngony). This number can be 81, 21, 13, 9, 5 and 2.

In this work classification of anisotropic linear elastic materials with residual deformations assuming that the material constants depend on the residual deformations is offered. Each generalised Hook's law is assigned some Lie point group, and this group in its turn is assigned Lie algebra. Each two-dimensional subalgebra of this algebra is matched with some experiment on combined loading which allows us to determine dependence of the material constants on residual deformations. This allows us to carry out classification of such materials and divide them into a considerably large

IOP Conf. Series: Materials Science and Engineering **822** (2020) 012031 doi:10.1088/1757-899X/822/1/012031

number of classes.

The plan of the article is as follows: 1) using the definition of Lie group and Lie algebra of continuous transformations to write Hook's law for materials with residual deformations and to verify that these laws comply with all the properties of the continuous transformations group, 2) as per these groups to build Lie algebras for different variants of anisotropic elasticity theory, 3) to describe possible experiments that allow us to classify elastic materials.

2. Necessary information from the theory of continuous groups of transformations

Let us remind necessary information from the theory of Lie groups of continuous transformations that is required for understanding of further presentation. Here we will adhere to the original source that is the recently translated into Russian Book of Sophus Lie [4].

Assume $x, x' \in R^n, a \in R^r$. Let us consider the transformations in the form of

$$x' = f(x;a) \tag{1}$$

where f – some smooth function, a – parameters that as a rule change in some neighbourhood of zero.

Assume $x'' = f(x';b)$, then this equation takes place

$$x'' = f(x';b) = f(f(x;a);b) = f(x;\varphi(a,b)) \tag{2}$$

Here Function $\varphi(a,b)$ is the law of multiplication of parameters which is assumed by the smooth function.

Let us demand that the continuous transformations (1) create a group. In this case Function $\varphi(a,b)$ must comply with the following properties

$$\begin{gathered}\varphi(0,a) = \varphi(a,0) = a \\ \varphi(a^{-1},a) = \varphi(a\ ,a^{-1}) = 0 \\ \varphi(a,\varphi(b,c)) = \varphi(\varphi(a,b),c)\,.\end{gathered} \tag{3}$$

If properties (3) are fulfilled in the neighbourhood of zero of Space R^r, then the group built is local and is called local Lie group. If $a \in R^1$, then such a group of transformations is called one-parameter, and if $a \in R^r$, then correspondingly – r – parameter. It may be shown that the r –parameter Lie group denoted as G_r is fully determined by its one-parameter subgroups G_1.

Each local one-dimensional subgroup G_1 is assigned a tangent vector according to the following rule

$$x_i' = f_i(x;a) \to \xi_i = \frac{df_i}{da}|_{a=0}, \tag{4}$$

and each tangent vector is assigned the operator in the form of

$$X_a = \xi_i \frac{\partial}{\partial x_i}. \tag{5}$$

Operators (4) for Group G_r create Lie algebra L_r, that can be classified as per their structure determined by the commutators in the form of

$$[X,Y] = [\xi_i \frac{\partial}{\partial x_i}, \eta_i \frac{\partial}{\partial x_i}] = (\xi_j \frac{\partial \eta_i}{\partial x_j}\frac{\partial}{\partial x_i} - \eta_j \frac{\partial \xi_i}{\partial x_j}\frac{\partial}{\partial x_i}).$$

Let us give a simple example that will allow us to in a simpler way carry out building of the groups and Lie algebras for the theory of elasticity with initial deformations.

Example [4].

Let us consider the three-parameter group of transformations of the one-dimensional space

$$x' = \frac{x + a_1}{a_2 x + a_3} \tag{6}$$

Then $x'' = \dfrac{x' + b_1}{b_2 x' + b_3}$

In this case Function $\varphi(a,b)$ is written as

$$\phi_1 = \frac{a_1 + b_1 a_3}{1 + b_1 a_2}, \phi_2 = \frac{b_2 + b_3 a_2}{1 + b_1 a_2}, \phi_3 = \frac{b_2 a_1 + b_3 a_3}{1 + b_1 a_2}.$$

It is not difficult to see that to the identity transformation there answer the values of $a_1 = 0, a_2 = 0, a_3 = 1$. To the backward transformations there correspond the parameters of $-a_1 / a_3, -a_2 / a_3, 1 / a_3$.

All the properties (3) are fulfilled that is why (6) is Group G_3. Let us build Lie algebra L_3 corresponding to this group. We have

$$\xi_1 = \frac{dx'}{da_1}\Big|_{(0,0,1)} = 1, \xi_2 = \frac{dx'}{da_2}\Big|_{(0,0,1)} = -x^2, \xi_3 = \frac{dx'}{da_3}\Big|_{(0,0,1)} = x.$$

Therefore Lie algebra L_3 is created by the operators

$$X_1 = \frac{\partial}{\partial x}, X_2 = x^2 \frac{\partial}{\partial x}, X_3 = x \frac{\partial}{\partial x}.$$

At that $[X_1, X_2] = X_3, [X_1, X_3] = X_1, [X_2, X_3] = X_2 - 2X_3$.

This example shows there are two ways to show that the transformations (6) create a group of continuous transformations: to check properties (3) or to build the corresponding Lie algebra and check its closeness relative to the operation of commutation.

Later on we will use both of the ways when we will be showing that the generalised Hook's laws built by us create groups of continuous transformations.

3. Classification of elastic bodies

The basis of the linear theory of elasticity of smaller deformations is Hook's law which for anisotropic materials is written as [4, 5]:

$$\sigma_{ij} = c_{ijkl} \varepsilon_{kl} \quad (i,j,k,l=1,2,3) \tag{7}$$

or

$$\varepsilon_{ij} = a_{ijkl} \sigma_{kl} \quad (i,j,k,l=1,2,3) \tag{8}$$

where σ_{ij} – components of symmetrical stress tensor, ε_{kl} components of symmetrical deformation tensor, c_{ijkl}, a_{ijkl} – mechanical material constants, for the repeated indices summarisation is carried out. Values c_{ijkl} create tensor that has 81 components. By virtue of the symmetry of Tensors σ_{ij}, ε_{kl} there will be no more than 21 of independent components. This number of constants correspond to triclinic crystal system.

Let us introduce the structure of Lie group into the theories of elasticity. Hook's law (7) in an even more general form can be written in two different ways [2]

$$\sigma_{ij} = f_{ij}(\varepsilon_{kl}) \quad (i,j,k,l=1,2,3) \tag{9}$$

$$\varepsilon_{ij} = g_{ij}(\sigma_{kl}) \quad (i,j,k,l=1,2,3) \tag{10}$$

This writing reminds (1), but there is a significant difference: in the left-hand part there is no σ_{ij}' or ε_{ij}'. Can they be introduced into there? They can be if we assume that in the elastic medium there are residual stresses $\sigma_{ij}{}^0$ (residual deformations $\varepsilon_{ij}{}^0$). Now if we remember that $\varepsilon_{kl} \in R^6$ ($\sigma_{kl} \in R^6$), then the relations (9)-(10) will be written as

$$\sigma_{ij}' = g_{ij}(\sigma_{ij}^0, \varepsilon_{kl}). \tag{11}$$

$$\varepsilon'_{ij} = f_{ij}(\varepsilon_{ij}^0, \sigma_{kl}) \tag{12}$$

The question arises: which of these formulas is better to be used to build the theory of elasticity, which one of them better corresponds to the group of continuous transformations (1.64)? In the author's opinion it is better to use Formulas (12), the reasons for that will be explained later. Let us highlight that elastic materials with initial stress condition were started to be looked at as early as by Cauchy, see [4] and the references available there, in recent time the properties of such materials were studied in [6]. The presence of Laws (11) or (12) mean that the constants included in Hook's law (7) or (8) depend on the initial stresses (deformations). This question if it is so can only be answered after performing experiments for the certain given material. Some of such experiments will be discussed below.

Now it is remaining to demand fulfilling of the properties (3). From (12) we will obtain the group determined by six parameters, in other words Group G_6.

$$\begin{gathered} \varepsilon_{ij}' = f_{ij}(\varepsilon_{ij}^0, \sigma_{kl}), \varepsilon_{ij}{}^0 = f_{ij}(\varepsilon_{ij}', \sigma_{kl}^{-1}), \varepsilon_{ij}{}^0 = f_{ij}(\varepsilon_{ij}^0, 0), \\ \varepsilon_{ij}'' = f_{ij}(\varepsilon_{ij}', \sigma_{rl}^1) = f_{ij}(f(\varepsilon_{ij}^0, \sigma_{rl}^1), \sigma_{rl}^0) = f_{ij}(\varepsilon_{ij}^0, \phi(\sigma_{rl}^1, \sigma_{kl})). \end{gathered} \tag{13}$$

These conditions are fulfilled by virtue of linearity of the equations of elasticity and the smallness of the deformation tensor components.

Of course we can check all the conditions of the continuous group as it is carried out in the example of S.Lie, but we can do it in another way: to calculate operators of Lie algebra and check if they create Lie algebra of dimension 6. If it is so then Transformations (12) create the group of continuous transformations.

Since $\varepsilon_{ij}, \varepsilon_{ij}^0$ are small, see [5, 7], then we expand $f_{ij}(\varepsilon_{ij}^0, \sigma_{kl})$ according to Taylor formula in the neighbourhood of zero. We have

$$f_{ij}(\varepsilon_{ij}^0, \sigma_{kl}) = a_{ijkl}(1 + \alpha_{nm}\varepsilon_{nm}^0)\sigma_{kl} = a_{ijkl}\alpha_{nm}\varepsilon_{nm}^0\sigma_{kl} + \varepsilon_{ij}^0 \tag{14}$$

Let us highlight that if $\varepsilon_{ij}^0 = 0$, then Relation (14) coincides with (8). Let us calculate the operator corresponding to Transformation (14), assuming σ_{kl} to be the parameters of the group of transformations. We have

$$X_{ij} = a_{ijkl}\alpha_{nm}\varepsilon_{nm}\frac{\partial}{\partial \varepsilon_{kl}}. \tag{15}$$

It is not difficult to see that the commutator of any two operators in the form of (15) is a linear combination of the operators in the form of (15). This depicts that the generalised Hook's law (14) in the case of small elastic deformations is a group of continuous transformations of deformation tensor components.

4. In this section we will consider some possible variants of the generalised Hook's law

As the first example of classification we will consider different variants of Hook's law that are used to describe elastic torsion. In this case only two components of stress tensor τ_{13}, τ_{23} will differ from zero.

4.1. Isotropic case

We have

$$\varepsilon_{13} = \varepsilon_{13}^0 + \mu\tau_{13}, \quad \varepsilon_{23} = \varepsilon_{23}^0 + \mu\tau_{23}, \tag{16}$$

where $\varepsilon_{13}^0, \varepsilon_{23}^0$ - initial stresses, μ - is function $\varepsilon_{13}^0, \varepsilon_{23}^0$.

We obtain

$$X_{13} = \mu\frac{\partial}{\partial\varepsilon_{13}}, X_{23} = \mu\frac{\partial}{\partial\varepsilon_{23}} \tag{17}$$

The commutator of operators (17) equals to

$$[X_{13}, X_{23}] = \mu\frac{\partial\ \mu}{\partial\varepsilon_{13}}\frac{\partial}{\partial\varepsilon_{23}} - \mu\frac{\partial\ \mu}{\partial\varepsilon_{23}}\frac{\partial}{\partial\varepsilon_{13}}.$$

Let us equate the commutator to the linear combination of operators (17) , we will see that in this case μ is a linear function $\mu = \alpha\varepsilon^0{}_{13} + \beta\varepsilon^0{}_{23} + \gamma$, $\alpha, \beta, \gamma - const$

4.2. Orthotropic symmetry

We have

$$\varepsilon_{13} = \varepsilon_{13}^0 + a_{44}\tau_{13}, \varepsilon_{23} = \varepsilon_{23}^0 + a_{55}\tau_{23}, \tag{18}$$

where $\varepsilon_{13}^0, \varepsilon_{23}^0$ – initial stresses, a_{44}, a_{55} – is function $\varepsilon_{13}^0, \varepsilon_{23}^0$.

We obtain

$$X_{13} = a_{44}\frac{\partial}{\partial\varepsilon_{13}}, X_{23} = a_{55}\frac{\partial}{\partial\varepsilon_{23}}. \tag{19}$$

The commutator of operators (19) equals to

$$[X_{13}, X_{23}] = a_{44}\frac{\partial\ a_{55}}{\partial\varepsilon_{13}}\frac{\partial}{\partial\varepsilon_{23}} - a_{55}\frac{\partial\ a_{44}}{\partial\varepsilon_{23}}\frac{\partial}{\partial\varepsilon_{13}}.$$

Let us equate it to the linear combination of operators (19) and we will see that in this case the coefficients satisfy the system of equations

$$a_{44}\frac{\partial\ a_{55}}{\partial\varepsilon_{13}} = Aa_{55}, a_{55}\frac{\partial\ a_{44}}{\partial\varepsilon_{23}} = Ba_{44}.$$

here *A*, *B* – are arbitrary constants.

4.3. Case of monoclinic crystal symmetry
We have

$$\varepsilon_{13} = \varepsilon_{13}^0 + a_{44}\tau_{13} + a_{45}\tau_{23}, \varepsilon_{23} = \varepsilon_{23}^0 + a_{45}\tau_{13} + a_{55}\tau_{23}, \tag{20}$$

where τ_{13}^0, τ_{23}^0 – initial stresses, a_{44}, a_{55}, a_{45} – is function $\varepsilon_{13}^0, \varepsilon_{23}^0$.
We obtain

$$X_{13} = a_{55}\frac{\partial}{\partial\varepsilon_{13}} + a_{45}\frac{\partial}{\partial\varepsilon_{23}} X_{23} = a_{55}\frac{\partial}{\partial\varepsilon_{23}} + a_{45}\frac{\partial}{\partial\varepsilon_{13}} \tag{21}$$

The commutator of operators (21) equals to

$$[X_{13}, X_{23}] = \left(a_{44}\frac{\partial\, a_{45}}{\partial\varepsilon_{13}} - + a_{45}\frac{\partial\, a_{45}}{\partial\varepsilon_{23}} - a_{55}\frac{\partial\, a_{44}}{\partial\varepsilon_{23}} - a_{45}\frac{\partial\, a_{44}}{\partial\varepsilon_{13}} \right)\frac{\partial}{\partial\varepsilon_{13}} +$$
$$+ \left(a_{44}\frac{\partial\, a_{55}}{\partial\varepsilon_{13}} + a_{45}\frac{\partial\, a_{55}}{\partial\varepsilon_{23}} - a_{55}\frac{\partial\, a_{45}}{\partial\varepsilon_{23}} - a_{45}\frac{\partial\, a_{45}}{\partial\varepsilon_{13}} \right)\frac{\partial}{\partial\varepsilon_{23}}$$

Let us equate it to the linear combination of operators X_{13}, X_{23} and we will see that in this case a_{44}, a_{55}, a_{45} satisfy the two differential equations:

$$\left(a_{44}\frac{\partial\, a_{45}}{\partial\varepsilon_{13}} + a_{45}\frac{\partial\, a_{45}}{\partial\varepsilon_{23}} - a_{55}\frac{\partial\, a_{44}}{\partial\varepsilon_{23}} - a_{45}\frac{\partial\, a_{44}}{\partial\varepsilon_{13}} \right) = C_1 a_{44} + C_2 a_{45}.$$
$$\left(a_{44}\frac{\partial\, a_{55}}{\partial\varepsilon_{13}} + a_{45}\frac{\partial\, a_{55}}{\partial\varepsilon_{23}} - a_{55}\frac{\partial\, a_{45}}{\partial\varepsilon_{23}} - a_{45}\frac{\partial\, a_{45}}{\partial\varepsilon_{13}} \right) = C_1 a_{45} + C_2 a_{55}.$$

Here C_i – arbitrary constants.

4.4. In conclusion of this section we will consider a special case of orthotropic medium
In conclusion of this section we will consider a special case of orthotropic medium [7].

$$\begin{aligned} \varepsilon_{11} &= a_{11}\sigma_{11} + a_{12}\sigma_{22} + a_{13}\sigma_{33}, \varepsilon_{22} = a_{12}\sigma_{11} + a_{11}\sigma_{22} + a_{23}\sigma_{33}, \\ \varepsilon_{33} &= a_{13}\sigma_{11} + a_{23}\sigma_{22} + a_{11}\sigma_{33}, \varepsilon_{12} = a_{44}\tau_{12}, \varepsilon_{23} = a_{55}\tau_{23}, \varepsilon_{13} = a_{66}\tau_{13}. \end{aligned} \tag{22}$$

After first loading the body obtains residual deformations

$$\begin{aligned} \varepsilon^0{}_{11} &= a_{11}\sigma_{11} + a_{12}\sigma_{22} + a_{13}\sigma_{33}, \varepsilon^0{}_{22} = a_{12}\sigma_{11} + a_{11}\sigma_{22} + a_{23}\sigma_{33}, \\ \varepsilon^0{}_{33} &= a_{13}\sigma_{11} + a_{23}\sigma_{22} + a_{11}\sigma_{33}, \varepsilon^0{}_{12} = a_{44}\tau_{12}, \varepsilon^0{}_{23} = a_{55}\tau_{23}, \varepsilon^0{}_{13} = a_{66}\tau_{13}. \end{aligned} \tag{23}$$

Assume expanding of coefficients according to Taylor formula is written as

$$a_{11} = a_{11}(1 + \alpha\varepsilon_{11}^0 + \beta\varepsilon_{22}^0 + \gamma\varepsilon_{33}^0), a_{44} = a_{44}(1 + \alpha_1\varepsilon_{12}^0), a_{66} = a_{66}(1 + \alpha_3\varepsilon_{13}^0), a_{55} = a_{55}(1 + \alpha_2\varepsilon_{23}^0). \tag{24}$$

For simplicity we consider other coefficients constant.
Then taking into account (23) we will get the following formulas

$$\begin{aligned} \varepsilon_{11} &= a_{11}(\alpha\varepsilon_{11}^0 + \beta\varepsilon_{22}^0 + \gamma\varepsilon_{33}^0)\sigma_{11} + \varepsilon_{11}^0, \varepsilon_{22} = a_{11}(\alpha\varepsilon_{11}^0 + \beta\varepsilon_{22}^0 + \gamma\varepsilon_{33}^0)\sigma_{22} + \varepsilon_{22}^0, \\ \varepsilon_{33} &= a_{11}(\alpha\varepsilon_{11}^0 + \beta\varepsilon_{22}^0 + \gamma\varepsilon_{33}^0)\sigma_{33} + \varepsilon_{33}^0, \varepsilon_{12} = a_{44}\alpha_1\varepsilon_{12}^0\sigma_{12} + \varepsilon_{12}^0, \varepsilon_{23} = a_{55}\alpha_1\varepsilon_{23}^0\sigma_{23} + \varepsilon_{23}^0, \varepsilon_{13} = a_{66}\alpha_1\varepsilon_{13}^0\sigma_{13} + \varepsilon_{13}^0. \end{aligned} \tag{25}$$

We build Lie algebra according to Group (25) we have

$$X_{11}=a_{11}(\alpha\varepsilon_{11}^0+\beta\varepsilon_{22}^0+\gamma\varepsilon_{33}^0)\frac{\partial}{\partial\varepsilon_{11}}, X_{22}=a_{11}(\alpha\varepsilon_{11}^0+\beta\varepsilon_{22}^0+\gamma\varepsilon_{33}^0)\frac{\partial}{\partial\varepsilon_{22}},$$
$$X_{33}=a_{11}(\alpha\varepsilon_{11}^0+\beta\varepsilon_{22}^0+\gamma\varepsilon_{33}^0)\frac{\partial}{\partial\varepsilon_{33}}, X_{12}=a_{12}\alpha_1\varepsilon_{12}^0\frac{\partial}{\partial\varepsilon_{12}}, X_{23}=a_{23}\alpha_2\varepsilon_{23}^0\frac{\partial}{\partial\varepsilon_{23}}, X_{13}=a_{13}\alpha_3\varepsilon_{13}^0\frac{\partial}{\partial\varepsilon_{13}}. \quad (26)$$

We calculate the commutators of Operators (25). We have

$$[X_{11},X_{22}]=\alpha X_2-\beta X_1,[X_{11},X_{33}]=\alpha X_3-\gamma X_1,[X_{22},X_{33}]=\beta X_3-\gamma X_2. \quad (27)$$

Other commutators are equal to zero.

Since Operators (26) create Lie algebra of dimension 6, then Hook's law in the form of (25) is a six-parameter group of continuous transformations.

5. Experimentation on the classification of materials

For simplicity we will consider mental experiments on combined loading that will allow us to determine the type of Hook's law.

We have

$$\varepsilon_{13}=(\alpha\varepsilon_{13}^0+\beta\varepsilon_{23}^0+\mu)\tau_{13}+\varepsilon_{13}^0, \varepsilon_{23}=(\alpha\varepsilon_{13}^0+\beta\varepsilon_{23}^0+\mu)\tau_{23}+\varepsilon_{23}^0. \quad (28)$$

5.1. Assume in the formulas (28) $\alpha=\beta=0$. In this case (28) is written as

$$\varepsilon_{13}=\mu\tau_{13}+\varepsilon_{13}^0, \varepsilon_{23}=\mu\tau_{23}+\varepsilon_{23}^0. \quad (29)$$

The experiment will consist of four steps. Assume the initial deformations $\varepsilon_{13}^0=\varepsilon_{23}^0=1$. This corresponds to Point A in figure 1.

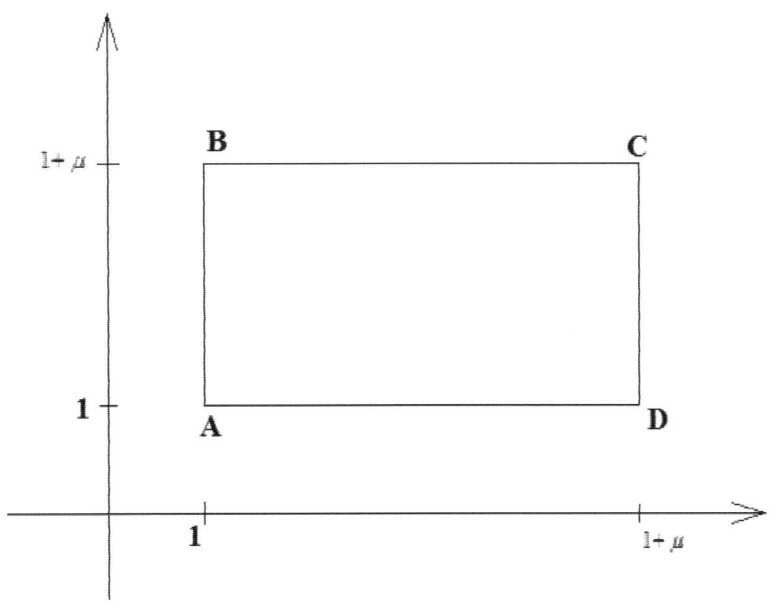

Figure 1. Experiment on combined loading with $\alpha=\beta=0$

Step 1. Assumes $\tau_{13}=1, \tau_{23}=0$. We have $\varepsilon'_{13}=1+\mu, \varepsilon'_{23}=\varepsilon_{23}^0=1$. This is Point B in figure 1.

Step 2. Assumes $\tau_{13}=1, \tau_{23}=1$. We have $\varepsilon''_{13}=1+\mu, \varepsilon''_{23}=1+\mu$. We get point C in figure 1.

Step 3. Assumes $\tau_{13}=-1, \tau_{23}=1$. We have $\varepsilon'''_{13}=1, \varepsilon'''_{23}=1+\mu$. We get Point D in figure 1.

Step 4. Assumes $\tau_{13}=-1, \tau_{23}=-1$. We have $\varepsilon^{1Y}_{13}=1, \varepsilon^{1Y}_{23}=1$. As a result we return to Point A in figure 1. We get a commutative experiment.

5.2. Assume in the formulas (28) $\alpha \neq 0, \beta=0, \mu \neq 0$. In this case (28) is written as

$$\varepsilon_{13}=(\alpha\varepsilon^0_{13}+\mu)\tau_{13}+\varepsilon^0_{13}, \varepsilon_{23}=(\alpha\varepsilon^0_{13}+\mu)\tau_{23}+\varepsilon^0_{23}. \quad (30)$$

Let us repeat the same steps as in the previous experiment

Assume the initial deformations $\varepsilon^0_{13}=\varepsilon^0_{23}=1$. This corresponds to Point A in figure 2.

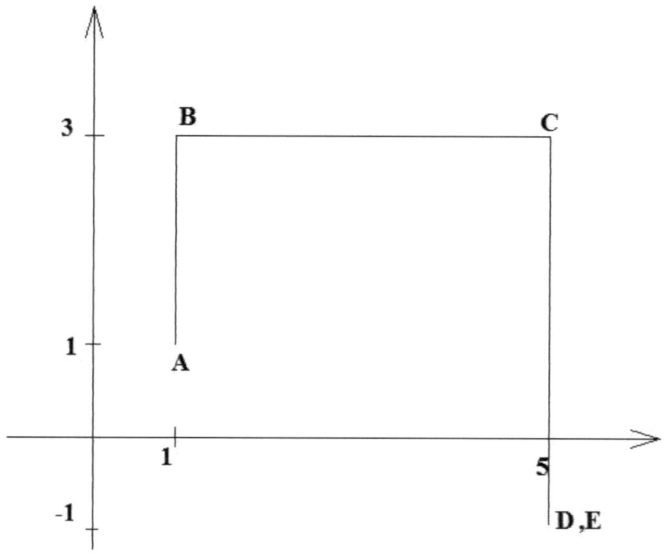

Figure 2. Experiment on combined loading with $\alpha \neq 0, \beta=0, \mu \neq 0$

Step 1. Assumes $\tau_{13}=1, \tau_{23}=0$. We have $\varepsilon'_{13}=1+\alpha+\mu, \varepsilon'_{23}=\varepsilon^0_{23}=1$.

Step 2. Assumes $\tau_{13}=1, \tau_{23}=1$. We have $\varepsilon''_{13}=1+(\alpha+\mu), \varepsilon''_{23}=1+\alpha(1+\alpha+\mu)+\mu$.

Step 3. Assumes $\tau_{13}=-1, \tau_{23}=1$. We have $\varepsilon'''_{13}=1+(\alpha+\mu)-(\alpha(1+\alpha+\mu)+\mu)$, $\varepsilon'''_{23}=1+\alpha(1+\alpha+\mu)+\mu$.

Step 4. Assumes $\tau_{13}=-1, \tau_{23}=-1$. We have $\varepsilon^{1Y}_{13}=\varepsilon'''_{13}, \varepsilon^{1Y}_{23}=\varepsilon'''_{23}-(\alpha\varepsilon'''_{13}+\mu)$.

For clarity we will assume $\alpha=1, \mu=1$. We get the following points in figure 2: *A(1,1), B(3,1), C(3,5), D(-1,5), E(-1,5).* We see that contrary to figure 1 we have not got the closed curve. Let us highlight that this result does not depend on the certain values of the parameters α, μ. Here all is determined by the commutator generated by Hook's law. In the first case it is equal to zero and in the second case it differs from zero (27).

It is known that Lie subalgebras of dimension two can only be of two types: either commutative or not. That is why elastic materials can only be divided into two classes with the use of one two-dimension experiment: «commutative» material or not.

If we take Hook's law for planar deformations then there we will have three parameters $\sigma_{11}, \sigma_{22}, \tau_{12}$. It is feasible to perform a three-dimension experiment but in practice it is very difficult to realise. It is simpler to perform a series of two-dimension experiments similar to the ones described above with each pair of the parameters: $(\sigma_{11}, \sigma_{22}), (\sigma_{11}, \tau_{12}), (\sigma_{22}, \tau_{12})$. With the use of such experiments it will be feasible to study most of the three-dimension subalgebras. And because their number is countless then this gives us vast opportunities for the classification of materials. In the same way it will be feasible to classify materials determined by 4, 5 or 6 components of stress tensor. This will allow us to classify most of the materials because all Lie algebras dimension 2-6 over the field of real numbers are described, see [2] and the articles referenced there.

References

[1] Lyav A 1935 *The Mathematical theory of elasticity* (Moscow; Leningrad) (in Russ)
[2] Senashov S I, Kiryakov P P and Jahno A N 2001 *Application of symmetry and conservation laws to the solution of differential equations* (Novosibirsk SB RAS) 192 p (in Russ)
[3] Senashov S. I. 1994 *The Lie's Groups and classification of elastic materials* DAN AN USSR vol 335 No 6 pp 712-715 (in Russ)
[4] Lie S 2011 *The theory of transformation groups. Part 1* (Moscow, Izhevsk) p 712 (in Russ)
[5] Lechnitsky S G 1977 *The Theory of elasticity of anisotropic bodies* (Moscow Nauka) p 416 (in Russ)
[6] Guz' A N 1983 *The Mechanics of brittle fracture of materials with initial stresses* (Kiev Nauk Dumka) p 296 (in Russ)
[7] Novatsky V 1975 *The Theory of elasticity* (Moscow Mir) p 860 (in Russ)

Analysis of the secondary emission current and X-ray radiation dependency on the beam's position along the joint during electron beam welding

V Ya Braverman, V V Bogdanov and L A Oborin

Reshetnev Siberian State University of Science and Technology, 31 Krasnoyarsky Rabochy Av., Krasnoyarsk, 660037, Russia

E-mail: braverman-vladimir@rambler.ru

Abstract. We analyzed secondary emission current and X-Ray radiation dependency on the beam's position along the joint during Electron Beam Welding of similar and dissimilar materials. We established that those characteristics were identical if the welding rim melting was absent. The characteristics are of extreme nature. During the welding of similar materials, the dependency minimum corresponds with the precise alignment of the beam's position and the joint. During the welding of dissimilar materials, the minimum of the secondary emission current shifts towards the material with the lowest coefficient of the secondary emission and the minimum of the X-ray radiation shifts towards the material with the lowest atom number. The similarity of the characteristics is explained by the fact that the appearance of the secondary emission and the X-ray radiation is the result of the electron beam's interaction with the material of the welded parts. Therefore, we can use the same mathematical tool to analyze the dependencies and joint tracking devices can be designed with unified hardware and software applications.

1. Introduction

Precise positioning of the beam along the joint of the welded parts is especially important when dealing with long joints during the welding of large parts. Acceptable error in positioning the beam with the joint usually does not exceed 0.2 mm. This precision requires the use of devices with automatic beam guidance along the joint.

Breaking X-ray radiation and secondary electron emission accompanying the Electron Beam Welding Process can be used as a source of information regarding beam's position at the joint [1-3]. Figure 1 provides schematic for a receipt of secondary and X-ray radiation [4]. Figure 2 provides examples for positioning of the secondary electron sensor (figure 1, a) and the X-ray radiation sensor (figure 1, b).

The current of secondary electrons I_{se} can be presented as follows: [5]:

$$I_{se}=\varphi I_b k_{se} k_{cse}, \qquad (1)$$

where I_b – the beam's current; k_{se} – coefficient that accounts for the number of secondary electrons reaching the sensor; φ – coefficient of the secondary emission of the welded materials; k_{cse}– coefficient that accounts for the influence of structural members on the current I_{se}.

Intensity of the X-ray radiation J is determined by the following equation [6]:

$$J=CZU^2 I_b k_{XR} k_{cXR}, \qquad (2)$$

where C – proportionality coefficient ($C\approx10^{-9}$, V^{1}); k_{XR} – coefficient that accounts for the share of radiation reaching the surface of the sensor; k_{cXR} – coefficient that accounts for the influence of structural members on the intensity J; U – accelerating voltage; Z – atom number of the welded materials.

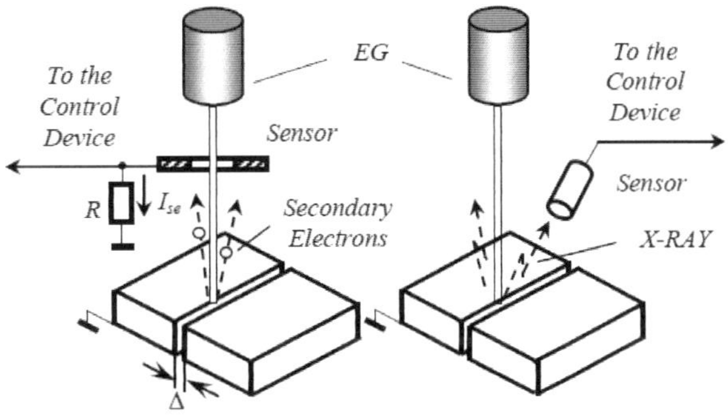

a) *b)*

Figure 1. Receiving signals.

Equations (1) and (2) show, that I_{se} and J with coefficient precision are determined by the beam current I_b. Once can assume that I_{se} and J depend on the beam's position along the joint and these dependencies are of a similar nature. Thus, one mathematical device can be used to study these dependencies.

a) *b)*

Figure 2. Arrangement of sensors.

2. Formulation of the objective. Current density in the electron beam is not the same in its cross section and has probabilistic nature. Therefore, the beam current I_b is expressed through the function of electron distribution along corresponding axes [7], and equations (1) and (2) can be presented as

$$I_{se} = k_{se} I_b \int_{-\infty}^{\infty} \int_{-\infty}^{\infty} k_{cse}(x, y) \varphi(x, y)\, j(x, y)\, d\,xdy,$$

$$J = Ck_{XR} U^2 I_b \int_{-\infty}^{\infty} \int_{-\infty}^{\infty} k_{cXR}(x, y) Z(x, y)\, j(x, y)\, d\,xdy,$$

where $j(x,y)$ – distribution density of electrons.

Density $j(x,y)$ is presented with a normal law of distribution [8]:

$$j(x, y) = \frac{1}{\sigma_x \sqrt{2\pi}} exp\left[-\frac{(x-\varepsilon_x)^2}{2\sigma_x^2}\right] \frac{1}{\sigma_y \sqrt{2\pi}} exp\left[-\frac{(y-\varepsilon_y)^2}{2\sigma_y^2}\right],$$

where σ_x, σ_y – standard deviation of electrons from the bunch axis along the corresponding axes; ε_x, ε_y – mathematical expectation – the beam's axis position in coordinates (figure 3).

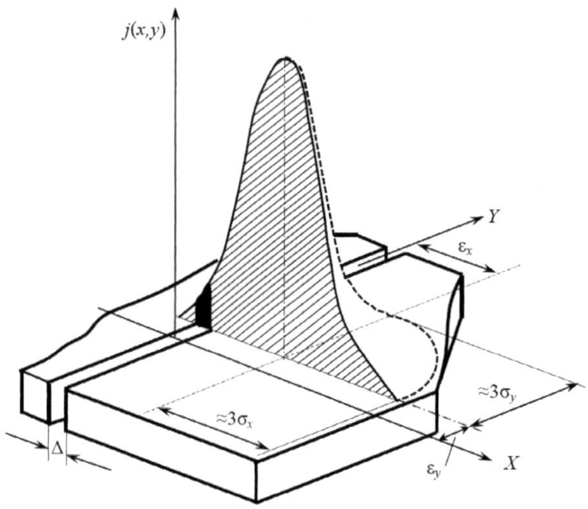

Figure 3. For calculating I_{se} and J.

The surface of the welded parts within the heating spot is considered homogeneous in along axis OY. Then I_{se} and J do not depend on y. Considering, that

$$\frac{1}{\sigma_y\sqrt{2\pi}}\int_{-\infty}^{\infty} exp\left[-\frac{(y-\varepsilon_y)^2}{2\sigma_y^2}\right]dy = 1,$$

as integral from density of distribution in infinity limits, we can provide equations for I_{se} and J:

$$I_{se} = \frac{k_{se}I_b}{\sigma\sqrt{2\pi}}\int_{-\infty}^{\infty} k_{cse}(x)\varphi(x)\, exp\left[-\frac{(x-\varepsilon)^2}{2\sigma^2}\right]dx,$$

$$J = \frac{Ck_{XR}U^2 I_b}{\sigma\sqrt{2\pi}}\int_{-\infty}^{\infty} k_{cXR}(x)Z(x)\, exp\left[-\frac{(x-\varepsilon)^2}{2\sigma^2}\right]dx,$$

where $\sigma = \sigma_x$, $\varepsilon = \varepsilon_x$.

Comparison of these equations demonstrates dependencies of the X-ray radiation intensity and the secondary emission current from the beam's position along the joint. The dependency is of a similar (extreme) nature. This provides an opportunity to use the same mathematical device to analyze the characteristics and application of well-known extremum seeking methods for achieving beam position control.

3. Specific Cases. During the Electron Beam Welding of parts shown on figure 3, there are no structural members interfering with the ability of the secondary electrons and the X-ray radiation to reach the corresponding sensors. In this case, k_{cse}=1 and k_{cXR}=1. Then

$$I_{se} = \frac{k_{se}I_b}{\sigma\sqrt{2\pi}}\int_{-\infty}^{\infty} \varphi(x)\, exp\left[-\frac{(x-\varepsilon)^2}{2\sigma^2}\right]dx,$$

$$J = \frac{Ck_{XR}U^2 I_b}{\sigma\sqrt{2\pi}}\int_{-\infty}^{\infty} Z(x)\, exp\left[-\frac{(x-\varepsilon)^2}{2\sigma^2}\right]dx.$$

If welding similar materials, then φ and Z do not depend on x and the equations become:

$$I_{se} = k_{se} I_b \varphi \left\{ 1 - \frac{1}{\sigma\sqrt{2\pi}} \int_{-\Delta/2}^{\Delta/2} exp\left[-\frac{(x-\varepsilon)^2}{2\sigma^2} \right] dx \right\},$$

$$J = Ck_{XR} U^2 I_b Z \left\{ 1 - \frac{1}{\sigma\sqrt{2\pi}} \int_{-\Delta/2}^{\Delta/2} exp\left[-\frac{(x-\varepsilon)^2}{2\sigma^2} \right] dx \right\},$$

where Δ - gap in the joint (figure 3).

Figure 4, shows dependency I_{se} и J (in relative units) on the movement ε of the beam along the joint during the Electron Beam Welding of similar materials. To calculate absolute values, it is sufficient to multiply coordinates of both functions by the corresponding coefficients:

$$I_{se}=k_{se}\varphi F,\ J=Ck_{XR}U^2ZF,$$

where

$$F = 1 - \frac{1}{\sigma\sqrt{2\pi}} \int_{-\Delta/2}^{\Delta/2} exp\left[-\frac{(x-\varepsilon)^2}{2\sigma^2} \right] dx.$$

Relative change of I_{se} и J with the constant gap in the joint grows with the decreasing σ (figure 4 b). This can be used along with a criterion of the electron beam focusing degree.

During the Electron Beam Welding of dissimilar materials, dependency of φ(*x*) on *Z*(*x*) needs to be taken into account. For example, during the welding of copper (Z_{Cu}=29) with steel (Z_{Fe}=26), the equation for J in relative units is as follows:

$$J = \frac{Z_{Cu}}{\sigma\sqrt{2\pi}} \int_{-\infty}^{-\Delta/2} exp\left[-\frac{(x-\varepsilon)^2}{2\sigma^2} \right] dx + \frac{Z_{Fe}}{\sigma\sqrt{2\pi}} \int_{\Delta/2}^{\infty} exp\left[-\frac{(x-\varepsilon)^2}{2\sigma^2} \right] dx.$$

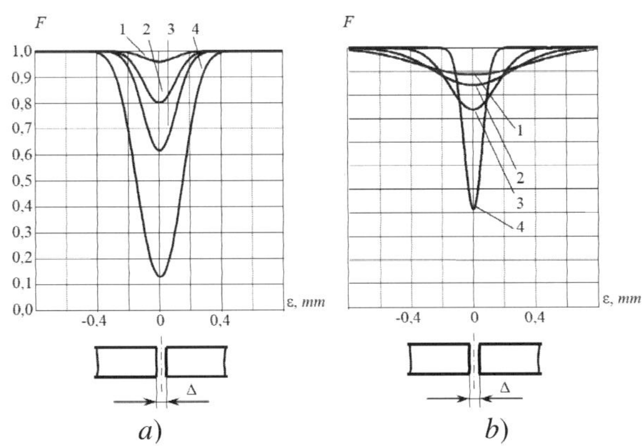

Figure 4. Calculation of dependency I_{se} и J on the beam position along the joint during electron beam welding of similar materials: *a*) – σ = *const* = 0,1 *mm*, Δ = *var*: 1 – Δ = 0,01 *mm*; 2 – Δ = 0,05 *mm*; 3 – Δ = 0,1 *mm*; 4 – Δ = 0,3 *mm*; *b*) – Δ = *const* = 0,1 *mm*, σ = *var*: 1 – σ = 0,35 *mm*; 2 – σ = 0,25 *mm*; 3 - σ = 0,15*mm*; 4 – σ = 0,05 *mm*.

Figure 5 *a* provides calculated values according to last equation. Figure 5 *b* shows dependency J/Z_{Cu}.

IOP Conf. Series: Materials Science and Engineering **822** (2020) 012032 doi:10.1088/1757-899X/822/1/012032

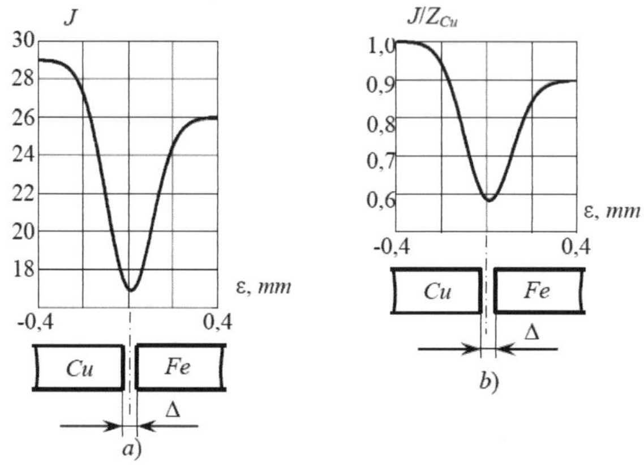

Figure 5. Dependency of J on the beam's position at the joint during electron beam welding of dissimilar materials.

Both graphs show that during electron beam welding of dissimilar materials slope angles for each characteristic are different. The metal with a higher atom number has a steeper angle (X-ray radiation intensity is proportionate with the atom number). Due to the same reason, the values of intensity saturation when the beam is fully positioned on one of the welded parts vary by a value that depends on the material atom number. In relative units, these values for *Cu* and *Fe* are equal to their atom numbers – 29 and 26, respectively (figure 5 a).

Another distinctive nature of these characteristics is that the extremum moves towards the material with the lower atom number. This is related to the fact that extremum coordinates are determined by the radiation parity from corresponding surfaces, which is proportionate to the electron bunch current and, therefore, to the area of the spot segment on the given metal. Since, $Z_{Cu}>Z_{Fe}$, for the above parity to happen, it is necessary for the spot area on the steel part to be bigger than the spot area on the copper part. Because of this, the beam is shifted towards the steel part.

Therefore, the joint position does not align with the position of the characteristic extremum, which needs to be taken into account during the electron beam welding of dissimilar materials.

The nature of extremum shift at various Δ is provided in figure 6 (graphs are based on the last equation).

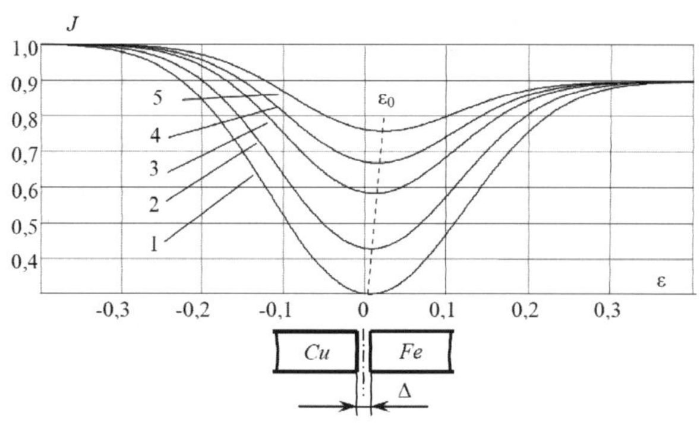

Figure 6. Value of $J(\varepsilon)$ at various Δ: $\sigma = 0{,}1$ *mm*; 1 - $\Delta = 0{,}2$ *mm*; 2 - $\Delta = 0{,}15$ *mm*; 3 - $\Delta = 0{,}1$ *mm*; 4 - $\Delta = 0{,}075$ *mm*; 5 - $\Delta = 0{,}05$ *mm.*

It is clear that noticeable position misalignment of characteristic extremum and the joint occur when $\Delta/\sigma \leq 1$. This is the most common case from practical standpoint and confirms the need to account for this misalignment.

By differentiating the last equation by ε and equating the derivative to zero, we can determine ε_0 – extremum abscissa and its dependency on the joint gap: $dJ/d\varepsilon=0$
Therefore,

$$\varepsilon_0 \approx \frac{0{,}0011}{\Delta}.$$

In other words, the extremum abscissa during the electron beam welding of dissimilar materials is related to the joint gap with the hyperbolic law (figure 7). When the gap is less than 0,01 mm, position misalignment of the extremum and the joint can be substantial and cause significant errors in beam positioning based on the position of the extremum of the considered dependencies.

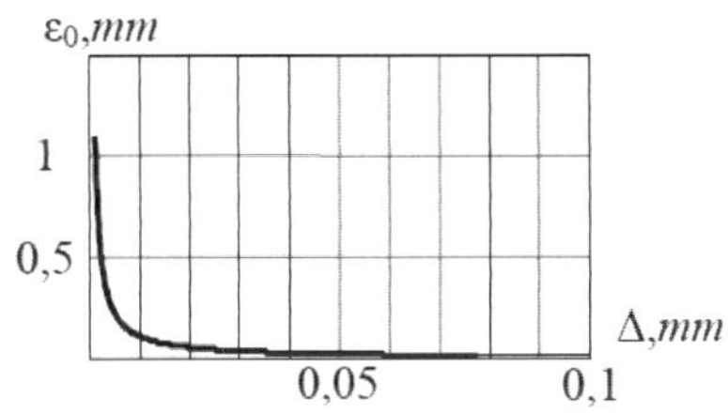

Figure 7. Dependence of extremum position on the joint gap.

Analysis of the dependency of the secondary emission current on the beam's position at the joint produces the same results.

4. Functional schematic for the joint tracking device. Extreme nature of the above described dependencies allows an opportunity to design devices that can track joints using an extremum seeking method. One of these methods is the synchronization signal detection. The method assumes that spectrum components are present in the sensor's signal and carry the information about the beam's position on the joint. These components appear, for example, when the joint is scanned with the beam. Mathematically, this can presented as follows:

$$\varepsilon=\varepsilon_0+\varepsilon_m sin\alpha,$$

where ε_0 – beam shift relative to joint; ε_m – searching beam shifting amplitude; $\alpha=\omega t$; ω – frequency, t – time. Then, expression F becomes:

$$F = 1 - \frac{1}{\sigma\sqrt{2\pi}} \int_{-\Delta/2}^{\Delta/2} exp\left[-\frac{(x - \varepsilon_0 - \varepsilon_m \sin\alpha)^2}{2\sigma^2} \right] dx.$$

Having periodical component in parameter ε, sensor output signal can be presented with Fourier series. Let us analyze the dependence of ε_0 on component b_1 with frequency ω. This component is defined as Fourier coefficient:

$$b_1(\varepsilon_0) = \frac{1}{\pi} \int_0^{2\pi} F \sin\alpha d\alpha$$

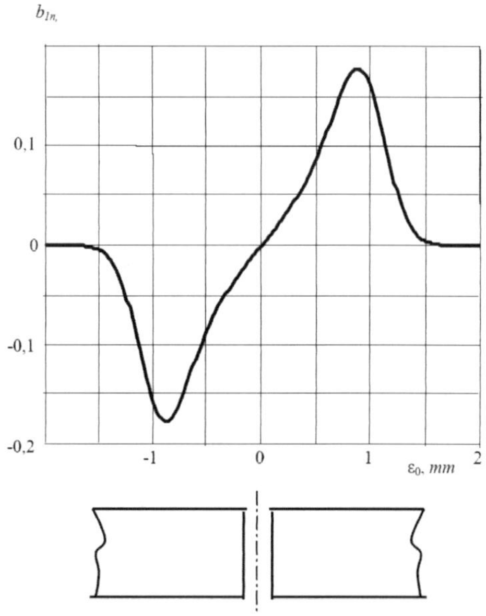

Figure 8. Dependence of the amplitude of sinusoidal component of the sensor signal with frequency ω on beam's position at the joint: σ=0.1*mm*; δ= 0.1*mm*; ε_m=1*mm*.

Figure 8 provides graphic representation of the function $b_1(\varepsilon_0)$ based on the above formula. It shows that, at the joint, $b_1(\varepsilon_0)$ is proportionate to the mismatch of the beam and joint positions and the sign determines the direction of the mismatch.

This fact provides an opportunity to use synchronous sensor signal detection to obtain information on the beam's position at the joint.

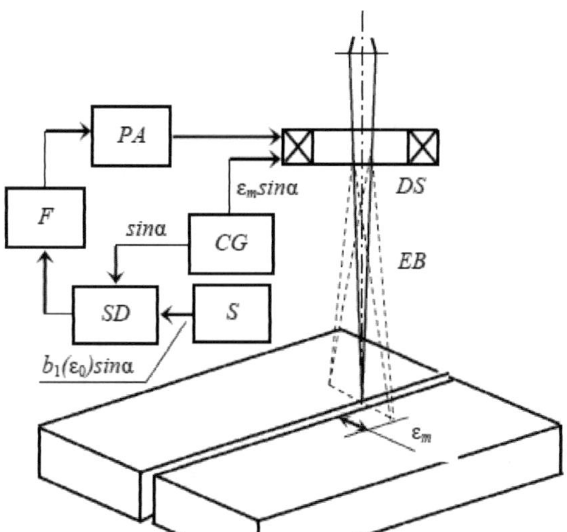

Figure 9. Functional diagram of the device for automatic beam control: *S* – sensor; *CG* – scanning beam generator; *SD* – synchronous detector; *F* – filter; *PA* – power amplifier; *DS* – deflection system.

Figure 9 provides functional schematic for the automatic joint tracking device. Generator *CG* helps joint scanning with the beam that has frequency ω and amplitude ε_m. At the same time, the generator *CG* creates voltage reference (multiplier *sin*α), which enters one of the synchronous detector *(SD)* inputs. A signal from the sensor *S* (the secondary electron sensor or X-ray sensor) enters the input *CD*. A signal forms at *CD* output, that is proportionate to the product of the input and reference signals:

$$[b_1(\varepsilon_0)sin\alpha]sin\alpha=[b_1(\varepsilon_0)]/2 - [b_1(\varepsilon_0)cos2\alpha]/2.$$

The signal from at the synchronous detector output is the sum of the constant component and the variable with the frequency of 2ω. Filter *F* shunts the high- frequency component, and the constant current signal proportionate to the beam deviation enters into the deviating system *DS* through the power amplifier *PS*. This helps correct the beam's position.

It is clear that, the signal form and its changes, depending on the beam's position at the joint, match the calculated characteristics.

Technical design of the device is not complicated. Collector of the secondary electrons is used as a secondary emission sensor. Scintillation sensor with electron or silicon photomultiplier can be used as an X-ray sensor.

5. Conclusions

1. The similarity of dependencies between the secondary emission current and the X-ray radiation on the beam's position at the joint is explained by the fact that the secondary emission and the X-ray radiation is the result of the interaction between the primary electron bunch and the material of the welded parts.

2. During the Electron Beam Welding of dissimilar materials, the minimum of the secondary emission current shifts towards the material with the lowest coefficient of the secondary emission and the minimum of the X-ray radiation shifts towards the material with the lowest atom number.

3. To reduce beam-joint alignment errors during the Electron Beam Welding of dissimilar materials, it is necessary to transmit an additional signal directed at the deflection system and that signal should be proportionate to the extremum shift of the reviewed characteristics.

References

[1] Reichmann A, Leffler D and Bartel R 2004 State of the art of a FEP control unit of beam scanning for electron beam equipment 6th Int. Conf. on Beam Technology Halle (Saale) pp 102-107

[2] Braverman V, Shabanov V, Bashenko V and Bajakin S 2004 Control over electron beam welding process by X-ray radiation from the zone of welding. 6th Int. Conf. on Beam Technology Halle (Saale) pp 68-74

[3] Braverman V 2006 The control of the process of electron-beam welding, Beam Technologies and Laser Application Proc. of the fifth int. scientific and technical conf. SPB: Publishing house SPbSpU pp 185 – 190

[4] Braverman V, Belozertsev V, Litvinov V and Rozanov O 2008 Issues relating to controlling seam formation process in electron beam welding *Vestnik SibSAU* **2** pp 148-152

[5] Bronshteyn N and Frayman B 1969 *Secondary Electron Emission* (Moscow: Nauka) p 468

[6] Haradza F 1966 *The general course of X-ray technique*, (Leningrad: Energiya) p 568

[7] Braverman V, Belozertsev V, Goryashin N and Lelekov A 2009 Mathematical models for the process of changing the x-ray radiation from the beam position relatively the seam during the electron beam welding *Vestnik SibSAU* **2** pp 247-251

[8] Rykalin N, Zuev I and Uglov A 1978 *Fundamentals of Electron Beam Processing of Materials* (Moscow: Mashinostroenie) p 239

IOP Conf. Series: Materials Science and Engineering **822** (2020) 012033 doi:10.1088/1757-899X/822/1/012033

Improving the process of anodic abrasive polishing of small-section channels

V I Trifanov, A V Sutyagin, O A Sukhanova, T V Strekaleva and I V Trifanov

Reshetnev Siberian State University of Science and Technology, 31 Krasnoyarsky Rabochy Av., Krasnoyarsk, 660037, Russia

E-mail: sibgau-uks@mail.ru

Abstract. The article studies the issues of increasing the quality of the surface and structure of a current-carrying skin layer in the small-section waveguide channels. We offer a promising technique of anodic abrasive polishing, which allows to remove anodic films from a machinable surface, activate and polish microroughnesses. We have developed the flow-chart of anodic abrasive polishing of small-section channels and undertaken a study. The process of anodic abrasive polishing can ensure the high quality of soft materials without impregnation of a machinable surface, as well as the high quality of hard materials with fine finish to Ra 0,04 micrometers keeping the original accuracy and structure of a current-carrying skin layer. We have presented a model and some recommendations how to perform anodic abrasive polishing.

Finishing techniques play a very important role in the context of maintaining the quality of machinery parts: they can be used to finish small-section channels, e.g. rectangular in section 7,2×3,4 mm and others, as well as round in section Ø 4,2÷2,0 mm.

It is known that performance characteristics of products are often determined by the parameters of the quality of the surface layer of parts (especially by surface roughness); for the channels of waveguides they are determined by the structure of a current-carrying skin layer as well [1]. The existing methods of surface machining are divided into mechanical (resurfacing), chemical, physical, electrochemical and combined [2, 3].

The most common are the methods of surface machining by resurfacing (mechanically) with the use of abrasive materials that allow to decrease surface roughness to Ra 0,08-0,2 micrometers. The task of a finishing operation is to remove any defects of previous working (grooves, scratches, small cavities, microasperities, etc.) in order to obtain a smooth surface capable to reflect light well or to decrease electric losses while electromagnetic waves are passing through the channels of the waveguides of antenna-feeder devices [4, 5].

At the time of mechanical polishing coarse irregularities are being removed successively, and at the completion coarse irregularities are removed and a mirrored shiny surface is obtained. Mechanical polishing can influence the structure of the skin layer of a current-conducting surface. While electropolishing microasperities are being removed in the first turn, and a shiny surface can be obtained when there is some considerable macroroughness. It occurs when the entire surface is not subject to the etching effect of an electrolyte. At the time of electrochemical polishing a machinable surface is not almost subject to a thermal and mechanical effect and it does not influence the structure of a current-carrying skin layer significantly.

In the paper [6] on the ground of the analysis of solution products it is concluded that smoothing of microroughnesses of a surface is connected to the exhaustion of chemical activity of an electrolyte in microdepressions of the machinable surface of a part that is an anode at electrochemical polishing.

The other authors [7] note that the process of electrochemical polishing is followed by the formation of passive films on the anodic surface of the workpiece being machined. The film can reach the thickness 0,07÷0,17mm. The work [8] notes that the process of electrochemical polishing may occur if the maximum current is reached, the value of its current density is in the range of 30-300 mA/cm^2. It is supposed that electrochemical polishing with the planarization of microrelief is performed by a film mechanism [9]. Smoothing of roughnesses occurs in consequence of different levels of passivation of bulges and depressions of microroughnesses. According to the theoretical analysis of the undertaken studies, we can conclude that the process of electrochemical polishing can be performed by a diffusive or film mechanism.

In the paper [10] the researchers note that electrochemical properties of a semiconductor oxide film decisively influence the mechanism of electrochemical polishing. The characteristics of this film depend on the composition and nature of an oxide, the composition and properties of an electrolyte being used in the diffuse layer. The dependence on the nature of an oxide is such that some metals are polished only at high current density within the conditions of oxygen release, when higher oxides of a machinable metal are formed, for example, iron and steel, aluminium and its alloys are subject to electrochemical polishing (in phosphoric-chromate or sulfur-phosphoric-chromate electrolytes, nickel in chlorhydric acid or sulphuric acid). Copper and its alloys are polished by this mechanism in phosphorus acid and phosphoric-chromate electrolyte. When performing electrochemical polishing, despite the formation of an oxide film on the machinable surface of a workpiece, during anodic dissolution of a metal it is necessary to make electric lines of force concentrate mainly on microirregularities. It will contribute to the augmentation of electrochemical polishing, at the same time decreasing the removal of metal and the time of machining in order to achieve the required surface roughness. The work [11] highlights that in some processes of electrochemical polishing the determining factor is not smoothing of the surface of a machinable part, but further appearance of microroughnesses due to the nonhomogeneous dissolution of an alloy being machined that is caused by its microstructure.

The electrochemical potentials of anodic dissolution of different metals in the same electrolyte and under the same conditions may be different [12].

Table 1 presents electrochemical potentials of dissolving iron and steels with the content of nickel and other elements in some electrolytes of neutral salts.

Table 1. Electrochemical potentials of dissolving iron and its alloys

	Metal or steel	Current density, A/cm^2	Electrolyte composition	Electrochemical potential of dissolution
			15% Na_2SO_4	+0,5
1	Iron	5	15% NaCl	+0,65
			15% $NaNO_3$	+1,15
2	H18N9T	5	15% NaCl	+1,7
	5ChNM		(15÷20)%$NaNO_3$	1,7÷2,8
3	EI958	5		
	5ChNV			

For homogenous dissolution of microroughnesses of a material being machined, it is necessary to choose an electrochemical potential ensuring the uniform dissolution of all the elements in the composition of a machinable material.

While studying a great number of metals it is discovered that the best quality of a surface is obtained with the certain range of potentials (figure 1), (Ea–Eb) is the area of voltammetric characteristics of a metal.

On the area Ea-Eb the current density hardly depends on the potential. In this area the anodic dissolution of a metal is controlled by mass transfer [13]. The maximum current corresponding to this range of the potentials characterizes the process of the formation on an anode of a passivating film that decelerates the etching effect of an electrolyte and burnishes the metallic surface. The thickness of a passivating film depends on the ratio of the speed of its formation and dissolution in an electrolyte.

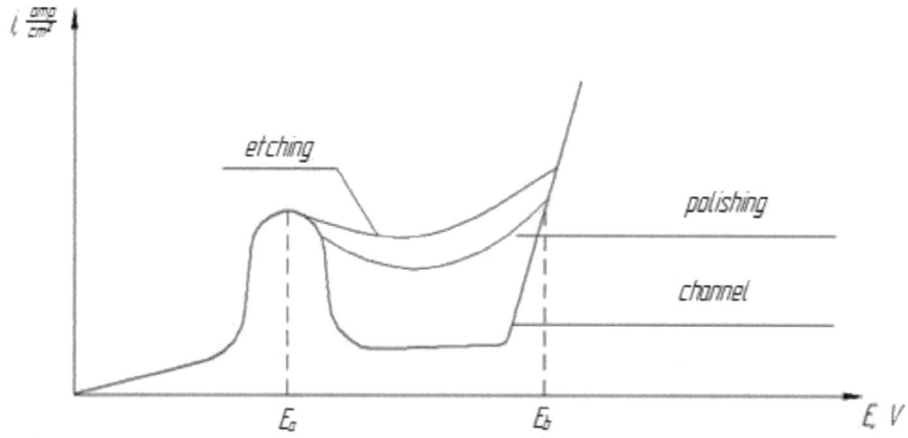

Figure 1. The schematic diagram of the anodic polarization of a metal

As it follows from the graph of anodic curves, the steel Ch18N9T being polarized with the high density anodic current (20-40 A/cm^2) in the solution of 15% NaCl at the rate of electrolytes 4 m/s is subject to the anodic dissolution in the active condition at the electrochemical potential φ_a=1,7 V. When the stream speed is decreased to 1 m/s, the anodic dissolution passes into the transpassive condition at φ_a=3 V [12]. In view of this, the speed of pumping of an electrolyte may have a great impact on the mechanism of electrochemical polishing in the neutral bath. The mechanism of electrochemical polishing of parts is influenced by original surface roughnesses as well, e.g. the current density on the bulges and depressions in the chloride electrolyte may differ by a factor of five, and in the nitrate electrolyte it may differ by a factor of approximately eight [14].

This difference may become greater if the potential of an anode increases. This difference is measured using the coefficient $K_i = \frac{i_{depr}}{i_{bulg}}$,

where i_{depr}, i_{bulg} – current density on bulges and depressions.

The increase of the current density on the bulges leads to the higher rate of the dissolution of microirregularities and to the augmentation of smoothing of a surface microprofile. The highest rate of smoothing is noticed while the electrochemical polishing of surface roughnesses with the narrow angle α at the peak of the depressions.

To ensure the reference accuracy and the uniform roughnesses of a surface while polishing small-section channels, it is offered to use a new combined method of anodic abrasive polishing [15] with an elastic abrasive tool that involves reciprocating motion and vibration of an elastic tool in relation to a workpiece. It ensures the pressure of an elastic tool to a machinable surface and simultaneous electrochemical anodic polishing of microroughnesses of a channel surface while supplying an omnipresent electrical or impulsive potential onto a workpiece being machined or a cathode. During anodic abrasive polishing specialists use the tool consisting of two abrasive elastic parts; a cathode is installed between these parts (figure 2).

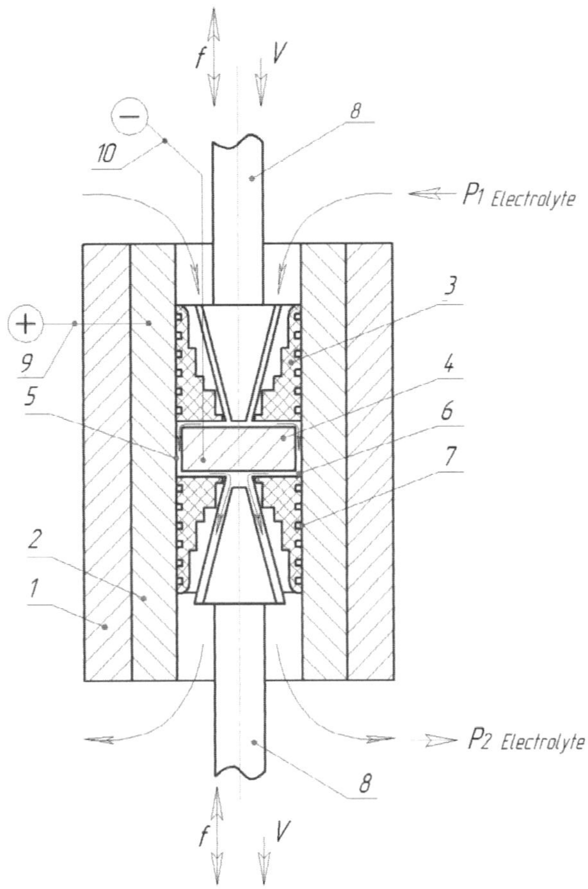

Figure 2. The model of abrasive anodic polishing of a small-section channel

The outer profiles of the above mentioned elastic parts and a cathode follow the shape of the channel being machined, provided that in both parts along a long axis cavities are performed in the shape of frustum of pyramid, their cross section follows the shape of the outer profile of an elastic tool and these cavities are directed towards each other by their small bases.

The cathode is installed in such a way that the channels are formed among the cathode, the surface being machined and the opposite butt ends of the elastic parts of the tool that provide electrolyte supply to the machinable surface. On the outside surface of the elastic parts of the tool there are two helical grooves to pump the electrolyte. The reciprocating motion of the elastic tool in the small-section channel is performed through the sequential feed of the electrolyte to the symmetric cavities of the elastic tool and with the help of a ram. The cathode integrated with the elastic tool may subject to the vibration from a mechanical drive with the definite frequency f_u.

At the elastic anodic abrasive polishing there is a decrease of specific pressure on an elastic abrasive tool. As a result, we have the scheme of sparking-out that ensures significant reduction of surface roughness. Herewith, the ratio of the amount of S shear to the radius of the blade edge of the abrasive grains r is defined by the frictional factor f_{fr} and it influences the formation of roughnesses due to the plastic deformation of microroughnesses of a material. The paper [17] shows that R_{max} depends on the diameter of V-abrasive, its mortality factor α, the coefficient of roughness excursion γ.

$$R_{max}=0{,}085B,\ a\ R_a=0{,}25,\ R_{max}=0{,}021B;\ R_{ai}=\alpha\gamma B.$$

The limit value of the ratio of shearing intensity S to the radius r of the truncation of the abrasive grain edge, at which the cutting process turns into the process of plastic deformation without the formation of metal chips, is defined by the index of friction f_{fr} of the abrasive grain with the material being machined.

The linear rate of electrochemical shearing of a metal layer from microroughnesses may be determined using the following formula:

$$V = \varepsilon_v \cdot \eta \cdot i \text{ cm/min}, \quad (1)$$

where ε_v– volumetric equivalent weight (mm^3/A·min); η – metal anode efficiency; i – current density, A/cm^2.

Current density may be calculated using the formula:

$$i = \frac{(U-\Delta U)\chi}{\delta_{ieg}}, \quad (2)$$

where U – electrode voltage, V; χ – electrolytic conductivity, cm·O^{-1}; ΔU – drop in voltage in cathodic and anodic layers, V; δ_{ieg} – the value of interelectrode gap, cm.

$$\Delta U = \varphi_a + \varphi_c, \quad (3)$$

where φ_a, φ_c – potentials on the anode and cathode, V.

The value of the metal layer being sheared using the electrochemical method can be determined using the following formula:

$$S = \varepsilon_v \eta i \tau = \frac{\varepsilon_v \eta (U-\Delta U)\chi 10^{-2}}{\delta_{ieg}}, \text{ cm} \quad (4)$$

where τ – time of the cycle of electrochemical polishing, min; 10^{-2} – the coefficient of value conversion.

The drop in voltage in near-electrode layers on bulges is considerably less than in depressions ($\Delta U_{bulg} < \Delta U_{depr}$) due to the activation of the combs of microroughnesses with an elastic abrasive tool approximately by 10-15 times. It leads to the higher rate of dissolving microroughnesses of the surface being machined compared to microdepressions.

Taking into account the cyclicity of machining, the surface roughness after each cycle of anodic abrasive polishing may be defined as:

$$R_{an} = R_{an-1} - (1 - n \cdot 10^{-1}) \left[\frac{\varepsilon_v \cdot \chi \eta (U - \Delta U_{bulg}) \cdot 10^{-2}}{\delta_{ieg}} + a \cdot \gamma \cdot B \right] \cdot \tau \quad (5)$$

where R_{an} – roughness of the next cycle, micrometers; R_{an-1} – roughness of the previous cycle, micrometers; n – sequence number of the cycle; α, γ, B – the influence of abrasive polishing on the decrease of surface roughness; α – the coefficient of the change of the diameter of an abrasive after each cycle of abrasion; γ – the coefficient of the change of roughness volume after each ith cycle of machining, B – the diameter of an abrasive grain, τ – time of the cycle of anodic abrasive polishing.

The time between the cycles τ is calculated using the formula [18]:

$$\tau_{bc} \geq \left(\frac{ZF}{2i}\right)^2 \pi D {C_s}^2 \quad (6)$$

where Z – ion charge; F – Faraday constant, kl/mole^{-1}, D – the coefficient of ion diffusion of a metal being machined, cm^2/s; C_s – ionic concentration on the surface of a resulting workpiece, mole/cm^3.

To calculate C_s we can use the known dependence [3]:

$$C_s = \frac{i \cdot \delta_{ieg}}{Z \cdot F \cdot D} \quad (7)$$

Putting the expression (7) into the formula 6 we get the following:

$$\tau_{bc} \geq \left(\frac{\pi {\delta_{ieg}}^2}{4D}\right) \quad (8)$$

The duration of the current impulse at anodic abrasive polishing can be calculated according to the formula[19]:

$$\tau_{imp} \leq \frac{\delta^2(\rho_e+\rho_{gl})}{2Uk} = \frac{\delta^2}{2ik} \quad (9)$$

where ρ_e and ρ_{gl} – volume resistivity of an electrolyte and gas liquid mixture; k – coefficient connecting the volume of flash gas with the thickness of gas liquid mixture.

As operating environment one uses the electrolyte that excludes intercrystalline etching of the surface being machined, e.g. 15% $NaNO_3$. Figure 2 shows the model of the process of anodic abrasive polishing.

In the vessel for polishing 1 the workpiece 2 is installed and the tool for abrasive anodic polishing is put into the port hole being machined. The tool is compound, it consists of two elastic parts 3 with the cathode 4 between them. The outside profiles of the elastic parts 3 and the cathode 4 follow the shape of the port hole being machined. In both elastic parts the symmetric cavities are performed in the shape of frustum of pyramid, their cross-section follows the shape of the outside profile of an elastic tool, these cavities are directed to each other by smaller bases. The cathode 4 is installed with the formation of the channel 5 between the cathode 4 and the surface of the workpiece being machined, and the channels 6 between the butt ends of the cathode and the butt ends of the elastic parts of a tool. To pump the electrolyte, on the outside surface of the elastic parts of a tool helical grooves 7 are made. An electrode is fed into the symmetric cavities of the parts of a tool through the conic stages of the stems 8, in turn under the pressure P_1 and the opposite pressure P_2 ensuring the contact of the tool with the surface being machined, then the electrolyte gets into the channels 5 and 6. The anodic dissolution between the cathode and the surface being machined is provided by the voltage from the supply equipment 9 connecting the positive bar to the workpiece 2 being machined, and the negative one 10 – to the cathode 4. Herewith, the removal of the combs of roughnesses on the surface being machined occurs. Resulting anodic and oxide films are removed using the abrasive grains of the elastic parts 3 of the tool performing reciprocating motion with vibrational oscillations from the stems 8.

The removal of the combs of roughnesses on the surface being machined is generally performed by means of anodic dissolution between the cathode of a tool and the surface being machined. The resulting anodic and oxide films are removed from the combs of microroughnesses using the abrasive grains of the abrasive-carrying elastic parts 3 of the tool performing reciprocating motion with simultaneous vibration. These both motions in the process of removing anodic and oxide films render the combs of microroughnesses active, it causes more intensive dissolution compared to the depressions that are covered with passivating films. It causes smoothing of the profile of the surface being machined as well. Machining products (slime, gas and destroyed particles of an abrasive) are scoured by the flow of an electrolyte. While supplying direct or impulse current onto a cathode and a machinable part we can reduce the roughness of a surface to Ra 0,16…0,04 micrometers.

The application of the pulse current with amplitude-time characteristics onto a cathode and a machinable part along the combs of microroughnesses can be defined using the formula:

$$\frac{\partial\varphi_a}{\partial t} = \int_0^{tu} \frac{\partial i}{\partial t} \leq \Delta\varphi_n \quad (10)$$

The increase and decrease of polarization allow to reduce the roughness of the workpiece being machined to Ra 0,02-0,04 micrometers [16].

In the formula (10) φ_a, φ_p – the potential of the anode in the process of active anodic dissolution on the combs of microroughnesses and the potential of passivation of the material of the surface being machined on the depressions of the hole microroughnesses respectively; i – current density; t – time unit; t_u – duration of current impulse (ms). The potential φ_a is provided during the anodic dissolution through the activation of microroughness combs with the abrasive particles of an elastic tool and the parameters of electric impulse.

During anodic abrasive polishing, thermal and mechanical forces do not influence significantly the surface being machined (the temperature of an electrolyte in the place of machining does not exceed

25°C), it causes higher resistance of the abrasive grains attached to the outside surface of an elastic tool, as well as it decreases abrasive layer wear and increases the operational life of a tool.

The structure of the current-carrying skin layer of a waveguide channel after treatment does not differ from the structure of a parent metal, there is no impregnation or burns, the roughness of the surface being machined is smooth, the operational life of elastic abrasive tools increases tenfold.

Figure 3 shows the graphs of measuring the roughness of the surface of the small-section channels made of the alloys 32NKD at anodic abrasive polishing in the electrolyte 15% $NaNO_3$ and at electrochemical polishing [20]. It is concluded that the decrease of surface roughness at anodic abrasive polishing occurs more intensely as a result of the removal of anodic films on the combs of microroughnesses.

Using the offered technique of polishing waveguide channels allows to enhance the quality of a surface being machined and the structure of a current-carrying skin-layer. It allows to increase the processing rate one-and-a-half times compared to the existing techniques of electrochemical polishing in a neutral bath.

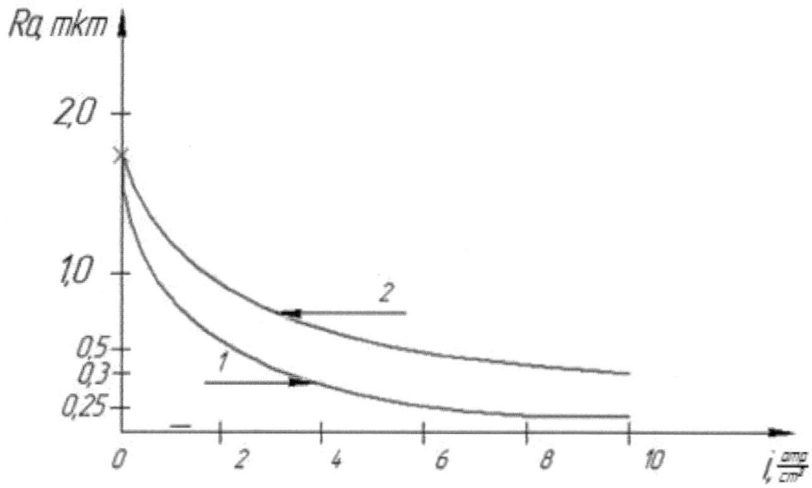

Figure 3. The dependence of surface roughness Ra on current density
1-Alloy 32NKD, electrolyte 15% $NaNO_3$ after anodic abrasive polishing.
2- Alloy 32NKD, electrolyte 15% $NaNO_3$ after electrochemical polishing.

Conclusion.

Following on from the theoretical studies of electrochemical polishing, we have:

– offered the technique of anodic abrasive polishing of small-section channels;

– developed the model of the process of anodic abrasive polishing, as well as the circuit schematic of a pilot unit for anodic abrasive polishing of parts and products;

– presented the results of the experimental studies of anodic abrasive polishing; these results confirm the efficiency of the offered technique of decreasing surface roughness keeping the original accuracy of small-section channels.

References

[1] Kuzovkin A V and Bes'ko A V 2004 Electrochemical finishing of metals and ensuring the quality of a surface coating (Moscow: Radio and Communication) pp 65-67

[2] Balmasov A V 2003 Materials and methods of improving the quality of a surface at electrochemical and chemical finishing of metals *Modern electrochemical technologies in mechanical engineering* (Russia: Ivanovo) pp 66-68

[3] Skorchelletti V V 1969 *Theoretical electrochemistry* (Moscow: Chemistry) p 608

[4] Trifanov I V 2006 *Technological maintenance of the quality while manufacturing a transmission line and the energy of antenna feeding devices* (Russia: Krasnoyarsk) p 242

[5] Amirhanova N A 2000 Electrochemical polishing of heat-resistant alloy nickel-chromium alloys HN45MVTYUBR i HN50VMTYUB *Metalworking.* **5/6** pp 29-33
[6] Batashev K P and Nikitin V P 1950 *Applied chemistry* **23.3** pp.120-122
[7] *Proceedings of UralNITOMASH* 1976 **1** pp 17-19
[8] Popilov L Ya 1978 *Electrochemical finishing of steel pipes.* (Moscow: Engineering) p 287
[9] Lipkin Ya N and Shtan'ko V M 1982 *Chemical and electrochemical finishing of steel pipes.* (Moscow: Metallurgy) p 256
[10] Valeev A Sh, Grechuhin T N and Gorbacheva A M 1965 On the mechanism of smoothing the surface of a metal at electrochemical dissolution *Journal of physical chemistry* **39** p 246
[11] Vozdvizhenskij G S 1953 Electrochemical finishing of cutting tools *(Kazan': TatargoPublisged)* p 72
[12] Shcherbak M V, Tolstaya M A et al 1981 *Fundamentals of the theory and practice of electrochemical finishing of metals and alloys* (Moscow: Engineering) p 263
[13] Dzhungurova G E 2014 *Electrochemical modification of the surface of metals using fluorine-containing ionic liquids* (Lomonosov Moscow State University) pp 20-21
[14] Atanasyanc A G *Electrochemical manufacturing of the parts of atomic furnaces* (Moscow: EnegroPublished) p 176
[15] *Method of anodic-abrasive polishing of holes* Author's license no 2588953 (In Russian)
[16] Oborina L I, Trifanov I V et al 2013 Improving the quality of the surface of the mirrors of beam mode waveguides by vibrational electrochemical honing *Siberian Journal of Science and Technology* **2** pp 223-229
[17] Novikov F V, Shkurupij V G et al 2014 Analytical definition of the conditionds of decreasing the roughnesses of a surface at abrasive polishing *Vestnik of Azov state technical University. Series: Technical Sciences* **28** pp 142-145
[18] Levich V G 1959 *Physico-chemical hydrodynamics* (Moscow: Fizmatgiz) p 699
[19] Moroz I I, Petrov Yu N et al 1977 *Basics of improving the accuracy of electrochemical shaping* Kishinev p 129
[20] Nadaraia C G, Babkina L A et al 2014 *Electrochemical polishing of metals in water solutions of neutral salt with electrode* (Krasnoyarsk: SibSau) p 117

Evaluation of the porous structure in polymer tooth implants and its effect of the transport processes

V R Pen[1], Y V Chijov[2], S I Levchenko[1], O V Pen[3] and L E Mascadynov[2]

[1] Reshetnev Siberian State University of Science and Technology, 31 Krasnoyarsky Rabochy Av., Krasnoyarsk, 660037, Russia
[2] Prof. Voyno-Yasenetsky's Krasnoyarsk State Medical University, 660022, 1 Partizana Zheleznyaka street, Krasnoyarsk, Russian Federation
[3] Wake Forest University, 1834 Wake Forest Rd, Winston-Salem, NC, USA, 27109

E-mail: 2507@inbox.ru

Abstract: the study focuses on the dynamics of the leaching of the low-molecular fragments from the volume of the polymer composition used in tooth implants. The saturation curves exhibit two characteristic time periods, leading to an assumption that the various transport mechanisms caused by the topology of the microporous structure might be the reason.

The content of the residual acrylic monomers in the polymer tooth implants is currently considered to be one of the most important quality assurance criteria [1, 2, 3]. The maximum allowed concentration in water is 0.01 mg/l [4], in the water extract of the medically-used plastic – 0.25 mg/l [5, 6]. Consequently, the laundering procedure over the length of time determined by the dynamics of the diffusion of the low-molecular fragments of the polymer composition from the product into the laundering solution becomes a necessary technological procedure for the production of the polymer dental structures. The following polymer compositions are studied:

- Sample 1 – Vertex;
- Sample 2 – Protakril
- Sample 3 – Belakril-M XO;
- Sample 4 – Etakril;
- Sample 5 – Ftorax .

They are used to produce the samples in the form of a block with 1 cm height and width and 8 cm length, weighing 8±0.5 g. Samples are prepared in accordance with the instructions for production of the polymer tooth implants and are identical to them in structure and composition. Samples are then submerged into distilled water volume of 42 ml. The experiment is conducted for a duration of 350 minutes.

The concentration of monomers in the solution is measured by conductometry [7]. The dynamics of the extract product accumulation are presented in figures 1-3.

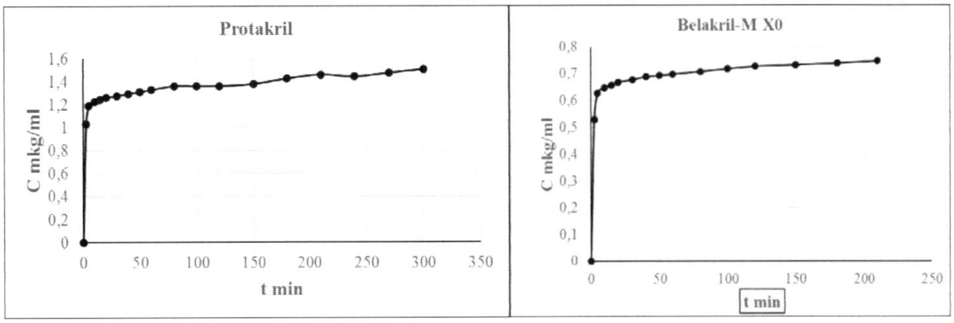

Figure 1. Protakril Belakril-M X0 Accumulation dynamics.

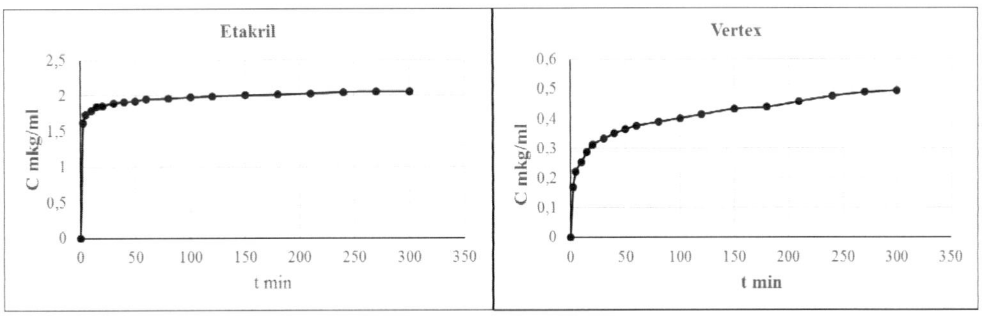

Figure 2. Etakril Vertex Accumulation dynamics.

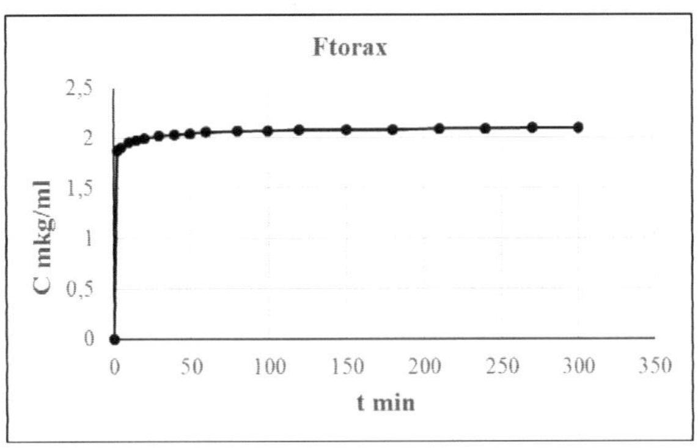

Figure 3. Ftorax Accumulation dynamics.

Two distinct characteristic time period regions can be observed on the curves. They correspond to the rapid change in the first region and slower process development in the later period. This behavior can be explained by two separate transport mechanisms. The rapid change is connected with mass transfer in the sub-surface layer, determined by the diffusion in the sub-surface layer porous structures and not connected to the transport mechanisms in the sample depth. The slower change region is correlated with the processes in the deeper porous layer of the sample.

The structure and size of the pores are determined by the method of raster electron microscopy [8].

Investigated samples were 10x10x10 mm in size, with chips running throughout the base of acrylic plastics. The samples were glued onto the metal base with 20 mm diameter with current-conducting tape. The working surface of the samples was covered with thin layers of carbon (10 nm depth) and gold (40 nm depth) with the high definition spraying system K550X (Hitachi, Japan). Prepared samples were studied under the electron microscope TM-1000 (Hitachi, Japan) with the accelerating voltage of 15 kV.

100X magnification factor has proved to be most suitable for the comparison of the characteristics of the pores, including their number, type, size, volume and surface. For separate groups of the pores, the magnification factors of 200X, 250X, 500X and 1000X have been used.

The porous coefficients were calculated with the JMicroVision program. The results are presented in figures 4-6 and table 1. A comparison with the accumulation curves allows to make a conclusion about the lack of correlation between the accumulation speed and the size of the macropores. An assumption can be made that the defining criteria is the degree of connection in the microporous system that is used to conduct the mass transfer of the low-molecular components of the polymer composition.

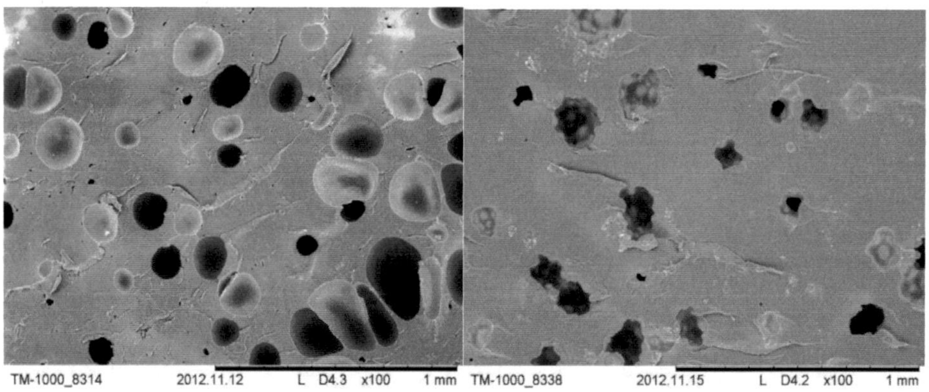

a) *b)*

Figure 4. Macroporous system, Vertex (a), Protakril (b).

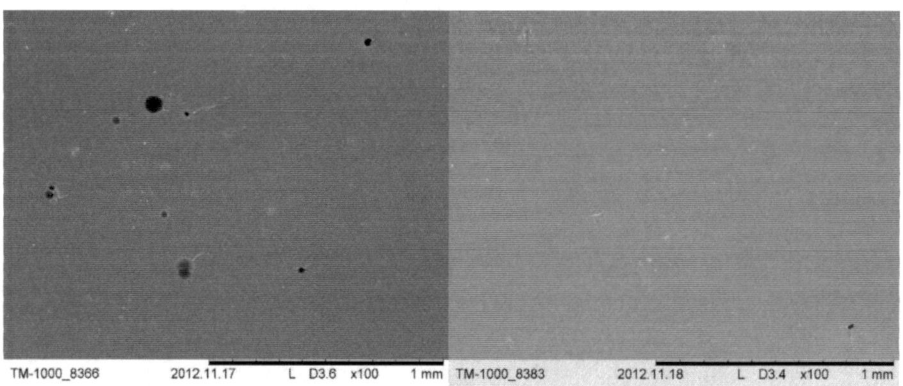

a) *b)*

Figure 5. Macroporous system, Belakril-M X0 (a), Etakril (b).

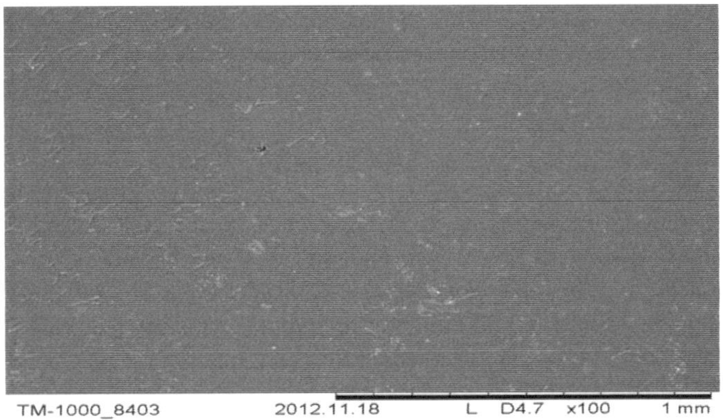

Figure 6. Macroporous system, Ftorax.

Table 1. A comparison in the polymer's porosity characteristics.

Plastic sample	Pore surface (mm^2) to the unit surface (100 mm^2)	Pore volume (mm^3) to the unit surface (100 mm^2)	Pore characteristic classification
Vertex	0.027 (23.04 %)	0.06 (4.43 %)	Granular, open – 50%, dead-end – 50 %
Protakril	0.46 (25.2 %)	1.1 (8.11 %)	Gasous, open – 50%, dead-end – 50 %
Belakril-M X0	0.008 (0.05 %)	0.0015 (0.05 %)	Granular, open – 40%, dead-end – 60 %
Etakril	0.0013 (0.008 %)	$4.3 \cdot 10^{-7}$ (0.009 %)	Granular, Compression pores – 100 %
Ftorax	0.0003 (0.001 %)	$3.7 \cdot 10^{-7}$ (0.008 %)	Granular, Compression pores – 100 %

Thus, a conclusion can be made that:

- The accumulation dynamics can be characterized by the two characteristic time regions defined by the different mechanisms of mass transfer
- The transport mechanism for the polymer matrix volume is assumingly defined by the topology of the microporous structure. The size of the micropores does not seem to be a decisive factor.

References

[1] Etaden M and Canston E E 1974 *Piastm Polim* Use of polymeric material in dentistry vol 41 Kq 153 pp 140-144
[2] Austin A T and Basker R M 1982 *Br. Dent J.* The level of residual monomer in acrylic denture base materials vol 153 pp 424 - 426

[3] Donovan T E, Hirst R G and Campagni W V 1985 *The Journal of Prosthetic Dentistry* Phisical properties of acrilic resin polemerized by four different techniques vol. 54 №4 pp 522- 524
[4] Grushko Y M 1982 *Reference book* Harmful chemical compounds in the industrial waste water (L. Chemical) p 483
[5] Tregubov I D, Boldyreva R I, Michaylenko L V, Malgakelidze V V and Tregubov S I 2007 *Use of the thermoplastic materials in dentistry Training manual* (Moscow Medical Press publishing house) p 140
[6] 1994 *Plastics used in orthopedic dentistry* Edited by S.E. (Joludyov: Yekaterinburg) p. 72
[7] Damaskin B B, Petriy G A and Cirlina O A *Electrochemistry* (Chemistry Moscow) p 672
[8] Zhou U 2013 *Raster electron microscopy for nanotechnology, methods and applications* (Binon: knowledge laboratory publishing house) p 611

Ecological issues of space industry

S Yu Piskorskaya

Reshetnev Siberian State University of Science and Technology, 31 Krasnoyarsky Rabochy Av., Krasnoyarsk, 660037, Russia

E-mail: piskorskaya1@rambler.ru

Abstract. The article deals with the negative impact on the ecological situation in the areas of rocket and space activities. It turns out that the fallen modules of carrier rockets and the remaining fuel in them (heptyl) which enters the environment by toxic dispersing in the atmosphere and gets in contact on the ground having a negative effect on the environment. The article presents a comparison of functioning spaceports, especially the advantages and disadvantages of the Baikonur and the Vostochny cosmodromes. As a proposal, a system of permanent monitoring is considered, which allows identifying the parameters of the environment in an open approach to reduce social stress in the areas of activity of the rocket and space industry.

1. Introduction

In recent years, the media have increasingly raised the issues of environmental safety, as well as the protection of the population concerning the use of liquid rocket fuel in the space industry. Information on the negative impact of rocket and space activities on the regional ecosystems leads to the certain social stress among people who live in the territories where the enterprises of the rocket and space industry are located. These people are, on the one hand, in a particular environmental situation, on the other hand, they distrust the official sources of information covering the consequences of the space industry activities.

It is obvious that the long-term activity of the rocket and space industry could not help but affect the environmental pollution of the areas, where the carrier rocket modules have fallen. The consequences of such intervention in the ecosystems of the areas depend on the degree of its impact. So the maximum load is experienced by the areas where the first stages of the carrier rocket have fallen. In places of their modules crash, the earth's cover is subject to mechanical, chemical and thermal damage. As a result, there are violations of soil and plant ground cover, compaction and clogging of soil, formation of craters and fires, soil damage by wheels of heavy equipment used to extinguish fires and evacuate fallen carrier rocket modules, changes in the chemical properties of the soil cover and. As a consequence there is reduction in the number and specific diversity of ecosystems, as well as the optimal ratio of their components.

The assessment of the consequences of such interference in the natural environment is one of the most important tasks of modern science, which, despite the half-century history of the practical exploration of outer space by humanity, is still being formed [1].

It should be noted that the impact of the rocket vehicle on the environmental situation in the areas of separating modules falling is complex. Together with the carrier rocket modules, the liquid rocket fuel enters the ground, the composition and degree of toxicity of which depends on the type of launch vehicle used. The most dangerous for the environment is heptyl. World Health Organization has included this substance in the list of especially dangerous chemical compounds that have a general toxic and skin-irritating effect on the human body. Heptyl is widely used in the space industry both in Russia and around the world, and therefore it can be argued that it affects the environmental safety of a number of countries. Used in propulsion systems of manned ships and automatic satellites, orbital stations and spacecraft, heptyl enters the environment by dispersing it in the atmosphere and in contact with the remains of fuel on the ground when the fuel tanks fall [2]. Untreated fuel residues and their combustion products enter the atmosphere and water, forming serious biochemical anomalies.

Considering that the space industry and its technological components are a strategically important sector of the economy for Russia, all its environmental consequences require constant monitoring of rocket and space activities, the launches of which are possible from six spaceports. They are the cosmodromes of Baikonur (Kazakhstan), Plesetsk (Arkhangelsk region), Kapustin Yar (Astrakhan region), Yasny (Orenburg region), Svobodny and Vostochny (Amur region). Currently, the most involved are the Baikonur and the Vostochny cosmodromes.

The Baikonur cosmodrome is the first and the largest spaceport in the world, located on the territory of Kazakhstan, created in the USSR in the mid-fifties of the twentieth century and leased by Russia until 2050. It is this cosmodrome from where the first space satellite has been launched, the first human flight into space has been carried out, as well as the first carrier rocket has launched.

Currently, the Russian presence on Baikonur remains an important part of the social, political and economic life of the city of the same name, which is home to more than 70 thousand people, of which almost a third are citizens of Russia. For the term of the lease, the Baikonur complex is endowed with the status corresponding to the cities of Federal importance of the Russian Federation, with a special regime of safe operation of facilities, enterprises and organizations, as well as the residence of citizens. Inside the city there is Russian legislation, as well as Kazakh laws are also applicable to Russian citizens.

Naturally, the functioning of the cosmodrome and the existence of the city now largely, if not completely, depend on Russia. In this regard, the reorientation of Russia to the Vostochny cosmodrome will mean a serious burden on the Kazakhstan budget. Russian experts estimate that now the Kazkosmos budget contributions to Baikonur facilities are not less than two million dollars. And if Kazkosmos cannot find alternative foreign partners who will provide the spaceport with investments and load it with work, the Kazakh government will face mass migration of citizens to other regions of the country [3, 4]. At the same time, this same social problem is also relevant for Russia itself: the authorities will also need to cope with the relocation of more than twenty thousand citizens.

Given the risks, the Russian and Kazakh parties are looking for ways to deepen mutual cooperation. So, back in the early 2000s, Presidents N. Nazarbayev and V. Putin agreed to build a new Baiterek (Topol) launch pad, from which an eco-friendly Angara launch vehicle could be started. This could become a new "growth point" for the spaceport. In July 2016, it became known that "Angara" will not launch from the "Baiterek" place. Therefore, the threat to the environment of the Kazakh steppe remains the main issue, as the simplified version of Proton, which is the workhorse of the Russian space Agency, most likely, will continue to fly on heptyl [5].

In this regard, the Kazakh authorities raise the issue of environmental safety and protection of people's health, especially after a number of accidents that have occurred, however, the economic benefits from cooperation with Russia still prevail. Therefore, the movement of local activists is increasing, demanding that the government of Kazakhstan take measures, including payment of compensation for environmental damage.

Environmentalists blame Roskosmos for non-compliance with its environmental obligations, as well as the fact that due to the influence of Roscosmos State Corporation and the particular status of the city of Baikonur, it is not possible to prove the damage to people's health and the environment, and, consequently, to achieve compensation. However, the elimination of environmental violations is necessary, as is the need for continuous monitoring of the state of ecosystems in the implementation of rocket and space activities at the Baikonur cosmodrome. Such monitoring is carried out at the regional (the area of the cosmodrome and the areas of fallen carrier rocket modules) and local (the spots of falling of these modules) levels.

Environmental monitoring involves monitoring the content of toxic pollutants and assessing the state of ecosystems in General and their individual components. Moreover, each monitoring level corresponds to specially developed observation programs that are formed according to the principle of choosing priority pollutants and integral characteristics reflecting a group of processes or changes in the composition of substances. The objects of research are more than 40 areas of fallen modules of carrier rocket launched from the territory of the Baikonur cosmodrome.

Twenty-one of these areas are located on the territory of the Russian Federation, their total area is about 35 thousand square kilometers in the territories of nine subjects of the Russian Federation (Novosibirsk, Tomsk, Omsk, Sverdlovsk regions, Altai and Perm regions, the Republic of Altai, Khakassia, Tuva), the rest are on the territory of the Republic of Kazakhstan [6].

In this regard, the need to implement space projects at the Vostochny cosmodrome, on the basis of which Russia will be able to implement a new project of the Angara heavy carrier rocket, more environmentally friendly than the Protons launched from Baikonur, which are charged with toxic heptil, is quite understandable [7].

It should be noted that Vostochny occupies a geographically more favorable location than the existing Russian cosmodromes of Plesetsk and Kapustin Yar, because it is closer to the equator, which provides an increase in launch vehicle speed due to the tangential launch point speed due to the daily rotation of the Earth. In this sense, Vostochniy is slightly inferior to Baikonur (the difference is a little more than six degrees in latitude), however, despite this, experts call them direct competitors. In addition, the location of the Vostochny spaceport in the Russian Far East creates an additional opportunity for the development of this region.

Analysis of the positive and negative aspects of Baikonur and Vostochny operation for Russia showed that the positive aspects of Baikonur should include its location, due to which the modules and debris fall in a deserted area, which does not affect the ecological situation in Russia, but in any case affects the ecology of Kazakhstan. The negative aspects of using Baikonur include the dependence on a foreign state and the cost of technology of the last century.

Let us turn to the analysis of the advantages and dis advantages of the use of the Vostochny cosmodrome. The positive aspects of functioning Vostochny include the advantages of the new technological wave, the lack of rent and freedom of action. The negative is the transfer of environmental burden to Russia, as the areas of falling modules cover the areas of active economic activity, including cities, navigable rivers and reserves. Moreover, it is almost impossible to evacuate the debris of rocket stages and emergency spacecraft from the taiga.

Currently, scientific and methodological approaches are being discussed how to create an environmental monitoring system for the Vostochny cosmodrome. It is noted that this system should become an integral part of the Roscosmos departmental environmental monitoring system, and recommendations have already been developed on the formation of a monitoring program. Studying the features of the impact of rocket and space activity on the environment of a territory, it should be noted that the Institute for Water and Environmental Problems of the Siberian Branch of the Russian Academy of Sciences has been conducting such studies since 1998.

They are talking about complex ecological-biogeochemical and medical-ecological studies of the impact of rocket and space activities on ecosystems and public health.

Since 2010, the named Institute has been involved in the environmental support of the construction of the Vostochny cosmodrome, including, since 2013, in the creation of an environmental monitoring system for its further operation. General analysis of research results 2013-2014 shows that the current ecological situation of the cosmodrome area corresponds to the category of "relatively satisfactory". The consequences of the negative impact of the construction processes of the cosmodrome facilities on the ecosystems of the region have not been identified. It is assumed that the environmental monitoring system of the Vostochny cosmodrome area should become a component of the Roscosmos departmental environmental monitoring system.

In the structure of Roscosmos, there is a Center for the operation of ground-based space infrastructure facilities, which is the head enterprise for environmental monitoring of the the Baikonur cosmodrome territory and the areas of booster debris landing and missiles launching spacecraft for various purposes from the Baikonur cosmodrome. In general, this center is responsible for conducting departmental environmental monitoring. To implement this task, the company has established the department of environmental safety of ground-based space infrastructure.

Within the framework of some objects of space infrastructure (for example, cosmodromes), systems of industrial environmental monitoring have been formed as a subsystem of departmental environmental monitoring. For example, there is a space center "Yuzhniy" at the cosmodrome Baikonur. It is also planned to create a similar structure at the Vostochny cosmodrome, which should be open enough so that in the future, as new objects are put into operation and their impact on natural environments and components are revealed, this information is available to the general public, removing the negative social assessment in advance. It will also enable interaction with other sources of information, databases, means of their production and processing [8].

The long-term experience of Russian scientists in this field suggests that in order to increase environmental safety in the territories of cosmodromes and the areas of carrier rocket modules falling, further development and improvement of the environmental monitoring system and improvement of the analytical bases and integral indicators of environmental impact are required. This is necessary to ensure the collection of reliable data sufficient to establish indicators on the basis of which it is expected to normalize their permissible impact on the environment.

It should be noted that in modern science, environmental monitoring is understood as a system of observations, assessment and forecasting of the state of the environment and the environmental situation on the basis of instrumental and other measurements of indicators of the state of objects of environmental monitoring.

Due to the heterogeneity of objects - territories exposed to rocket and space activities - the degree of impact on their components is different. Therefore, the environmental monitoring system pays special attention to the issue of territorial location of observation points. Monitoring spots cover the maximum diversity of areas exposed to space-rocket activities with different levels of impact on them, which helps to assess the degree of change in ecosystems and their components.

One approach to determining the location of monitoring spots can be the zoning of the territories of cosmodromes and areas of impact of the carrier rocket separating modules on the stability of ecosystems and their components by different types and scales of impact. For example, the factors responsible for the resistance of ecosystems to chemical effects can be integrated into three groups.

The first group includes indicators determining the intensity of self-purification of the environment: the degree of dissection of the surface, features of the structure of the territory, characteristics of the surface layer, features of the hydrogeological conditions of the area, meteorological indicators (rainfall, wind speed, etc.).

The second group is the indicators responsible for the possible forms and speed of transformation: soil climate, biochemical activity, the total amount of solar radiation and other factors.

The third group of characteristics is responsible for the possible strength of the fixing of polluting products and includes indicators of soil and rock capacity that can absorb pollutants, redox conditions, mineralogical composition, elemental composition of soils and soils, as well as characteristics of geochemical barriers and biogeochemical specialization of plant communities.

The fourth group are natural factors that contribute to the stability of ecosystems to mechanical damage, these include the strength properties of soils and the nature of the vegetation cover of the territory.

Identifying the ecological parameters of assessing the state of ecosystems, it should be noted that the primary task of environmental monitoring is collecting, analyzing and systematizing information that allows not only to diagnose changes in ecosystems, but also to determine in time the approximation of a certain critical level in their functioning.

In this sense, the optimal selection of indicators and the precise determination of their critical values, in combination with the monitoring time limit, will allow for the timely implementation of all necessary protective and preventive measures [6].

When determining the environmental parameters for assessing the impact of rocket and space activity on the properties of ecosystem components, it is necessary to use generally accepted theoretical approaches, as well as practical results of field and laboratory experiments on modeling pollution of ecosystems with rocket fuel components and their transformation products.

2. Conclusion

Thus, it is too early to say anything about the prospects for environmental safety at the Baikonur and Vostochny cosmodromes with some confidence, since the launch of objects into space must meet, on the one hand, the strategic economic plans for the development of the government, and on the other hand, the environmental safety of all humanity.

However, already now it is possible to state the need and the possibility of organizing environmental monitoring of the territories of cosmodromes and areas of carrier rocket modules falling. This monitoring allows reducing social stress regarding the availability to the general public the information on the environmental problems of the space industry and ways to solve them.

References

[1] Krechetov P P, Koroleva T V, Chernitsova O V, Zhdanov D F and Modenova E I 2009 Ecological consequences of the operation and disposal of rocket and space technology *Natural and Technical Sciences* **5** p 116-122

[2] Panin L E and Perova A Yu 2006 Medico-social and environmental problems of using liquid-fuel rockets (heptyl) *Bulletin of the Russian Academy of Medical Sciences* Vol **1 (119)** p 124-131

[3] Vorotilov A 2013 If Russia leaves Forbes Kazakhstan Business Journal **24**, available at: http://forbes.kz/process/probing/kogda_rossiya_uydet (access date 01/04/2019)
[4] John C K 2014 Daly is a non-resident Fellow at the Johns Hopkins Central Asia Caucasus Institute in Washington DC, available at: http://www.silkroadreporters.com/2014/09/10/kazakhstans-space-launching-cosmodrome-pressure-russia/ (access date 01/04/2019)
[5] German A Baikonur and Vostochny: cooperation vs competition *Sputnik Kyrgyzstan,* available at: http://ru.sputnik.kg/analytics/20160429/1024868188.html (access date 21/04/2019)
[6] Koroleva T V, Krechetov P P, Chernitsova O V and Sharapova A V 2013 Ecological monitoring of the Baikonur cosmodrome *Natural and Technical Sciences* **6** p 221-4
[7] Sabitov D 2016 Kazakhstan space: reality and prospects (Astana – Almaty: The Institute of World Economy and Politics (IWEP) under the Foundation of the First President of the Republic of Kazakhstan, the Leader of the Nation.) p 7
[8] Puzanov A V, Bezmaternykh D M, Kirillov V V and Vdovina O N 2015 Assessment of modern ecological state of the position area at the spaceport "Vostochny" including the areas of falling carrier rocket modules as the basis for the environmental monitoring establishment *Regional Environmental Issies* **2** p 97-102

IOP Conf. Series: Materials Science and Engineering **822** (2020) 012036 doi:10.1088/1757-899X/822/1/012036

Demand for satellite reconfigurability

A S Andreev[1], A V Leonova[2], A M Popov[1]

[1] Reshetnev Siberian State University of Science and Technology, 31 Krasnoyarsky Rabochy Av., Krasnoyarsk, 660037, Russia
[2] JSC Information Satellite Systems Reshetnev, Lenina str., 52, Zeleznogorsk, Krasnoyarsk krai, 662972, Russia

E-mail: mail.4o@ya.ru

Abstract. This article analyzes several approaches to in-orbit satellite reconfiguration: rapid launch, reconfigurable satellite constellations, modular satellites and software defined radio. A number of significant projects are studied under each category. Software defined satellites is selected as the most promising technology.

1. Introduction
Design life of modern satellites can reach up to 15 years compared to 3-5 years a decade ago. This important achievement accelerated the growth in the number of active satellites. But it had a negative effect in the long run: payload of such satellites becomes deprecated. A vivid example of this tendency resides with communication satellites, which become deprecated due to swift evolution of communication standards and digital signal processing algorithms.

2. Approaches to solving the problem of satellite payload deprecation
A number of approaches to the problem of satellite payload deprecation exists:
1. Launching new satellites when the need for service improvement arises (rapid launch);
2. Reconfigurable satellite constellations;
3. Modular satellite architecture;
4. Designing the payload with flexibility and reconfigurability in mind.

3. Rapid launch approach
First approach mostly finds its use in satellite surveillance systems. An example is Air Force's Operationally Responsive Space (OSR) program, which is aimed at enabling temporary improvement of surveillance coverage quality for military applications [1, 2]. This program assumes an existence of a pre-fabricated, ready to launch satellite pool and sets strict limits for launch procedures duration. Two week time frame is set for a satellite to be tested, launched and reach the destination orbit position in the earlier work [1]. The later work [2] sets even more severe requirements: 24 hours for launch vehicle integration and another 24 hours for reaching the destination orbital position. Another related program is DARPA SEEM [3], aimed at developing a low-cost satellites with design life of 60 to 90 days.

This approach has several disadvantages as a means of solving a problem of communication satellite payload deprecation. First of all, the effect of this approach is constrained by launch complex resources and technologies, which could be limited. Secondly, current effort is focused on the launch procedures and does not mitigate the long development cycle for new payload. Lastly, this approach reduces the cost of satellites through the use of cheaper and/or commercial grade components, but does not affect the high cost of launching services. Those arguments indicate that rapid launch approach is ambiguous at solving a problem of payload deprecation. This conclusion is partially supported by the fact that rapid launch approach was developed for specialized military applications [4].

4. Reconfigurable satellite constellation

Second approach consist in forming a group of remotely controlled satellites (constellation), which are able to change their orbit promptly (reconfiguration) to reach desired coverage. This approach is aimed at increasing the link capacity for a particular area of land, e.g. world-grade sports or cultural event (olympic games, political forums, etc.).

The ReCon [4] is an example of satellite constellation project. This work describes to distinctive modes of operation for a group of satellites: global observation mode (GOM) as a basic constellation state and regional observation mode (ROM) aimed at increasing the quality of communication service for a particular region (figure 1). Later work [5] applies genetic algorithms and simulated annealing to facilitate constellation reconfiguration. The authors prove that this approach can applicable for radiolocation and radio occultation. Both works use model simulation and are not supported by field tests.

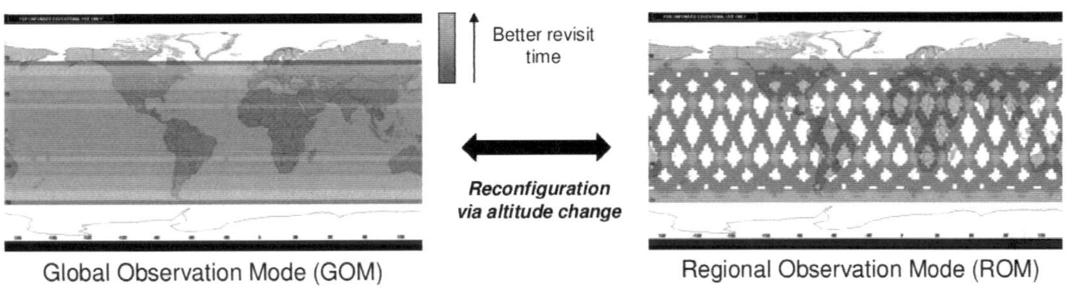

Figure 1. ReCON modes of operation

Modelling [5, 6] indicates, that reconfigurable constellations with large number of small spacecraft are more efficient. Reconfiguration efficiency increases with the number of satellites, while light class of the proposed satellites improves their maneuverability. On the other hand, the complexity of reconfiguration plan increases with the number of spacecraft involved as well. Traditional ground-based control model may be inadequate in terms of cost and complexity. This issue is addressed in [6] with the use of autonomous model of reconfiguration. The letter algorithm is only tested for a constellation of a single ring topology on GEO.

Economically efficient deployment is another problem that reconfigurable constellations face. Works [7, 8, 9] provide analysis for this problem with [9] focusing on light-class spacecraft. The letter work states, that launch and deployment for light-class satellites, that are preferable for constellations, suffer penalties compared to heavier spacecraft.

5. Modular satellite approach

Modular satellite is a spacecraft consisting from interchangeable modules partially or completely. The key principle for this approach resides with satellite upgrade in-orbit by replacing it's modules with new ones implementing required functions. This approach is characterized with the following list of advantages:

- building block unification becomes economically efficient with mass production;

- design life of modular satellite management becomes viable through modules replacement;
- new satellite development time is reduced with an opportunity to use "off-the-shelf" modules.

Along with benefits this approach may provide, it also brings up a set of non-trivial problems:

- define module composition in terms of space, durability, complexity, etc;
- define module interface which would not place complex constraints on module assembly process. This interface should ensure satellite flexibility as well as heat, energy transfer and high bandwidth;
- develop a flexible and reliable networking standard with possible redundancy for inter-module communication;
- develop a distributed application layer operating over a satellite with arbitrary composition;
- define deployment and reconfiguration procedures.

Reconfigurable satellites approach is being developed by "Intelligent Building Blocks for On-Orbit Satellite Servicing" (iBOSS) project [10] (figure 2). Most of the abovementioned problems were solved in a ground laboratory environment.

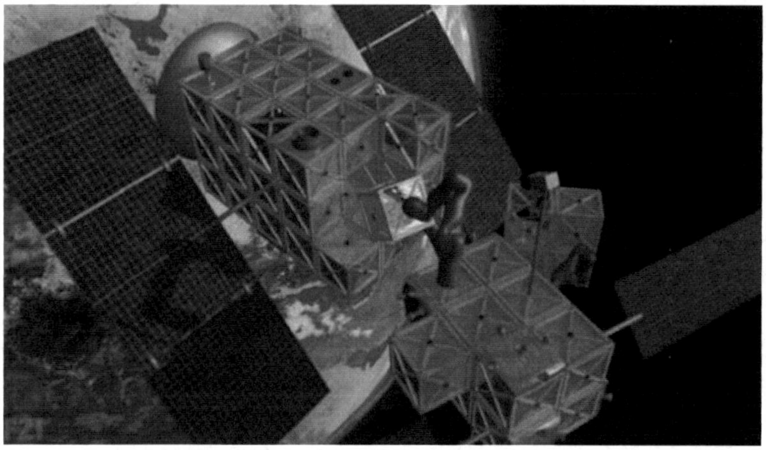

Figure 2. iBOSS satellite model (iSAT)

Under this project, a standard building block, iBLOCK, was developed [11] (figure 3). Each of such blocks consists of on-board computer (OBC), power control unit (PCU) and a router. TTEthernet (with RSTP and SNMP support) was selected as a final candidate for networking layer instead of SpaceFiber in earlier project papers. PCU of an iBLOCK is controlled by OBC thus enabling dynamic power scaling.

Figure 3. iBOSS module (iBLOCK)

The unified block interface standard [12] for connecting iBlocks to each other or to other components such as solar panels, additional payloads was developed under iBOSS project. Mechanical layer of this interface is an androgynous connector (figure 4). Fuel piping uses separate connectors with similar interface.

Figure 4. iBOSS physical interface

iBOSS project utilises distributed computing model for software application. Two distinctive types programming modules: one tied to specific iBLOCK, the other is iBLOCK independent, which could be executed on any available module. RSTP protocol is used to track changes in satellite composition. This information is handled by special device - monitoring and control node, which composes routing tables with redundant paths and configures routers over SNMP protocol in each iBLOCK. A special IDE is proposed for satellite construction and testing automatization.

iBOSS project authors report that they are working on in-orbit reconfiguration procedures for their modular spacecraft and a library of pre-tested functional blocks. It is worth noting that iBOSS project carried out successful testing of several laboratory prototypes and is still to carry out in-orbit tests for their reconfigurable spacecraft.

It is important to note that there are several projects focusing on non-reconfigurable modular satellites. ARaMiS project [13] proposes standard block designs, which could be reused for CubeSat-class spacecraft. Authors state, that this particular initiative is intended to lower new satellite development and fabrication cost to make them affordable for universities and small businesses. Such projects fall under rapid launch approach as they are aimed at decreasing development time for new spacecraft rather then upgrade existing ones.

6. Reconfigurable payload approach

This approach is also known as software defined satellite (SDS), or a satellite, which functionality could be changed by software rather then hardware. The hardware of such a satellite is a redundant set of reconfigurable components like processors, FPGA, FPAA or microcontrollers. Such components do not define application directly, but provide an ability to set application by a set of configuration files and change it on demand. It worths noting that mass and size of such satellite has to be increased to house redundant set of components as a trade-off for reconfigurability. In-orbit reconfigurability could be utilized to manage payload deprecation.

SDS technology was addressed on World Satellite Business Week (WSBW) conference [14] and was acknowledged as perspective satellite architecture. Thales Alenia Space president declared, that under their SDS initiative, the company will focus on processor-controlled phased-array antennas. It was stated that medium-class satellite architecture is to benefit the most from adapting SDS approach. Airbus CEO declared that over 15 airbus SDS satellites are already operating or ready to launch. Lockheed Martin Space Systems plans to provide financial support for start-ups developing SDS technology. Maxar's SSL CEO stated that the company is aiming for integrating programmably-defined payloads with ground 5G segment.

Several project exist that put effort into SDS technology development. GalacticSky works on a microsatellite that would be subject to in-orbit reconfiguration in accord with particular customer demands. Proposed payload technology is a virtualized platform under GalacticOS (figure 5) hypervisor. SDK, API and cloud simulation service is provided for developers. The core hardware is Intel Atom or Intel i7/Xeon processors on a platform providing standard interfaces: RS232, USB, PCIe, SATA and 1Gbit LAN. Client software is executed on guest machines with virtual processor, memory and sensors for multiple customers simultaneously. Citrix virtualization solution was chosen as hypervisor software.

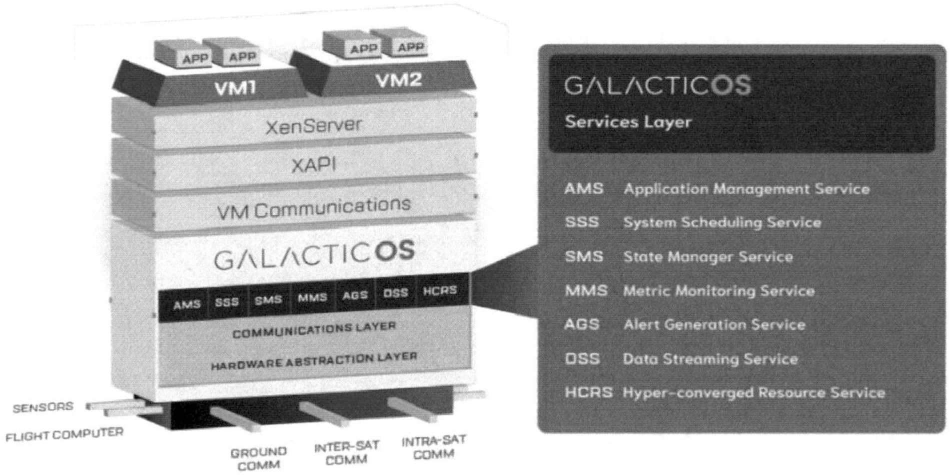

Figure 5. SDS protocol stack solution from GalacticSky

GalacticSky and University of Southern California's Space Engineering Research Center (SERC) developed GSky-1 (figure 6) spacecraft to test proposed technology. This satellite is to be launched in 2019 [15]. Payloads for this spacecraft were developed by NASA and Air Force Research Laboratory (AFRL).

Figure 6. GSky-1 satellite

Quantum project (figures 7, 8) by Eutelsat, Arbus and ESA is the most developed SDS, scheduled to be launched in 2019 [16]. This is the only heavy satellite design utilizing SDS technology. It is has reconfigurable coverage pattern, signal power and frequencies due to phased array antenna being used onboard.

Figure 7. Quantum satellite platform

Specialized software on Quantum satellite monitors resources utilization efficiency, predicts the need for characteristic change and controls reconfiguration process. Also reconfiguration could be initiated with control center commands. On top of that Quantum spacecraft will be able to perform orbital position change.

Figure 8. Quantum satellite model

Eutelsat representatives state, that they already secured contracts for 50% satellite resources capacity, which indicates, that there's demand for SDS technology.

7. Conclusion
Software defined satellite technology is the most efficient at managing payload deprecation out of the list of approaches analized in this article. While being beneficial in some spheres, rapid launch approach is the most expensive one. Reconfigurable constellation approach is not universal and is only effective on a limited range of satellite weight classes. Modular satellites approach is the second to top, but there is a number of problems that have be solved in order for it to be practically applicable.

Demand for SDS technology is stated by major aerospace-related companies. But current projects either intended for micro-class satellites, or restricted to radio transceiver reconfigurability.

All the presented information proves that software defined satellite technology is a promising course in spacecraft design phere and should attract attention of Russian space agency developers.

At the time of writing, set of software tools, reconfiguration procedure specifications and SDS hardware platform is being developed by the group of authors as core parts of SDS technology. Current work consists in:

1. developing specification for SDS hardware platform;
2. automating DSP circuit development process and incorporate it into specialized IDE;
3. developing algorithms for digital circuit design segmentation and mapping such segments onto SDS platform resources;
4. developing specifications for reconfiguration process.

More information on SDS concept being developed as well as obtained results will be provided in future papers.

References

[1]. Siddiqi A 2006 Reconfigurability in space systems: architecting framework and case studies *Doctoral dissertation* (MIT, Cambridge, USA)

[2]. Knight F 2006 Concept of operations for operationally responsive space *Proc.* of the *4th Responsive Space Conf.* (Los Angeles)

[3]. Barnhart D 2012 SeeMe *Broad Agency Announcement* (Arlingron, USA)

[4]. Legge Jr. R S 2014 Optimization and Valuation of Reconfigurable Satellite Constellations Under Uncertainty *Doctoral dissertation* (MIT, Cambridge, USA)

[5]. Paek S W, Kim S and de Weck, O L 2019 Optimization of Reconfigurable Satellite Constellations Using Simulated Annealing and Genetic Algorithm *J. Sensors* **19**

[6]. Mushet G, Colombo C and McInnes C 2012 Autonomous control of a reconfigurable constellation of satellites on geostationary orbit with artificial potential fields *Proc. of the 23rd International Symposium on Space Flight Dynamics* (Pasadena, California, USA)

[7]. McGrath C and Macdonald M 2015 Design of a reconfigurable satellite constellation *Proc. of the 66th International Astronautical Congress* (Jerusalim, Israel)

[8]. Appel L, Guelman M and Mishne D 2014 Optimization of satellite constellation reconfiguration maneuvers *J Acta Astronautica* **99** 166-174

[9]. Crisp N, Smith K and Hollingsworth P 2015 Launch and deployment of distributed small satellite systems *J Acta Astronautica* **114** 65-78

[10]. Oberländer J, Uhl K, Pfotzer L, Göller M, Roennau A and Dillmann R 2012 Management and Manipulation of Modular and Reconfigurable Satellites *Proc. of the 7th German Conference on Robotics* (Munich, German)

[11]. Kortmann M, Schervan T, Schmidt H, Ruehl S, Weise J and Kreisel J 2015 Building Block-based "iBOSS" Approach: Fully Modular Systems with Standard Interface to Enhance Future Satellites *Proc. of the 66th International Astronautical Congress* (Jerusalim, Israel)

[12]. Kortmann M, Meinert T, Dafnis A and Schröder K U 2017 Multifunctional Interface for Modular Satellite Systems with Robotic Servicing Capabilities *Proc. of the 68th International Astronautical Congress* (Adelaide, Australia)

[13]. Mughal M, Ali A and Reyneri L 2014 Plug-and-play design approach to smart harness for modular small satellites *J Acta Astronautica* **94** 754–764

[14]. Holmes M 2018 Major Satellite Manufacturers Predict a Software-Defined Future https://www.satellitetoday.com/innovation/2018/09/17/major-satellite-manufacturers-predict-a-software-defined-future/http://www.galacticsky.net/

[15]. GSky-1 Satellite Developed with USC's Space Engineering Research Center Ready for Launch Later This Year, 2019 available at https://www.prnewswire.com/news-releases/vectors-

galacticsky-gsky-1-satellite-developed-with-uscs-space-engineering-research-center-ready-for-launch-later-this-year-300817368.html?tc=eml_cleartime

[16]. Four things you should know about EUTELSAT QUANTUM, 2018 available at http://news.eutelsat.com/blog_posts/four-things-you-should-know-about-eutelsat-quantum-69078

Influence of Graft and Stock on the Stability of Grafted Pinus Sibirica Trees

Iu E Scherba, R N Matveeva, O F Butorova, N P Bratilova and V S Martynov

Reshetnev Siberian State University of Science and Technology, 31 Krasnoyarsky Rabochy Av., Krasnoyarsk, 660037, Russia

E-mail: selekcia@sibgu.ru

Abstract. The growth characteristics of a 30-year-old Pinus sibirica grafted with cuttings from plus trees and a 36-year-old one of different geographical origin on the rootstock of Pinus sylvestris and Pinus sibirica, the results of their studying are presented in the article. The analysis of accretion of grafted components depending on the type, age of the rootstock was done. It has been established that 36-year-old grafted trees using cuttings of Pinus Sibirica of different geographical origin (Altai, Biryusa, Sverdlovsk, Khanty-Mansi) on the Pinus sylvestris tree differ in the growth of scion with rootstock. Ramets with good accretion (without exceeding the diameter of the stem of the scion over the stock) with maximum incompatibility (13.1 cm) were selected. The growth of 30-year-old grafted trees depends on the genotype of plus trees. The most intensive growth of ramet is noted at clone 91/55 in comparison with 100/64 and 94/58. The grafted trees of one clone (91/55) on the Pinus sylvestris had an average graft stem excess of 6.2 cm; there is practically no excess on the Pinus sibirica stock. It is recommended to use the selected specimens without sagging in the places of accretion of grafted components as uterine trees for their further reproduction and the creation of target plantations for the accelerated cultivation and use of industrial wood from Pinus sibirica.

1. Introduction

Wood of Pinus sibirica (Pinus sibirica Du Tour) is light, soft, well-processed, valued for its straightness, beautiful texture and is widely used in the construction of houses, furniture, musical instruments, various handicrafts, in pencil production. Due to the prohibition of cutting of cedar plantations, target plantations are created for the accelerated cultivation of Pinus sibirica to produce valuable wood. Grafting of Pinus sibirica from positive trees certified on stem productivity will make it possible to obtain high-quality wood quickly [1].

Much attention is paid to reproduction of Pinus sibirica due to the fact that this method allows to accelerate not only the growth, but also the beginning of seed production, to obtain increased harvests of seeds for further reproduction of ramets selected on stem productivity [2-5].

The growth of scions in height depends on the individual characteristics of uterine trees, the physiological conformity of the scion and rootstock, environmental conditions [6-10].

When grafting Pinus sibirica, both Pinus sibirica and Pinus sylvestris are used as a stock. A.I. Severova [11] planted Pinus sylvestris, mountain, Rumelian, Weymouth, Pinus sibirica, Korean pine on Pinus sylvestris. Similar experimental studies were also carried out by N.F. Khramova [12], O.P.Olisova and et al. [13], R.N. Matveeva and et al. [14], E.V. Titov [15] and others.

Pinus sylvestris, as a stock for Pinus sibirica, has a number of positive properties: it grows quickly, has low soil requirements, is resistant to climatic conditions, has a developed root system. However, the Pinus sibirica scion in diameter initially catches up with the Pinus sylvestris stock, and later begins to outgo it in most cases, and it sometimes breaks off in places where the grafted components grow together under the weight of the expanded crown, what leads to plant death [16-19].

The different rate of growth of conductive tissues in the graft and rootstock complicates the passage of nutrients. As a result, the graft gradually outgrows the stock, slowing down the radial growth of the stock [20-23]. The rejection is due to the physiological and biochemical heterogeneity of the stock and graft [24].

In the experiments done by A.I. Severova [11], the successful accretion of the Pinus sibirica graft with the pine stock was observed in 80-90% of cases in the first years. According to G.V. Kuznetsova [25] the percentage of 35-year-old vaccinations with incompatibility ranged from 11 to 95 in different ecotypes.

R.N. Matveeva et al. [26] showed that the Pinus sibirica graft grows much more slowly on the stock of the pine tree than on the Pinus sylvestris, but forms individuals of a better growth of the grafted components, and therefore their durability is noted.

Great influence on the accretion of grafted components has the type and the age of the rootstock. To identify the cause of death of heteroplastic vaccinations, the radial growth of the stock of Pinus sylvestris and the scion of Pinus sibirica with a varying degree of incompatibility was investigated. It has been shown that the interspecific incompatibility of vaccinations can be overcome by selecting the components of the vaccination. In the case of heteroplastic grafts, the phenomenon of anatomical and morphological incompatibility is observed, with different intensity of growth of the graft components by diameter [27].

Analysis of the radial growth and climatic response of heteroplastic grafts of Pinus sibirica and European cedar of three climatypes grafted on Pinus sylvestris stock showed that the radial growth of Pinus sibirica grafts compared to European cedar was 23% higher, which is a consequence of their species characteristics. The radial growth of the Pinus sylvestris depends on the type of graft: the growth of Pinus sylvestris (rootstock for Pinus sibirica) is higher than the growth of control trees by 20% for Surgut climate type and by 15% for Yemelyanovo climate type, while the increment in pine diameter (for European cedar) is 10% less than that of control unvaccinated trees [28].

An active work to create the clonal plantations of various tree species is done abroad. Thus, Li Bailian et al. [29] noted that the increase in the volume of incense pine wood was higher by 12% when rejecting the worst clones on the plantations of the first generation and 30% of the second generation. Clone plantations of Ladanna pine have been created in the USA [30], Calabrian pine, black and Pinus sylvestris in Turkey [31]. Xu Jin et al. [32] studied the variability of Masson pine on the plantation of 140 clones, Y. Moriguchi et al. [35] - Japanese cryptomeria, A.D. Yanchuk et al. [34] studied Sithoi fir tree in Canada, where the variability of wood stocks in depending on clone affiliation was established.

2. Materials and methods

Our research was carried out on the inoculative hybrid and seed plantation of Pinus sibirica in the Karaulny forest area of the Training and Experimental Forestry of SibSU (suburban zone of Krasnoyarsk).

When creating a plantation, cuttings for grafts were harvested from Pinus sibirica plants - offspring of populations of different geographical origin, growing in the arboretum of SibSU, and plus trees certified in the Novosibirsk region. Vaccinations were carried out according to the method of E.P. Prokazina [35] in 1982 in the undergrowth of Pinus sylvestris of 5–10 years old, in 1988 — on 6-year-old seedlings of Pinus sibirica and 20-year-old growth of Pinus sylvestris. The trees were measured in height, diameter of the stem of the stock and scion in the places of accretion, diameter, length of the crown, length of the needles according to generally accepted methods [36]. The obtained results were statistically processed using software packages (Microsoft Office, Excel).

3. Results and discussion

As a result of the research, it was established that the death of grafted trees that have a significant excess of the diameter of the graft trunk over the rootstock occurs as a result of snowbreak, windfall and other factors. It is established that the biometric indicators of grafted trees at 36 years of age range from low to very high levels in accordance with the S.A. Mamaev scale [37]. A very high variation (67%) was noted in excess of the diameter of the scion above the stock (table 1).

Table 1. Variability of indicators of grafted Pinus sibirica trees on Pinus sylvestris.

Indicator	min	max	X av.	±m	V, %	P, %
Height, m	10.8	16.4	13.6	0.14	8.7	1.0
Diameter of stock, cm	14.0	50.6	35.5	0.92	21.7	2.7
Diameter of scion, cm	20.0	64.3	42.6	1.10	21.6	2.6
Excess, cm	0.0	22.6	7.1	0.57	67.0	8.0
Diameter of crone, m	3.8	9.6	6.9	0.14	17.7	2.0
Crown length, m	3.8	15.5	11.9	0.29	20.7	2.4
Needles length, cm	6.5	14.0	10.0	0.19	15.8	1.9

When using cuttings of different geographic origin, it was found that there is a significant variability in excess of the diameter of the scion above the stock diameter (from 0 to 13.1 cm), depending on the geographical origin of the scion and the individual characteristics of uterine trees (figure 1).

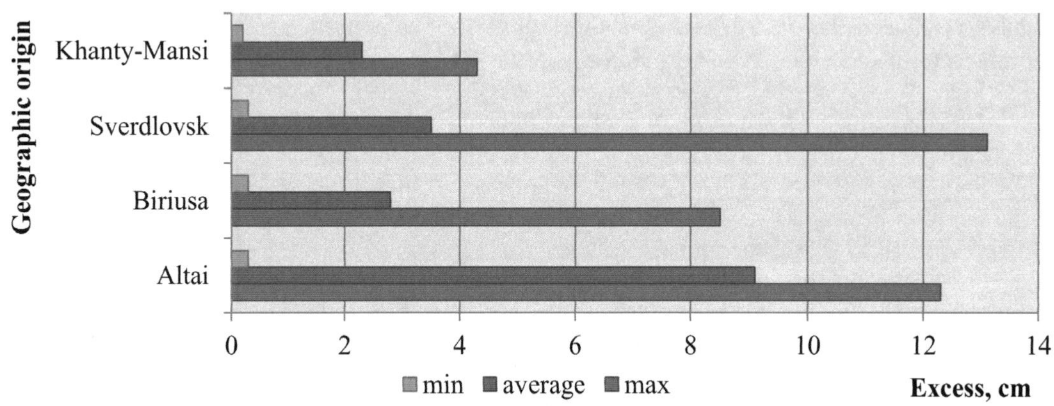

Figure 1. Excess of diameter of the scion above the rootstock in grafted trees of different geographical origin.

Some ramets of Altai and Sverdlovsk origins had the greatest influxes in the places of accretion.

The individual characteristics of uterine trees have great importance for the quality of accretion of graft and rootstock. Good fusion was observed in individual trees of tested origins (table 2, figure 2).

Table 2. Biometric indicators of trees characterized by good graft accretion with rootstock.

Geographic origin	Ramet	Height, m	Diameter of stock, cm	Diameter of scion, cm	Excess, cm	Diameter of crone, m	Live crown length, m	Needles length, cm
Altai	7-11	12.8	38.8	38.8	0.0	7.6	11.7	11.0
Biryusa	9-2	12.4	26.7	26.7	0.0	4.8	10.9	12.7
Sverdlovsk	14-9	14.8	40.7	40.7	0.0	6.4	13.9	10.3
	16-8	14.0	33.7	33.7	0.0	6.0	12.7	12.3
Khanty-Mansi	16-9	14.9	41.7	41.7	0.0	6.6	13.6	8.6

Some specimens at the age of 36 years old with a height of 8.3-16.6 m, with the greatest excess of the scion over the stock, have survived (table 3).

Table 3. Ramets of different geographical origin with the greatest excess of diameter of the scion over the stock.

Geographic origin	Ramet	Height, m	Diameter, cm		
			stock	scion	excess
Altai	15-8	16.6	38.2	50.6	12.4
Biryusa	8-10	8.3	24.2	32.8	8.6
Sverdlovsk	6-13	15.1	42.6	55.7	13.1
Khanty-Mansi	4-15	13.3	39.8	44.2	4.4

During reproduction of Pinus sibirica by grafting in 1988 using cuttings harvested from plus trees No. 91/55, 94/58, 100/64, to 6-year-old Pinus sibirica seedlings, a good accretion of the grafted components was found (Table 4).

Table 4. Indicators of 30-year-old grafted trees on the rootstock of 6 years old Pinus sibirica.

Plus tree	Height, m	Diameter of stem, cm			Diameter of crone, m	Live crown length, m	Needles length,cm
		stock	scion	excess			
100/64	9,1	26.9	24.9	-2.0	5.3	8.7	10.1
94/58	8,4	28.6	28.6	0.0	6.1	8.1	10.0
91/55	9,9	27.9	28.0	0.1	5.6	9.6	9.9

There is a variability of graft specimens of Pinus sibirica depending on the genotype of uterine trees from which the cuttings were harvested. So, the highest growth figures were in ramets from plus tree № 91/55. The ramet of the tree 94/58 is lagging behind in height by 17.8% in comparison with the clonal progeny of the positive tree 91/55 and by 8.8% - in the tree 100/64. The diameter of the crown had the greatest value in the ramet tree 94/58. K.G. Zatzepina et al. [38] note that when creating clone plantations it is necessary to certify trees, taking into account the establishment of links between biometric indicators and genetically marker features.

Figure 2. Good graft healing with stock (a - graft - Pinus sibirica; b - stock of Pinus sylvestris)

A comparison of the growth rates of 30-year-old grafted trees using a graft of the same genotype (plus tree No. 91/55) on a 20-year old pine tree stock and a 6-year-old Pinus sibirica stock was made (table 5).

Table 5. Indicators of the offspring of plus tree 91/55 when grafting on stocks of different types and ages.

Type of stock	Age of stock, years	Height, m	Diameter of trunk, cm stock	scion	excess	Diameter of crone, m	Needles length, cm
Pinus sylvestris	20	9.4	28.2	33.4	6.2	6.0	10.2
Pinus sibirica	6	9.9	27.9	28.0	0.1	5.6	9.9

The excess of the diameter of the trunk of 30-year-old grafted trees in the places of accretion of grafted components is observed when a stock of Pinus sylvestris of 20 years age is used. Other indicators, such as height, diameter of the stock stem, diameter of the crown, the length of the needles vary within small limits.

4. Conclusion

As a result of the research, the possibility of accelerated cultivation of Pinus sibirica specimens resistant to negative natural factors was established. The type of stock, growth characteristics of uterine trees and the ability of cuttings harvested from them for good fusion with the stock are taken into account. The selected grafted specimens without sagging in the places of accretion of grafted components are recommended to be used as uterine plants for their further reproduction and creation of target plantations for the accelerated cultivation and use of industrial wood from Pinus sibirica.

5. References

[1] Matveeva R N, Butorova O F and Bratilova N P 2003 *Poleznyye svoistva i metody rasmnozheniya kedra sibirskogo* [Useful properties and methods of reproduction of Pinus sibirica] (Krasnoyarsk: SibGTU) p 154

[2] Kolegova N F 1997 Geograficheskiye privivochnuye plantatsii sosny i kedra v Krasnoyarskoy lesostepi [Geographical grafting plantations of pine and cedar in the Krasnoyarsk forest-steppe] *Geographical cultures and plantations of conifers in Siberia* (Novosibirsk: Science) pp 154-166

[3] Martynov V S, Matveeva R N and Butorova O F 2017 Individual'naya izmenchivost radial'nogo prorosta privitykh derev'yev sosny kedrovoy sibirskoy na plantatsii Uchebno-opytnogo leskhoza SibGU im.M.F.Reshetneva [Individual variability of the radial growth of grafted Pinus sibirica trees on the plantation of the Training and Experimental Forestry of SibSU named after M.F.Reshetnev] *Coniferous boreal zone* **35** pp 38-41

[4] Tvelenev M V 1975 Razmnozheniye privivkoy khozyastvenno-tzennykh derev'yev kedra sibirskogo [Reproduction by grafting of economically valuable trees of Pinus sibirica] // *Genetics, selection, seed production and introduction of forest species* (Moscow: Science) pp 76-82

[5] Titov Ye V 2016 Otbor tsennykh genotipov – klonov kedra sibirskogo na privivochnykh plantatsiyakh [Selection of valuable genotypes - clones of Pinus sibirica on grafting plantations] *Coniferous boreal zone* **5-6** pp 284-289

[6] Rabtsun A S 2013 Metody identifikatsii klonov sosny obyknovennoy na lesosemennykh plantatsiyakh [Methods for the identification of Pinus sylvestris clones on forest seed plantations] *MGUL Bulletin. Forest messenger [EE]* **3** pp 7-10

[7] Drozdov I I and Bryncev V A 1989 Problems of forming seed base for Siberian cedar (Pinus sibirica Du Tour) in the region of species introduction *Proc. Int. Simp. Forest Genetics, Breeding and Physiology of Woody Plants* (Moscow: CRIFGaB) pp 205-206

[8] Liu Guogang, Xiao Jie and Yan Lijun 2004 Physiological indicators and regulation of the growth of Korean pine with intragenital graft *Dongbei linye daxue xuebao* (J. North-East Forest. Univ.) **32** pp 12-13

[9] Mutke S 2005 Cone yield characterization of a stone pine (*Pinus pinea* L.) clone bank *Silvae genet* **54** pp 189-197

[10] Schmidtling R C 1983 Rootstock influences flowering, growth, and survival of loblolly pine grafts *For. Sci.* **29** pp117-124

[11] Severova A I 1958 *Vegetativnoye razmnozheniye khvoynykh drevesnykh porod* [Vegetative reproduction of coniferous trees] (Moscow-Leningrad: Goslesbumizdat) p 143

[12] Khramova N F 1964 Privivki kak metod sozdaniya semennykh uchastkov kedra i kedrovykh sadov v Novosibirskoy oblasti (Vaccinations as a method of creating seed plots of cedar and cedar gardens in the Novosibirsk region) *Renewal and improvement of forests* (Novosibirsk: Science) **8** pp 139-144

[13] Olisova O P, Larionova N A and Luzganov A G 1966 Ritm rosta kedra sibirskogo v geograficheskikh kul'turakh pod Krasnoyarskom (Growth rhythm of Pinus sibirica in geographical cultures near Krasnoyarsk) *Proc. report to conf. following the results of n. works for 1965* (Krasnoyarsk: STI) pp 78-79

[14] Matveeva R N 1982 *Vegetativnoye razmnozheniye sosny kedrovoy sibirskoy* [Vegetative reproduction of Pinus sibirica] (Krasnoyarsk: KPI) p 72

[15] Titov Ye V 1977 *Geograficheskiye privivki kak selektsionnyy priyem razvedeniya kedronykh sosen* [Geographical vaccinations as a selection method of cultivation of cedar pines] (Voronezh: CSILGiS) pp 49-52

[16] Dokuchaeva M I 1967 *Vegetativnoye razmnozeniye khvoynykh* [Vegetative reproduction of conifers] (Moscow: Lesn. prom-st) p 103

[17] Ibe A A, Chubugina I V, Lozitskaya G M, Dygalo I P, Shaprun E N and Belyaev V V 2012 Otsenka sostoyaniya arkhiva klonov sosny kedrovoy sibirskoy (Pinus sibirica) v Zapadno-Sayanskom OLKH [Estimation of the archive of Siberian cedar pine clones (Pinus sibirica) in the West Sayan EF] *Coniferous boreal zone* **30** pp 77-79

[18] Kuznetsova G V 1998 Opyt sozdaniya geograficheskoy privivochnoy plantatsii kedra sibirskogo v Krasnoyarskoy lesostepi [The experience of creating a geographic graft plantation of Siberian cedar in the Krasnoyarsk forest-steppe] *Forest science* **6** pp 63-70

[19] Kuznetsova G V 2016 Privivki kedrovukh sosen v Krasnoyarskoy lesostepi [Grafts of Pinus sibirica in the Krasnoyarsk forest-steppe] *Intensification of forestry in Russia* pp 122-123

[20] Krystev M T and Protas S A 2012 Anatomiya privivki nekotorykh khvoynykh rasteniy, vypolnennoy sposobom vrasshchep [Anatomy of grafting of some coniferous plants, made by splitting] *Byul. GBS RAS* **198** pp 64-67

[21] Fedotov A V 2000 Osobennosti rosta privitykh klonov kedrovykh sosen v Ivanteyevskom dendrosadu [Peculiarities of growth of grafted clones of Pinus sibirica in the Ivanteevsky dendrogarden] *Scientific. tr. Mosk. state un-t of forestry* **303** pp 191-195

[22] Jayawickrama K J S 1991 Rootstock effects in grafted conifers: A review *New firests* **5** pp 157-173

[23] Krakowski J and El-Kassaby Y A 2003 Effects of stratification and simulated aging on germination of Douglas-fir seed from a clonal seed orchard *Forest Genet.* **10** pp 65-70

[24] Titov Ye V 1995 Klonovyye ispytaniya kedrovykh sosen [Clone testing of cedar pines] *Forestry* **6** pp 25–26

[25] Kuznetsova G V and Savva Yu V 2004 K voprosu o mezvidovoy nesovmestimosti privivok kedra sibirskogo na sosnu obuknovennuyu [On the issue of interspecific incompatibility of Pinus sibirica inoculations on Pinus sylvestris] *Tomsk Bulletin. state un-tate* **10** pp 45-46

[26] Matveeva R N, Butorova O F and Shcherba Iu E 2016 *Semennoye i vegetativnoye razmnozheniye otselektirovannykh derev'yev sosny kedrovoy sibirskoy* [Seed and vegetative reproduction of selected Pinus sibirica trees] (Krasnoyarsk: SibGTU) p 206

[27] Shcherba Iu E and Grishlova M V 2015 Pokazateli odnoletnikh gomoplasticheskikh i geteroplasticheskikh privivok kedrovykh sosen [Indicators of annual homoplastic and heteroplastic grafts of Pinus sibirica] *Coniferous boreal zone* **33** pp 248-252

[28] Darikova Yu A, Vaganov E A, Kuznetsova G V and Grachev A M 2013 Radial'nyy rost privivok kedrovykh sosen (Pinaceae) v usloviakh Krasnoyarskoy lesostepi [Radial growth of pine grafts (Pinaceae) in conditions of the Krasnoyarsk forest-steppe] *Journal of the Siberian Federal University. Biology* **6** pp 3-17

[29] Li Bailian, McKeand Steve and Weir Robert 2000 Impact of forest genetics on sustainable forestry - results from two cycles of loblolly pine breeding in the U.S. *J. Sustainable Forest.* **10** pp 79-85
[30] McKean S E and Raley (Fred) E M 2000 Interstock effects on strobilus initiation in topgrafted loblolly pine *Forest Genet.* **7** pp 179-182
[31] Bilir N, Kang K S and Ozturk H 2002 Fertility variation and gene diversity in clonal seed orchards of Pinus brutia, Pinus nigra and Pinus sylvestris in Turkey *Silvae genet.* **51** pp 112-115
[32] Xu Jin, Wang Zhangrong, Chen Yabin and Qiu Jinqing 2004 Analysis of genetic indicators, signs of seeds and cones, as well as the productivity of clone cones growing in the seed nursery Pinus massoniana *Linye kexue = Sci. silv. sin.* **40** pp 201-205
[33] Moriguchi Y, Tsuchiya S, Iwata H, Itoo S, Tani N, Taira H and Tsumura Y 2007 Factors influencing male reproductive success in a Cryptomeria japonica seed orchard revealed by microsatellite marker analysis *Silvae genet.* **56** pp 207-214
[34] Yanchuk A D, Bishir J, Russell J H and Polsson K R 2006 Variation in volume production through clonal deployment: Results from a simulaion model to minimize risk for both a currently known and unknown future pest *Silvae genet.* **55** pp 25-37
[35] Prokazin Ye P 1960 Novyy metod privivki khvoinykh dlya sozdaniya lesosemennykh uchastkov [A new method of grafting conifers to create forest seed plots] *Forestry* **5** pp 22-28
[36] Molchanov A A and Smirnov V V 1967 *Metodika opredeleniya prirosta drevesnykh rasteniy* [Method of determining the growth of woody plants] (Moscow: Science) p 27
[37] Mamaev S A 1972 Formy vnutrividovoy izmenchivosti drevesnykh rasteniy (na primere Pinaceae na Urale) [Forms of intraspecific variability of woody plants (using the example of Pinaceae in the Urals)] (Moscow: Nauka) p 282
[38] Zatsepina K G, Eckart A K and Tarakanov V V 2012 Genotipirovaniye derev'yev na klonovykh plantatsiyakh khvoynykh lesoobrazuyushikh vidov v Zapadnoy Sibiri [Genotyping of trees on clone plantations of coniferous forest-forming species in Western Siberia] *Coniferous boreal zone* **30** pp 67-71

IOP Conf. Series: Materials Science and Engineering **822** (2020) 012038 doi:10.1088/1757-899X/822/1/012038

Using the Raspberry Pi single-board computer for solving problems in MATLAB when developing an underwater vehicle

D A Volkov, A A Kuznetsov, A V Sayapin
Reshetnev Siberian State University of Science and Technology
31, Krasnoyarsky Rabochy Av., Krasnoyarsk, 660037, Russian Federation

E-mail: vlkden@yandex.ru

Abstract. This article represents description of the developed concept of an underwater remotely controlled vehicle (URCV). Explains the steps required to establish a connection between the MATLAB computational environment and the Raspberry Pi single-board computer. The ways of using the Raspberry Pi hardware in MATLAB are considered. Using the Raspberry Pi hardware resources allows you to establish a connection between the computer environment and the outside world.

1. Introduction

In the 2018 the underwater remotely controlled vehicle (URCV) (fig. 1), developed by a team consisted of students of Reshetnev Siberian State University of Science and Technology and an associate professor of the university, Alexander Sayapin, was presented. The basic principle of the vehicle is similar to the one of multirotor unmanned aerial vehicles. The vehicle was intended to perform underwater archaeological studies of lower reaches of the Yenisei River.

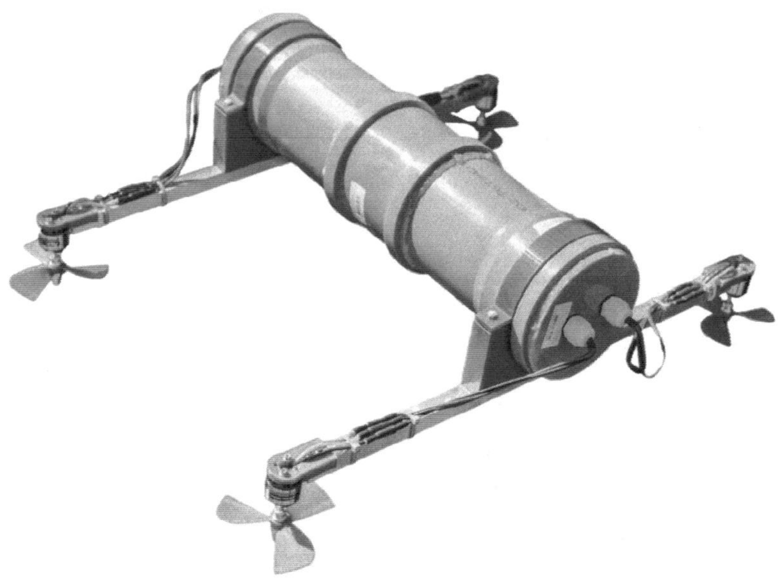

Figure 1. General view of the underwater vehicle.

The hardware of the vehicle consists of the vehicle frame, sealed case, onboard electronics, power source, engines and screws. As the main onboard computer, the Orange Pi has been chosen. It can be used as a host for a messaging server software and to provide online video translation from the vehicle. The computer is compatible with most of the hardware modules developed for the Raspberry Pi computer, which allows to debug and use most of the modules both with Raspberry Pi and Orange Pi.

The main engines of the vehicle are 4 brushless motors which is powered and controlled by electronic speed controllers (ESCs) combined with BECs. To control ESCs the MultiWii NanoWii module is used. It features 6-axis IMU and ATmega32U4 MCU. The MCU and the onboard computer are connected with an USB interface.

To control the vehicle an Ethernet network cable is used. It connects the onboard computer and a wireless router which is placed atop the water surface. To provide the video translation the vehicle is equipped with a USB web-camera in a sealed box which is connected to the onboard computer. The more detailed description of the URCV construction, its hardware and connectivity you can find in [1].

The onboard computer software detailed description is provided in [2]. At the moment the development of a software complex including the android application for the remote control of the vehicle is in progress.

The development of the system that includes both URCV and its software is a hard task which consists of simulation of the sealed case behavior under water pressure, simulation the vehicle movements and interaction with the environment, obstacles avoidance and so on and so forth.

To control the system and to debug the software complex of the URCV the developers decided to use MATLAB computational environment.

The MATLAB environment includes extension packages that allows to perform data analysis, visualization, specific types of computation and so on. There are special extension packages for interacting with Raspberry Pi microcomputer which allows to use different kinds of sensors and other peripheral devices of Raspberry Pi [3].

Work [4] suggests that following hardware of Raspberry Pi can be used:

- Raspberry Pi LED;
- General Purpose Input-Output pins;
- Devices connected through UART, I2C, SPI;
- Raspberry Pi camera module as well as USB web-camera;
- PWM outputs;
- Sense HAT IO board;
- Linux system shell.

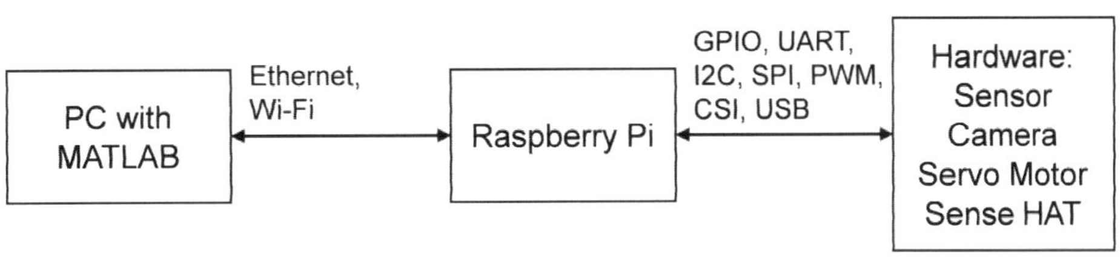

Figure 2. A scheme of an interaction between MATLAB and Raspberry Pi hardware.

Figure 2 shows scheme of an interaction between personal computer (PC) with MATLAB environment and Raspberry Pi hardware.

IOP Conf. Series: Materials Science and Engineering **822** (2020) 012038 doi:10.1088/1757-899X/822/1/012038

2. Getting started

MATLAB itself cannot be used to interact with Raspberry Pi. To provide such kind of functionality, special steps are to be performed:

MATLAB Support Package for Raspberry Pi Hardware [5] must be installed. It is necessary to note that that package is compatible with MATLAB R2014a and newer and supports only Raspberry Pi 1 B+ and newer.

Also, special image for Raspberry Pi operating system (OS) must be downloaded and installed.

To establish an interaction between MATLAB and Raspberry Pi a special object in MATLAB environment has to be created. It can be done by using special command:

```
mypi = raspi('ipaddress','username','password')
```

where "mypi" is the name of the newly created object, "ipaddress" is the address of the Raspberry Pi you are connected, "username" and "password" are credentials of a user registered in the Raspberry Pi Linux operating system. If the connection is successful, the user will see an appropriate information in the MATLAB command window.

3. Hardware

The hardware of Raspberry Pi can be accessed from MATLAB environment. The onboard green LED near the power indicator can be switched on or off using the following command:

```
writeLED(mypi,'led0',value)
```

where "value" is a LED state (0 – the LED is switched off, 1 – the LED is switched on).

It is also possible to get an information from different sensors connected to Raspberry Pi. To get a state of the URCV an Bosch Sensortec BMP/BME pressure sensor [6] is supposed to be used.

To get a spatial position of the vehicle the so called MPU (Motion Processor Unit) sensors can be used. A typical MPU combines accelerometers to measure linear movements together with gyroscopic sensors supposed to measure rotation of a sensor (and so the vehicle) [7]. This kind of sensors are typically connected to Raspberry Pi using I2C interface.

MATLAB environment has particular functions designed to get information through the I2C bus. To access I2C sensors data it is necessary to know the I2C addresses of the sensor and its internal registers structure.

To get a data from the sensor, the following functions can be used [8]:

1. AvailableI2CBuses – display the names of the available I2C buses on the Raspberry Pi hardware;
2. scanI2CBus – scan I2C bus device addresses;
3. i2cdev – connection to device on Raspberry Pi hardware;
4. readRegister – read from register on I2C device.

The data we get from the sensors can be used to show diagrams of the state of the vehicle in time or to control the URCV after corresponding data processing.

MATLAB environment also can be used to interact with Raspberry Pi GPIO pins. The state of the pins can read by MATLAB as well as it can be set in it.

There is packet in MATLAB that allows to work with a video stream from Raspberry Pi. The video stream or single frames can be captured and processed, using «Computer Vision System Toolbox» and «Image Processing Toolbox» packets. Both provide high performance and wide range of means to process images and to control the vehicle using the video stream. These MATLAB packets in some way overcome popular OpenCV library [9].

To use a USB web-camera connected to Raspberry Pi the following steps must be performed:

1. Physically connect the camera to the Raspberry Pi;
2. Create the connection between Raspberry Pi and MATLAB environment;
3. Create the connection between the camera and MATLAB using command webcam.

After that the camera is ready for the streaming.

An interesting mean to use in URCV is also the Sense HAT shield. It features accelerometers, gyroscopes, magnetometer and LED matrix for visualization.

4. Simulink

Simulink is a graphical programming environment for modeling, simulating and analyzing multi-domain dynamical systems, integrated with MATLAB. It utilizes a graphical programming language allowing to create a pipeline of data processing procedures and packages, installed in MATLAB.

Simulink includes a special packet named «Simulink Support Package for Raspberry Pi Hardware» [10] supposed to be used to connect Simulink to Raspberry Pi to use its input-output ports.

This packet allows to:

– Audio-video input-output capabilities;
– Sending-receiving TCP/UDP packet over the network;
– Input-output using Raspberry Pi GPIO;
– Sending-receiving MQTT messages;
– Digital signal processing of Raspberry Pi audio and video data.

This packet can be used to process the data stream Raspberry Pi produces, to build systems for augmented or virtual reality or to build a computer vision system to avoid obstacles.

This packet can be used also to develop a URCV control system.

5. Conclusions

Connection between Raspberry Pi and MATLAB environment allows to build a system for underwater remotely controlled vehicle which is more powerful than one build using Raspberry Pi resources only.

This approach also allows to build a virtual environment to test and debug algorithms for URCV control which reduces the number of real-life tests.

References

[1] Volkov D A, Sayapin A V 2018 Remotely operated underwater vehicle for an archaeological survey *Robotics and Artificial Intelligence* Pp 21-25 [in Russian]

[2] Sayapin A V, Volkov D A 2018 Software of remotely operated underwater vehicle for an archaeological survey *Robotics and Artificial Intelligence* Pp 72-77 [in Russian]

[3] Volkov D A 2019 Realization of library determination of distance to objects in the water environment *Digital Signal Processing and its Applications (DSPA-2019)* Pp 470-475 [in Russian]

[4] Davidov B 2014 Import and export MATLAB data through Raspberry Pi *Portal of scientific and practical publications* [in Russian]

[5] Raspberry Pi Support from MATLAB https://www.mathworks.com/hardware-support/raspberry-pi-matlab.html (accessed on 23 April 2019)

[6] Bosch Sensortec GmbH 2018 Bosch Sensortec MEMS sensors and solutions

[7] Zhmud' V A, Kuznetsov K A, Kondrat'ev N O, Trubin V G, Trubin M V 2018 Accelerometer and Gyroscope MPU6050: first switch on the STM32 and study the readings in statics *Nauchno-tekhnicheskiy zhurnal Avtomatika i programmnaya inzheneriya* **25** Pp 9-22 [in Russian]

[8] MATLAB Support Package for Raspberry Pi Hardware Documentation https://www.mathworks.com/help/supportpkg/raspberrypiio/ (accessed on 19 April 2019)

[9] Volosatova T M, Yablokov V E 2015 Tracking selected objects throughout the video sequence *Elektronnyy nauchnyy-tekhnicheskiy zhurnal Inzhenernyy vestnik Dona* **7** Pp 518-531 [in Russian]

[10] Raspberry Pi Support from Simulink https://www.mathworks.com/hardware-support/raspberry-pi-simulink.html (accessed on 23 April 2019)

IOP Conf. Series: Materials Science and Engineering **822** (2020) 012039 doi:10.1088/1757-899X/822/1/012039

Large deflection of a cantilever beam under transverse loading. A modification of linear theory.

D M Zuev, Yu V Zakharov

Reshetnev Siberian State University of Science and Technology

E-mail: ZuevDmitriy93@yandex.ru

Abstract. The paper provides a modification method for a classical linear theory of strength of materials in case of a cantilever under transverse loading. The modification uses a condition of saving arc-length of a cantilever. A sagging deflection of the modified linear theory has accuracy comparable with exact solutions in elliptic integrals and functions. The modification expand loading interval of applicability of the linear theory significantly. A comparison with an exact solution was made. We also considered a cantilever under combined loading: concentrated transverse load and uniformly distributed load (an own weight of a cantilever). A comparison with experimental data confirms accuracy of the proposed modification method.

1. Introduction

There are two approaches of solving bending beams problems: small and large deflection. The small deflection approach is of use in strength of materials theory and uses linearized equations. It is believed that an interval of applicability of this approach is limited by the ratio of 3-4% maximum deflection to the length of a beam [1]. The small deflection approach is sufficient for most engineering and construction problems, because deflections are restricted in these problems.

However, many modern problems of science and industry use long elastic rods that are subject to large deflections. For example, these problems include: development of deployable rods construction of small satellites [2], robotic flexible manipulators [3], polymer actuators [4], lamina emergent mechanism and compliant mechanisms [5]. Linearized equations can't be used for solving of these problems. Solutions for these problems have high complexity and use elliptic integrals and functions. Numeric methods can be used, but it has another difficulties. Therefore development of analytic solutions for different cases of loading is important. It is especially important to obtain simple analytical formulas for practical use.

In the paper we consider a case of thin elastic cantilever under transverse concentrated loading P on the free end of a cantilever (Figure 1).

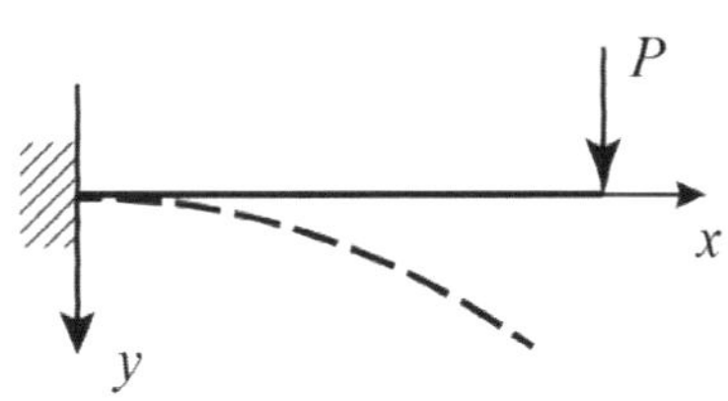

Figure 1. A cantilever under transverse concentrated loading on the free end

To solve a bending problem is required to solve Euler-Bernoulli equation (1) in angular or Cartesian coordinates:

$$\frac{d\theta}{ds}=\frac{d^2y}{dx^2}[1+(\frac{dy}{dx})^2]^{-3/2}=\frac{M(x)}{EJ}, \tag{1}$$

where $M\ (x)$ – a bending moment, EJ – bending stiffness, θ – an angle of the tangent at a point as it moves along coordinate s (the distance measured along the curve itself).

Solving of equation (1) with different one-point and two-point conditions lead up to equations in elliptic integrals and functions. For example, the problem was solved at 1999 in [6]. In this work eigenvalues problem and obtained characteristic critical forces for bending modes was considered. This solution is a special case of a general solution obtained in [7]. The above solutions use elliptic integrals 1st and 2nd kinds. In work [8] solution for transverse loading with the use elliptic integrals of 1st and 3rd kind was obtained by using one-point conditions.

2. Linear theory

Linearized Euler-Bernoulli equation (2) describes bending of a beam in classical strength of materials theory [1]:

$$\frac{d^2y}{dx^2}=\frac{M(x)}{EJ} \tag{2}$$

In case of transverse loading on the free end of a cantilever $M(x) = Px$. After integrating of (2) with initial conditions $y(L) = 0$, $y'(L) = 0$ a formula (3) for a bending shape is obtained. The formula in dimensionless coordinates shown below:

$$\eta=\eta(\xi,\mu)=\frac{\mu}{3}(2-3\xi+\xi^3), \tag{3}$$

where $\eta = y/L$ and $\xi = x/L$ – dimensionless the ordinate and the abscisse respectively, $\mu = PL^2/2EJ$ a dimensionless parameter of load, P – an acting force, L – a length of a cantilever. The parameter can be written in Euler force units λ: $\mu = \pi^2 \lambda/8$ ($\approx 1.23\ \lambda$). A sagging deflection in the linear theory in the considered case of bending is determined at the point of intersection of the ordinate axis by the bending form (4). It corresponds to the condition of saving a projection of the length of a cantilever on the abscisse axis:

$$f=f(\mu)=\eta(0,\mu)=\frac{2}{3}\mu. \tag{4}$$

It necessary to note that this expression can be obtained with a power series expansion of the sagging deflection of the exact solution [6] with respect to the loading parameter [9].

3. A modification of the linear theory

This paper presents the modification of the linear theory for sagging deflection under transverse loading. The modification uses a condition of saving a cantilever arc-length which equals one in dimensionless coordinates. The modification method keeps track of the maximum deflection point on the free end of a cantilever ξ_k. In this way a sagging deflection formula can be found from formula below:

$$f=\eta(\xi_k,\mu), \tag{5}$$

where ξ_k calculated with solving equation obtained from of one (that is the length of a cantilever)to an arc-length integral with limits of integration from ξ_k to an abscisse point of the fixed end [10]:

$$\int_{\xi_k}^{1}\left\{1+[\frac{d\eta(\xi,\mu)}{d\xi}]^2\right\}^{1/2} d\xi = 1. \qquad (6)$$

Root of equation (6) was obtained through numerical computations by the bisection method.

Graphs of sagging deflection of the exact solution [6], the linear theory and the modified linear there was plotted for a comparison in loading interval $0 < \mu < 5$ (Figure 2). Maximum deflection f is on the ordinate axis, loading parameter μ is on abscissa axis. Also, deviation graph of the modified linear theory from the exact solution was plotted (Figure 3).

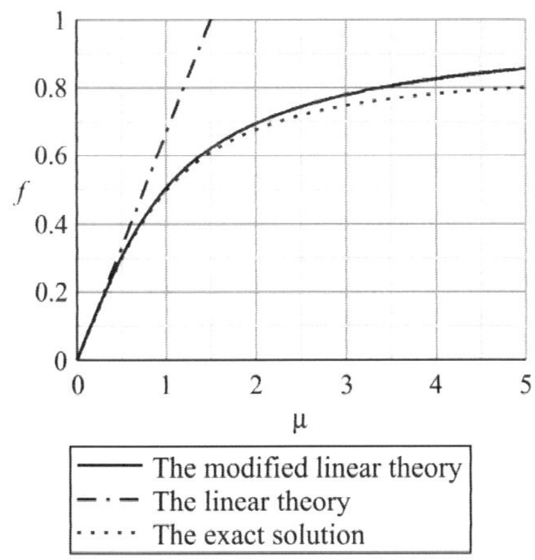

Figure 2. Sagging deflection plots for a cantilever under transverse loading

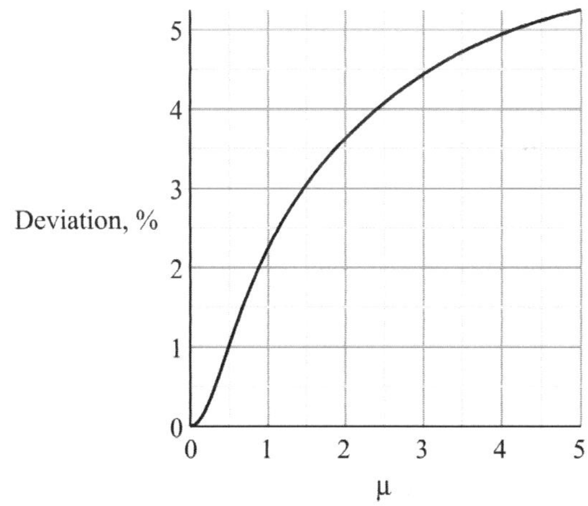

Figure 3. Deviation graph of the modified linear theory from the exact solution

We can see on Figure 2 that all solution match in interval $0 < \mu < 0.05$, which corresponds to the range of applicability of the linear theory. The modified linear theory is in good agreement with the exact solution in elliptic integrals over the whole range of load $0 < \mu < 5$. According Figure 3 deviation of the modified linear theory from the exact solution is no more than 6% over the interval $0 < \mu < 5$. These preliminary results are interesting and deserves further investigation because this interval of loads corresponds to the large deflection bending of a cantilever. We can say that the modification expand loading interval of applicability of the linear theory significantly [10].

4. Experimental verification

A comparison with experimental data from [11] was made for experimental verification. In the above paper a demonstration experiment of thin elastic steel cantilever under transverse loading on the free end and cantilever's own weight was made. The length of the beam is L = 0.40 m and it has a uniform rectangular cross-section of width b = 0.025 m and height h = 0.0004 m. The weight of the beam and the value of the load uniformly distributed over its entire length are W = 0.3032 N and $w = W/L$ = 0.758 N/m, respectively [11].

From equation (2), by analogy with the previous case, we obtain the expression of the cantilever under combined load: transverse loading on the free end and uniformly distributed load over the whole length. In this case, the bending moment:

$$M(x) = Px + \frac{1}{2}wx^2 .$$

By using initial conditions $y(L) = 0$, $y'(L) = 0$ obtain:

$$\eta = \frac{L^2}{24EJ}[W\xi^4 + 4P\xi^3 - (12P + 4W)\xi + 8P + 3W]. \quad (7)$$

In the linear theory, the expression for the sagging deflection is given by expression (7) with $\xi = 0$:

$$f = \eta(0) = \frac{L^2}{24EJ}(8P + 3W). \quad (8)$$

After differentiate of (7) and substitution to (6) we obtain the following equation for ξ_k parameter:

$$\int_{\xi_k}^{1} \left\{ 1 + \frac{L^4}{36(EJ)^2}[3P(\xi^2 - 1) + W(\xi^3 - 1)]^2 \right\}^{1/2} \mathrm{d}\xi = 1. \quad (9)$$

The combined load sagging deflection plots for the linear theory, the modified linear theory in comparison with experimental data shown on Figure 4. Also, we add the sagging deflection of the exact solution for transversal load for clarity.

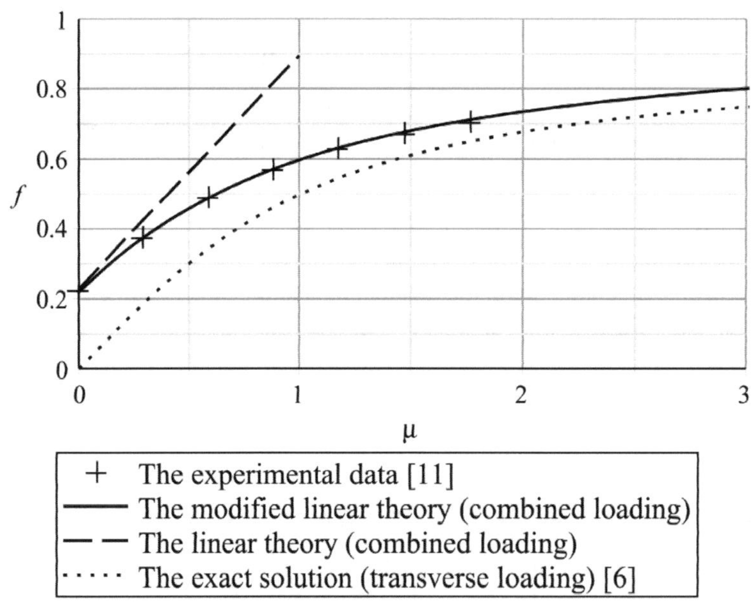

Figure 4. Sagging deflections for case of combined load.

As we can see on Figure 4, the sagging deflection of the linear theory is in agree with experimental data only on the initial part of the curve corresponding to small parameters of load. The sagging deflection of the modified linear theory is fully consistent with experimental data for the entire load range $0 < \mu < 1.75$. For the exact solution it is clear that the shape has the same form as the experimental data.

IOP Conf. Series: Materials Science and Engineering **822** (2020) 012039 doi:10.1088/1757-899X/822/1/012039

5. Conclusions

The classical linear theory uses a condition of saving a projection of the length of a cantilever on the abscisse axis. It works well for small deflections. But in the case of large deflection the end of a beam significantly deviates from an equilibrium position and it is impossible to use this approach. It is possible to use a condition of arc-length saving.

In this work the modification method for the linear theory in case of a cantilever under transverse loading was suggested. The sagging deflection of the modified linear theory was validated with a comparison with experimental data. In the experimental comparison we considered case of combined load: transverse loading on the free end and uniformly distributed loading (own weight of a cantilever). The suggested method significantly expands loading range of applicability of the linear theory. A sagging deflection of the modified linear theory has accuracy comparable with exact solutions in elliptic integrals and functions.

References

[1] Gere J M 2004 *Mechanics of materials* (Brooks/Cole)

[2] Nakamura Y, Eishima T, Nagai M, Funase R, Enokuchi A, Nakada K, Cheng Y, Takei E T, Funane T, Sasaki F, Nojiri Y, Yamamoto T, Nagayama E and Nakasuka S 2004 University of Tokyo's Ongoing Student-Lead Pico-Satellite Projects-CubeSat XI and PRISM *55th Int. Astronaut. Congr.*Vol **4** no 4.06 pp 4-8

[3] Ang M H, Wei W, Teck-Seng L 1993 On the estimation of the large deflection of a cantilever beam *Proc. IECON '93 - 19th Annu. Conf. IEEE Ind. Electron.* pp 1604-9

[4] Alici G 2009 An effective modelling approach to estimate nonlinear bending behaviour of cantilever type conducting polymer actuators *Sensors Actuators, B Chem.* Vol **141** no 1 pp 284-92

[5] Jacobsen J O, Winder B G, Howell L L and Magleby S P 2010 Lamina Emergent Mechanisms and Their Basic Elements *J. Mech. Robot.* Vol **2** no 1 pp 011003

[6] Zakharov Yu V, Zakharenko A A 1999 Dinamicheskaja poterja ustojchivosti v nelinejnoj zadache o konsoli [Dynamic instability at a nonlinear problem of a cantilever (in Russ.)] *Vychislitel'nye tehnologii. [Computational technologies]* Vol **4** no 1 pp 48-54

[7] Zakharov Yu V, Okhotkin K G 2002 Nonlinear bending of thin elastic rods *J. Appl. Mech. Tech. Phys.* Vol **43** no 5 pp 739-44

[8] Scarpello G M, Ritelli D 2011 Exact solutions of nonlinear equation of rod deflections involving the Lauricella hypergeometric functions *Int. J. Math. Math. Sci.* Vol **2011**

[9] Zakharov Yu V, Okhotkin K G, Vlasov A Yu 2002 Approximate formulas for sagging deflection of an elastic rod under transverse loading *J. Appl. Mech. Tech. Phys.* Vol **43** no 5 pp 745-7

[10] Zuev D M Zakharov Yu V 2016 Sagging deflection of cantilever loaded with a transversal load. Modification of linear theory (in Russ.) *Proc. of Reshetnev Readings* Krasnoyarsk Vol **2** no 20 pp 131–3

[11] Beléndez T, Neipp C and Beléndez A 2003 Numerical and Experimental Analysis of a Cantilever Beam: A Laboratory Project to Introduce Geometric Nonlinearity in Mechanics of Materials *Int. J. Engng Ed.* Vol **19** no 6 pp 885-92

IOP Conf. Series: Materials Science and Engineering **822** (2020) 012040 doi:10.1088/1757-899X/822/1/012040

A GIS for the Yenisei: an overview of maritime industrial archaeology on the Yenisei River and prospects for its development

A E Goncharov
Reshetnev University of Science and Technology, Krasnoyarsk, 660037, Russian Federation

E-mail: goncharovae@sibsau.ru

Abstract. In 2018, Reshetnev Siberian State University of Science and Technology launched the Geographic Information System Yenisei – Arctic Project with the objective of creating a digital cartographic system of the Yenisei River and its basin. Among the primary tasks is visualizing the riverbed using side-scan sonar and locating and cataloging submerged man-made objects, including those with certain historical value. In this report, the author addresses some issues relevant to maritime archaeology on the Yenisei River in connection with the project.

1. Introduction
In 2018, Reshetnev Siberian State University of Science and Technology (Reshetnev University), Krasnoyarsk, Russia initiated work on an interdisciplinary project titled the 'Geographic Information System Yenisei – Arctic'. The main objective of the project, as follows from its title, is the construction of a specialized Geographic Information System (GIS) of the Yenisei River basin (Fig. 1). Along with collecting updated geographic and environmental data (Fig. 1), it is planned to include a large historical content into the system's framework.

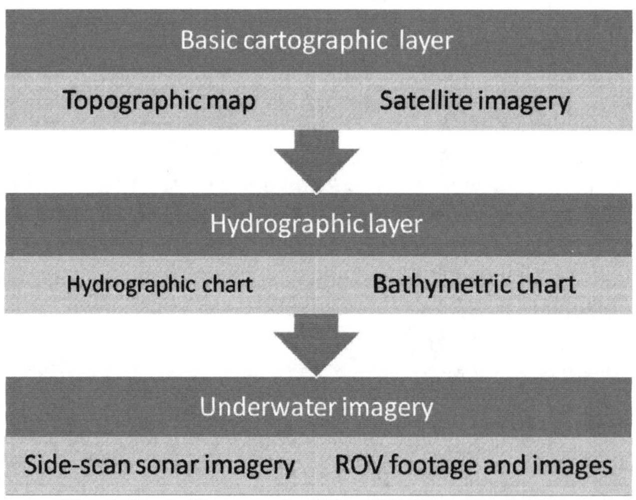

Fig. 1. Basic framework of the GIS Yenisei – Arctic.

2. Project background
The river Yenisei and its largest tributaries (Fig. 2) boast a rich history of shipping and industrial activity that have deeply impacted the river and its basin. The history of shipping on the Yenisei dates back to the 17th century and is firmly connected to the arrival of the first Russian explorers. During the 17th and 18th centuries, multiple unsuccessful attempts had been made to establish a sea route from this river to European ports, a feat that was first accomplished only in the 1870s. Kara Sea shipping evolved with a particular intensity during the 1920s–30s; and in 1929, the first permanent seaport – Igarka was constructed on the Yenisei. Ever since, this Siberian river has been a key destination for seagoing vessels traversing the Northern Sea Route. Auxiliary inland river navigation also saw intense development during the Soviet era (1920–1991). The remnants of this activity are scattered along the riverbanks and in the waters of the Yenisei and its tributaries.

Fig. 2. Map of the lower reaches of the Yenisei River.

Until recently, there have been few cases of maritime archeological investigations in Siberia. The main reasons for this are, firstly, a settled notion that since the region is mostly landlocked and situated far inland, there is little potential for underwater archaeological research. Secondly, knowledge on the history of riverine and maritime navigation on the Yenisei especially during its development in the 18th, 19th, and 20th centuries is chiefly restricted to factual data, with only a handful of studies addressing the bulk of human interactions with the river. Overall, despite the existence of numerous investigations on the development of the Northern Sea Route and Arctic exploration, few of them can be classified as complex studies on the maritime history of Siberia. For instance, the authors of many standard references [e.g., 1; 2] focus more on the economic and political aspects of shipping along the Northern Sea Route and, to a lesser extent, the great rivers of Siberia – the Ob', Yenisei, and Lena. Such topics as the history of local shipbuilding, navigation techniques, and seamanship are often overlooked despite their vital importance to understanding the history of human activity on the major waterways of Siberia. To our knowledge there are no up-to-date comprehensive investigations into the material world of the shipping industry of the 19th and particularly the 20th century, both as desk studies and *in situ* research, a practice known as industrial archaeology.

This brings us to the third reason for the slow development of underwater archaeology in Siberia, which is the lack of adequate and readily available knowledge on the existence of local maritime historical monuments. Furthermore, such vital archaeological practices as underwater research are aggravated by the deficit of necessary equipment, which often has a high cost, thereby deterring research institutions from engaging in relevant studies. To add to this, proper training is required for the operating personnel of hydroacoustic systems and remotely operated underwater vehicles (ROVs), along with an adequate expertise in scuba diving.

Thus, the above-mentioned factors have restrained a full-scale emergence of marine and riverine archaeology in Siberia. At the same time, the massive investments into various industrial projects, such as the development of oil and gas mining, transportation systems, and other projects related to Siberia and the Arctic have resulted in the intensification of construction. In compliance with Russian legislation, particularly the Federal Law No. 73-FZ (25 June 2002), it is mandatory to perform thorough archeological investigation at projected construction sites prior to the building activities. This includes underwater archaeological sites.

During the last two decades, there has been a significant increase of interest towards the exploration of underwater archaeological and historical sites in the Russian Federation [3]. Exploration is conducted by both professional and (often illegally) amateur divers. However, the geography of Russian maritime archaeology is chiefly limited to the Black Sea, the Baltic Sea, and the Far East. Most underwater investigations are performed using traditional methods, such a scuba diving.

Recently, there has been a worldwide boom in the usage of underwater acoustic imaging systems, both professional and recreational, e.g. fishfinders with integrated side-scan sonar (SSS) modules, and various ROVs. However, as aforementioned, this practice has found very limited application in Siberia. To our knowledge, there have been only a few cases of maritime archaeological investigation in Siberia and its coastal seas. In 2015, an expedition was organized into the Kara Sea under the auspices of the Administration of Yamalo-Nenets Autonomous Okrug and the Russian Geographical Society. Using advanced sonar technology and ROVs the expedition revealed the remains of two Second World War Arctic convoys, including the wrecks of the ships *Marina Raskova* (sunk by a German U-boat in 1944) and *Aleksandr Sibiriakov* (sunk by the German battleship *Admiral Sheer* in 1942) [4]. Unfortunately, there is no knowledge of further follow-up research in connection with this project.

An ongoing project was begun in 2015 by a team of enthusiastic polar historians from Reshetnev University and the Krasnoyarsk Branch of the Russian Geographical Society. The objective of the project was to explore and document existing historical monuments related to the history of Arctic exploration, the Northern Sea Route, and shipping on the Yenisei. Without any funding, the team traveled almost 500 km by kayak from the town of Eniseisk to the village of Bor on the Yenisei River. Besides gathering extensive historical and ethnographic materials, the expedition established the probable location of the steamer *Phoenix,* which sunk on the Yenisei in 1892 after hitting an uncharted shoal in the rapids of the Osinivskii Porog. This steamer was delivered to the Yenisei in 1887 by Captain Joseph Wiggins (1832–1905) and was used by the Anglo-Siberian Trading Syndicate until being sold to a local Russian steamship company. The discovery was largely based on historical records and a report by the local hydrographic service of a shipwrecked vessel buried beneath the sands of the Feniksova Kosa island (Fig. 3). According to this report, fragments of the vessel's superstructure are visible during low water levels. Unfortunately, none of the team members were able to revisit the island when the conditions for visualizing the wreck were fitting. Any archaeological investigations beyond determine the vessel's location and implementing nondestructive methods were ruled out considering that a special permit is required by Russian legislation when the monument in question is over one hundred years old.

a

b

Fig. 3. The island Feniksova Kosa in high water in 2016 (a); an unidentified shipwrecked vessel off the head of a nearby island (2015).

Further field investigations included the discovery of the steam schooner *Thames* in 2016. This vessel came to the Yenisei by sea in 1876. It sank in 1878 in the mouth of the river Sal'naia Kur'ia. Using archival data and secondary historical sources, the location of the wreck was determined. A series of dives revealed some ship details, which were first taken to be part of the superstructure, but after a follow-up study in 2018, showed that they were part of the hull that had been severely damaged during earlier dredging works.

The results of the field investigations in 2015 and 2016 convinced the team that in order to continue the research, it was necessary to obtain underwater imaging equipment that would allow collecting new data on the underwater historical heritage of the Yenisei River, its tributaries, and the coastal waters of the Kara Sea. Later archival research revealed that there were many more undiscovered shipwrecked vessels in these waters. It was decided that the most suitable imaging technique was the side scan sonar (SSS). The operating principles and exploitation of this device are exhaustively described in existing literature, among which [5] can be singled out as the most distinguished investigation. Strong currents and high sediment levels restrict the use of ROVs equipped with optical imaging devices. Besides, low temperatures, fast currents, and underwater obstacles make northern rivers such as the Yenisei difficult and dangerous places for human divers. Unlike ROVs and human divers, SSS allows covering large areas of the riverbed and determining the location of submerged objects, along with assessing their initial condition. This enables the researchers to determine further investigation techniques and assess the potential of the overall study of a given historical monument.

In 2018, after a long history of persuading various university authorities and funding organizations, the research team received its first sonar equipment including a StarFish 990F™ high-resolution side scan sonar manufactured by Tritech International Ltd for operation in shallow water, and a Humminbird ONIX 10 cxi SI Combo™ fishfinder. Due to the fact that Reshetnev University is, primarily, a technical educational and research institution, it was necessary to expand the overall objective of the project and transform it into a multidisciplinary investigation. For this purpose the research team recruited specialists working in the fields of remote sensing and geographical information systems, geodesists, mathematicians, and programmers. The project was now aimed at building a GIS of the Yenisei River for the purpose of aiding small craft navigation (local river transportation companies already had their own information system for navigation) and determining the dynamics of changes in the river topography and benthic environment. The task of studying underwater archaeological monuments remained. However, it was expanded to locating, mapping, and determining the potential risk of submerged manmade objects, including shipwrecked vessels, land vehicles, aircraft, and various structures.

3. Field investigations in 2018.

In 2018, the first series of field investigations was launched. Traveling aboard an inflatable two-tonner catamaran, Taimyr-600 (the same watercraft used in 2016 (Fig. 4)), two members of the research team, including the author of this paper traveled about 600 km from the village of Turukhansk on the Yenisei River to the city of Dudinka. Another 150 km north were traversed without making any underwater investigations. En route, the researchers studied three major areas: (1) the mouth of the river Sal'naia Kur'ia (Fig. 5) (the unusual geomorphology of this site was noted in 2016 [6] and, therefore, required further investigation); (2) the Igarskai Protoka (this was the site of the former seaport of Igarka and it was rightfully anticipated that the site contained large quantities of anthropogenic objects); (3) the area of 30-meter depths at the village of Potapovo. The collected data is currently being processed in order to be incorporated into the GIS.

Fig. 4. The Taimyr-600 catamaran used by the research team in 2016 and 2018.

Besides collecting geographic data, the expedition of 2018 had the objective of understanding how to operate the new equipment in the waters of the lower Yenisei. Among the primary problems was maintaining steady movement and constant speed of the towed transducer. The river currents are unstable and, therefore, greatly affect the behavior of the transducer. Thereby, in shallower areas it is preferable to use a hull-mounted transducer or mount the towed transducer on a metallic pole or pipe, which is firmly attached to the vessel. This option, however, requires calm waters and a low intensity of river traffic as the wake of bypassing vessels and river waves cause an intense roll motion of the pole-mounted transducer. In areas with many motorboats, such as the Igarskia Protoka, it was necessary to tow the transducer. Here local fishermen, often curious of the work done by the team, would pass within ten meters of the catamaran thereby generating severe wakes. However, despite its greater stability in rough seas, towing is more labor intensive and negatively affects image quality if the towing speed is abruptly changed by currents. Furthermore, underwater currents have different speeds and unpredictable movement patterns, making it difficult to maintain stable transducer movement. A stationary mount, on the other hand, is preferable when the study area has a constant depth. In this case the transducer is to a lesser degree affected by the current. However, it might be difficult to retrieve the transducer in case there is a threat of colliding with an obstacle.

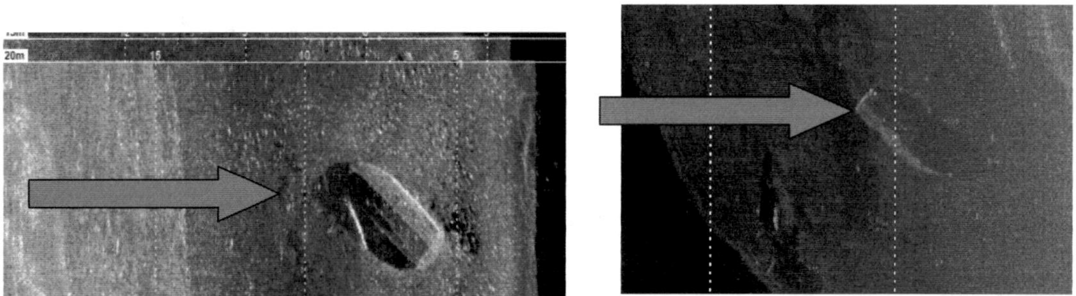

a b

Fig. 5. Sunken motorboat in Sal'naia Kur' as seen on SSS. Port (a) and starboard (b) image.

Fig. 6. Examples of shipwrecked vessels on the banks of the lower reaches of the Yenisei.

Another serious problem was sun glare. The monitor manufactured by the Russian Tetis Pro Company for the Starfish SSS did not have anti-glare protection, despite being designed for operating in outdoor conditions. This made it necessary to place the monitor under a canopy, or work only during cloudy weather or when the sun was low. The latter may be a serious problem when the fieldwork takes place during the Arctic summer.

Overall, the sonar data gathered in 2018 provides interesting insights into the underwater world of the Yenisei. In fact, these are the first known images of the Yenisei's riverbed. In another report included in this same volume, the author presents more detailed results of the investigation of the Igarskaia Protoka, which was thoroughly surveyed in 2018. In later publications, it is planned to cover other areas of the Yenisei.

4. Conclusion

In conclusion, it is necessary to stress on the importance of future maritime archaeological investigations of the Yenisei and its basin. There are still many more historical objects and sites that await their discovery and study, such as the remains of shipwrecked vessels shown in Fig. 6. The proposed Yenisei – Arctic GIS is expected to have a special layer that will cover these monuments and show their location and contemporary state. This investigation is essential to our understanding of the history of both shipping on the Yenisei, which lay at the heart of the economic and social development of this region, and shipping on the Northern Sea Route.

5. Acknowledgements.

This research was completed thanks to the assistance of Andrei Shapovalov during the field work in 2018 and Svetlana Fedorovna Titova during the archival work at Igarka in 2016. The research was supported by the Reshetnev Siberian State University and Technology Geographic Information System Yenisei – Arctic Project.

References

[1] Belov M I *Sovetskoe arkticheskoe moreplavanie 1917–1932* [Soviet Arctic Navigation, 1917–32] (Leningrad: Morskoi transport 1959).

[2] Pinkhenson, D M *Problema Severnogo morskogo puti v epokhu kapitalizma* [The Problem of the Northern Sea Route during the Epoch of Capitalism] (Leningrad: Morskoi transport 1962)

[3] Fazlullin S M 2017 Podvodnoe kul'turnoe nasledie i problem ego muzeefikatsii, *Voprosy museologii* **1(15)**, 3–16.

[4] Il'asov R M 2016 Po rezul'tatam nauchno-memorial'noi Karskoi Ekspeditsii 2015 goda, *The North and the Arctic in the New Global Development Paradigm. Luzin Readings – 2016 (Proceedings of VIII international scientific and practical conference)* 149–156

[5] Blondel P (2009) *The Handbook of Sidescan Sonar* (Springer Praxis Books. Springer, Berlin, Heidelberg 2009)

[6] Goncharov A E, Mednikov D M, Karelin N M, Nasyrov I R 2016 Prospects for using sonar for underwater archeology on the Yenisei: exploring a 19th century shipwreck, *IOP Conference Series: Materials Science and Engineering*, **155**, doi:10.1088/1757-899X/155/1/012037

An assessment of the condition of the Igarskaia Protoka on the Yenisei in 2018: side-scan sonar imagery interpretations and bathymetry

A E Goncharov
Reshetnev University of Science and Technology, Krasnoyarsk, 660037, Russian Federation

E-mail: goncharovae@sibsau.ru

Abstract. For almost eighty years, the town of Igarka on the Yenisei River had been an important timber industry center and seaport on the Northern Sea Route. Since the early 2000s, however, the town and its economy fell into decline resulting in the termination of timber export and, consequently, maritime navigation. In September 2018, a team of researchers from Reshetnev Siberian State University of Science and Technology, headed by the author of this paper conducted a side-scan survey of the Igarskaia Protoka, which had been the port's harbor, in order to access the current state of its aquatic environment. This paper describes some preliminary results of this investigation, which is the first study of the underwater environment of this particular body of water.

1. Introduction

In 2018, side-scan sonar and bathymetric data was collected within the Yenisei River in central Siberia as part of a multidisciplinary scientific project under Reshetnev Siberian State University of Science and Technology (Krasnoyarsk, Russia) to provide information on the state of the riverbed under the Geographic Information System Yenisei – Arctic Project. The purpose of the project is to build a complex system of digital models and maps of the Yenisei and its tributaries that will include and encompass data on riverbed topography, sedimentology, hydrology, river and bank archaeology, and the location of significant anthropogenic underwater objects, e.g. shipwrecked vessels, sunken vehicles, and clusters of submerged timbers. The project is designed to help better understand the Yenisei River and the relationship between the river's limnology and its exploitation through historical and modern themes. While conceived a few years earlier, the project received official support from the university administration only in 2018, allowing the project team to secure necessary scientific instruments and arrange fieldwork. It is panned that during the course of the project, a series of expeditions will be staged to the lower reaches of the Yenisei, the river's delta, and estuary to collect new geographic data. Further, the gathered information will be processed using a high-level performance computer for compiling the GIS.

During the last decades, the use of side-scan sonar (SSS) for river research has become a widespread practice – a phenomena attributed to the availability of various acoustic imaging devices and their comparatively low price. Despite being designed for recreational purposes, many low-cost side-scan sonar modals have been effectively used to study the hydrographic, benthic (biological) geomorphologic and archaeology of riverbeds [1]. Sonar modals with a fixed transducer can produce effective imagery of the riverbed when operated in shallow environments; however depths of over 10 meters limit the effective range of these devices as the acoustic shadow increases. To investigate the riverbed at depths exceeding the effective range for low-cost side scan sonar modals, it was necessary to use a towed transducer, commonly referred to as a fish. As the fish is towed behind the vessel, it dives and moves closer to the riverbed allowing, thus to reduce the acoustic shadow and visualize the submarine environment.

2. Study area and methods.

The town of Igarka (67°28′N, 86°34′E), situated within the Turukhanskii Raion municipal district of the Krasnoiarskii Krai, had until recently been among the largest and timber exporting centers of Russia and one of the most developed seaports on the Northern Sea Route. Since the dissolution of the Soviet Union, Igarka has been in decline, particularly after the Siberian timber trade shifted towards the Chinese market. The last seagoing vessel sailed from Igarka in 2006, thus putting an end to a history of maritime shipping which had begun in 1928. Today, there are less than 6,000 inhabitants in this town, which had once boasted a population of over 18,000 (1989 census [2]).

The study area for this research encompasses the lower nine kilometers of the Igarskaia Protoka, from its confluence with the main channel of the Yenisei at Mys [cape] Karmakuly to a point near a high voltage power line spanning between Igarka and Ostrov Igarkskii [Igarka Island] (Fig. 1). Overall during the investigation, 3,750 m^2 of the survey area were covered with the riverbed surface visualized by the side-scan sonar.

Fig. 1. Igarskaya Protoka. Satellite image from Yandex Maps web mapping service with central track of study area. Map legend: (1) Mys Karmakuly; (2) Ostrov Igarskii; (3) Site of the original location of Igarka and its sawmills.

The Igarskaia Protoka (historically known as Samoedskaia Protoka) is an approximately 15-kilometer side channel of the Yenisei situated between the right bank of the Yenisei and Ostrov Igarskii. Unlike other side channels of the lower Yenisei, Igarskaia Protoka is deep in its lower part (~ 13–15 m) and sheltered from northern, eastern, and southern winds, making the site an ideal location for harboring both riverine and seagoing vessels.

Due to the fact that September this region on the Yenisei is characterized by severe north winds, which generate 1.5–2.5 m waves on the Yenisei, the Igarskaia Protoka was the perfect location for investigating the performance of side-scan sonar on the lower Yenisei and developing techniques for acquiring quality imagery of the riverbed. What is more, the channel is known to be full of sunken timber and woody debris, shipwrecked vessels, sunken boats, and other traces of human activity, which made it an ideal location to not only assess the pollutedness of the riverbed, but to investigate the industrial history of this site.

For the purpose of our investigation, the Nautilus-50 Underwater Search Complex manufactured by the Russian Tetis-Pro Company had been employed. The complex is largely built on the StarFish SSS by Tritech International Ltd. with the difference of incorporating the sonar equipment, GPS module and processor into one unit, which includes a water and shatterproof computer.

In Table 1 are some of StarFish 990F towed SSS specifications:

Towing cable length	25 m
Maximum towing speed	5 kn
Maximum towing depth	8 m
Transducer vertical beam	60° nominal width
Transducer horizontal beam	0.3° nominal width
Frequency	1 MHz
Acoustic range	1–35 m on each channel
Mode	CHIRP pulse compression
Pulse length	100μs

Table 1. Specifications of StarFish 990F (compiled from product data sheet [3]).

3. Desk study information.

There is little published information on the Igarskaia Protoka, despite the fact that the site had been thoroughly mapped by Soviet and later Russian authorities during the time when Igarka was an important port on the Northern Sea Route. It is understood that the most recent bathymetric chart of the area was produced in 2006 and is largely a corrected edition of charts produced during the 1980s [4]. Nautical charts containing historic bathymetric soundings of the Igarskaia Protoka taken during 1936–37, 1956, 1964, 1981, 1987, and 2006 were obtained for the desk study performed prior to the field work. From the accumulated information, it was concluded that the depth of the channel range from 10 to 28 meters; greater depths were found at Mys Karmakuly – at the mouth of the channel, whereas the average depth of the channel between Ostrov Igarskii and the mainland ranged from 10 to 15 m. Depths decreased further southeast into the channel, ranging from 10 to 6 m along the centerline of the waterway.

The original port complex occupied approximately160 hectares of land and water along 3.5 km of waterfront. Overall, approximately 10 km of the channel had been exploited for various purposes: (1) a sheltered wintering harbor was established in the mouth of the River Chernaia; (2) coal was unloaded between Lisii Log and Volchii Log; (3) timber rafts were floated through the channel to the sawmills (1 km upriver from the wharves of the seaport); (4) a roadstead for seagoing vessels awaiting transshipment was located opposite of the wharves and Medvezhii Log; (5) wharves for riverine vessels were situated to the southwest of the passenger floating pier; (6) a roadstead for riverine vessels was located opposite of the wharves for riverine vessels along the southwestern tip of Ostrov Igarskii; (7) the pier of the fuel installations was the last of the industrial facilities in the channels and was situated on the mainland. The above information arrives from the chart of the channel published in 2006 (the last time Igarka was used as a seaport) [4].

Fig. 2. A view of the port of Igarka and the Igarskaia Protoka (looking east) during the 1970s. Courtesy of the Municipal Archive of the City of Igarka (file: A1/38/19/2).

A ferry service has been operating across the Igarskaia Protoka since the 1930s, linking Ostrov Igarskii and the mainland. Presently, the ferry provides communication between Igarka and the airport, which is situated on the island. During the Soviet era, Ostrov Igarskii was the location of a number of lesser industries, such as the state-run agricultural farm 'Poliarnyi', which is currently abandoned.

The desk study amassed information on the scale and intensity of the timber industry and shipping at Igarka. As anticipated, there was a steady development of both of these activities throughout the Soviet era. The table below illustrates the scale of timber exports through the port of Igarka from 1928 to 2005:

Year	Amount of timber exported in m^3	Exported timber produced in Igarka in m^3
1928	9,400	–
1932	123,900	83,800
1940	178,700	97,400
1957	307,000	167,000
1965	899,900	249,100
1978	1,322,500	383,500

1985	1,217,900	370,500
1991	750,400	275,600
2005	81,500	65,500

Table 2. Amount of timber exported from Igarka, 1928–2005. Compiled from [5].

As seen from the aforementioned table, wood production and export peaked during the 1970–80s and rapidly declined after the collapse of the Soviet Union. Igarka produced a line of wood products and yielded an average of 35.5% of the exports from the port. The remaining products were delivered to Igarka from the sawmills at Lesosibirsk.

It was expected that this massive amount of processed and unprocessed timber, which came to Igarka for almost eight decades had great affect on the aquatic environment of the Igarskaia Protoka. As said above, the city's timber production rate was intense and followed a simple scheme. Timber rafts were floated or towed to Igarka from the upper reaches of the Yenisei, typically, from the basin of the Angara River, which was heavily exploited due to the presence of valuable tree species such as the Angara pine, an ecospecies of the Scots pine (*Pinus sylvestris*), which is prized for its density. After passing through the Igarskaia Protoka, the rafts were unchained and transferred to the log booms; from here the logs were collected and transported to the sawmills (Fig. 3).

Fig. 3. Timber rafts being dissembled in the Igarskaia Protoka at log booms number 3 and 4 (1987). Courtesy of the Municipal Archive of the City of Igarka (file: 227).

The fact that river currents allowed floating timber rafts to Igarka without the need to employ riverine vessels (this was especially important during the 1920–30s when there was a shortage of riverine vessels on the Yenisei) was among the key factors for selecting Igarkaia Protoka as a location for a timber exporting center.

4. Results of side-scan sonar investigations.

Side-scan sonar investigations of the Igarskaia Protoka were carried out using a 6-m inflatable catamaran during September 2018. The working conditions were quite difficult as they had to be performed from an open deck, exposing the operators and the equipment to temperatures around 5°C and constant precipitation. Intense motorboat traffic on the waterway also posed problems because closely passing watercraft generated interference and waves that negatively affected the sonar imagery. Another problem was the need for the operator to simultaneously focus on the side-scan sonar monitor and a tablet computer in order to follow a steady track. Despite a more or less stable bathymetry of the study area, there occurred numerous situations when the towfish had to be abruptly extracted due to threat of snagging. Nonetheless, despite these difficulties, the results of the investigation produced some fascinating results, including the first known underwater images of this part of the

Yenisei. More importantly, it is now possible to assess the amount of sunken timber in the channel and estimate its impact on the aquatic environment of the Yenisei.

Microsoft Paint was used to construct a mosaic of the side-scan sonar imagery (Fig. 4); a similar method is described by Kaeser and Litts [6]. The reviewed images enable us to clearly identify both individual timbers and their clusters, many of which are composed of hundreds of logs and cut boards (Fig. 5). These log caches originated from logs that absorbed access water, described by Kaeser and Litts as dead-head logs [5] and boards that had been dropped into the water during ship loading operations. Other large woody debris originated from port and ship structures, such as ramps, railings, cargo cranes, etc.

Fig. 4. Side-scan sonar mosaic fragment showing a shipwrecked vessel (upper center), layers of sunken timber (middle left), and a sudden rise of the riverbed (lower center), caused by piled sunken logs.

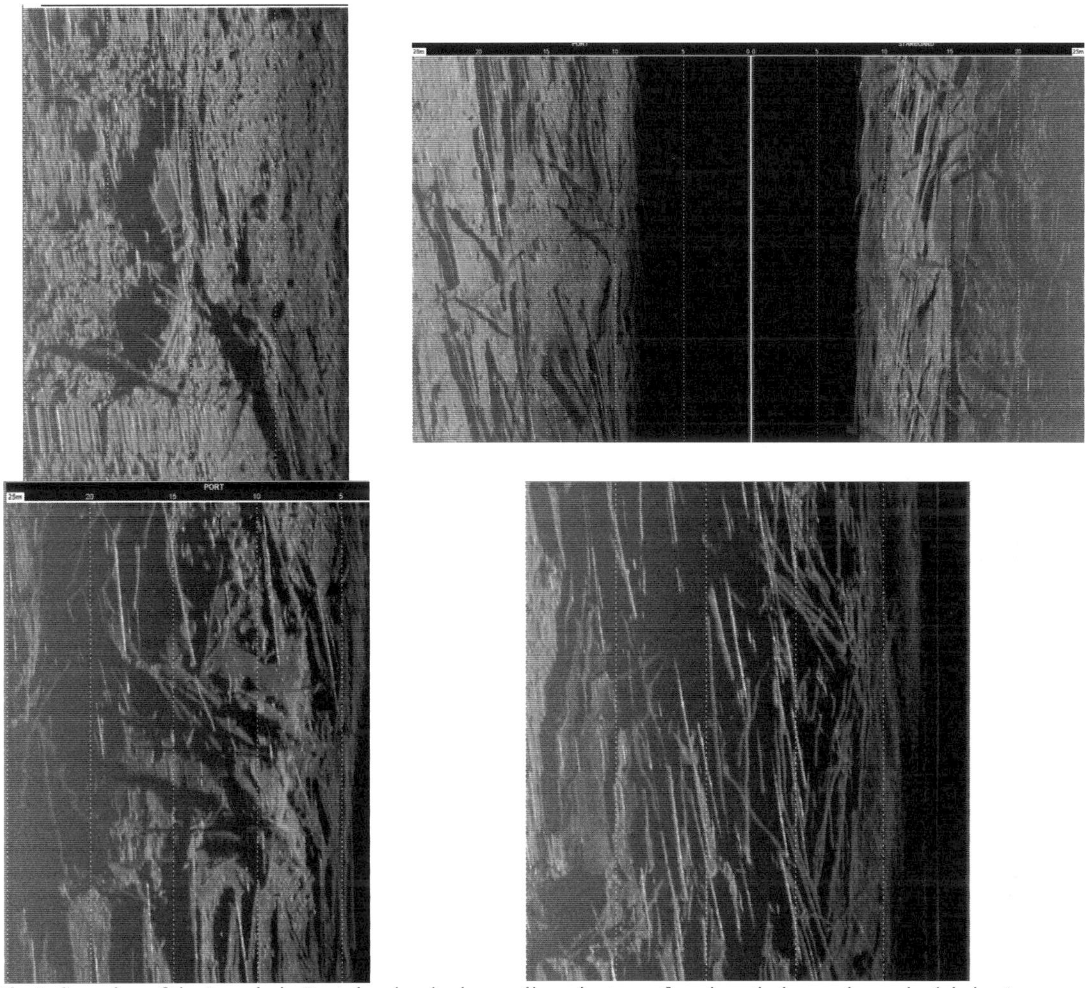

Fig.5. Samples of the Igarskaia Protoka riverbed revealing clusters of sunken timber and woody debris. Average depth at 10–15 m. Sonar acoustic range is set at 25 m on each channel.

Along with vast amounts of sunken timbers, the investigation revealed the presence of at least five shipwrecked vessels, the largest of which are 100-meter barges, at least ten sunken motorboats, and large amounts of miscellaneous river debris (steel cables, vehicle tires, containers, etc.) (Fig. 6). Most of the litter entered the environment after being dropped (accidentally or intentionally) from vessels or from the river ice after it had melted away.

IOP Conf. Series: Materials Science and Engineering **822** (2020) 012041 doi:10.1088/1757-899X/822/1/012041

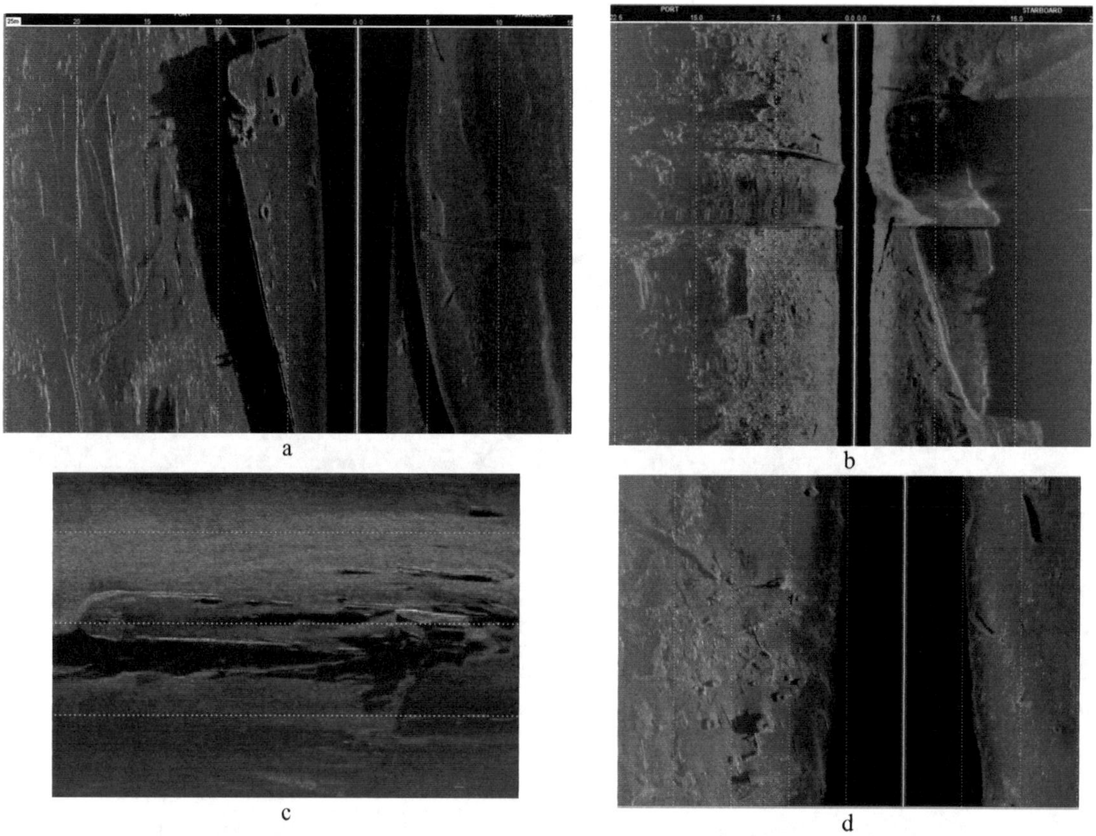

Fig. 6. SSS imagery revealing shipwrecked vessels (a,b,c) and a cache of vehicle tires (d)

5. Conclusion.

This study is the first to reveal the current state of the underwater world of the Igarskaia Protoka, which had once been the site of the busiest and most important port of the Northern Sea Route. The investigation provided information for estimating the amount of sunken timber and its location (it is planned to count how many logs, boards, and woody debris are situated within the studied area in the nearest future). It was possible to locate and identify at least five large shipwrecked vessels in the Igarskaia Protoka along with at least twenty sunken small craft. Applying SSS revealed large amounts of river debris left over from shipping operations and human activity in this part of the Arctic.

The Igarskaia Protoka and the remains of the port of Igarka are potential candidates for conducting further environmental studies and research on the history of the Arctic. Since the potential of the first field was touched upon in this paper and has, in terms of SSS application, been widely discussed in scientific literature [e.g., 1, 6, 7], it is worth to briefly address the second. Igarka vanished as an industrial center in Russia's Arctic frontline almost as quickly as it had once appeared. Its destiny was, on the one hand, bound to the Soviet economic model, which ceased existing soon after the collapse of the Soviet Union. This inevitably signaled the demise of Igarka, which itself was a creation of the socialist economy. On the other hand, Igarka was a product of what historian Michael Wood refers to as the "modern, Western scientific civilization" [8, p. 204], the fundamental principal of which was the exploitation of nature for the benefit of humanity. SSS has offered us a glimpse into this materialistic past of Igarka – a benchmark of Soviet Arctic development. Like many other industrial projects of the era, little attention was paid to certain environmental issues. Today, this careless treatment of the environment has returned to haunt the town's residents. It is known that Igarka's tap water arrives from the Igarskaia Protoka. During 2015–18, there have been numerous media reports on the topic of low water quality in Igarka. According to them, it is hazardous to not only to drink local tap water, but to even to bath in it [9]. It seems very likely that this is the result of the heavy pollutedness of the Igarskaia Protoka. A complex historical investigation of Igarka's industrial development, including the application of industrial archaeological methods, is therefore required to obtain a full picture of the ecological problem we are facing today.

6. Acknowledgements.

This research was completed thanks to the assistance of Andrei Shapovalov during the field work in 2018 and Svetlana Fedorovna Titova during the archival work at Igarka in 2016. The research was supported by the Reshetnev Siberian State University and Technology Geographic Information System Yenisei – Arctic Project.

References

[1] Kaeser A J, Litts T L, Tracy T W (2012) Using low-cost side-scan sonar for benthic mapping throughout the Lower Flint River, Georgia, USA, *River Research and Applications* 29: 634-644. doi:10.1002/rra.2556

[2] Vsesoiuznaia perepis' naseleniia 1989 goda [All-Union Population Census, 1989] www.demoscope.ru

[3] StarFish 990F product data sheet, available at: https://www.trichech.co.uk

[4] Karta reki Enisei ot ust'ia reki Podkanennaia Tunguska do porta Igarki [A chart of the river Yenisei from the confluence of the Podkanennaia Tunguska River to the port of Igarka] (2006) St. Petersburg: "Russo-Balt"

[5] Lapin G (2012) Letopis' Igarskogo LPK za 80 let [The 80-year history of the Igarka Wood Processing Complex] Krasnoyarsk

[6] Kaeser A J, Litts T L An assessment of deadhead logs and large woody debris using side scan sonar and field surveys in streams of southwest Georgia (2008), *Fisheries* 33(12): 589–597

[7] Tian W M (2011) Side-scan sonar techniques for the characterization of physical properties of artificial benthic habitats, *Brazilian Journal of Oceanography* 59(special issue CARAH) pp. 77–90

[8] Wood M (1993) Legacy: the origins of civilization. Network Books. Retrieved from: https://books.google.ru/books/about/Legacy.html?id=00sWAQAAIAAJ&redir_esc=y

[9] Zhiteli Igarki zhaluiutsia na otvratitel'noe kachestvo vodoprovodnoi vody [The inhabitants of Igarka complain about horrible tap water], 13/06/2018 available at: https://www.enisey.tv/news/post-9917/

IOP Conf. Series: Materials Science and Engineering **822** (2020) 012042 doi:10.1088/1757-899X/822/1/012042

An underwater research drone for the Yenisei

A M Derben, A A Kuznetsov

Reshetnev University of Science and Technology, Krasnoyarsk, 660037, Russian Federation

E-mail: derben@sibsau.ru

Abstract. The study of the rivers and lakes of Eastern Siberia has its own peculiarities. Fast currents, small depths, silt, algae, and the abundance of small non-ferrous artifacts on the riverbed are all factors, which make such investigations complicated. In this report, the authors discuss a software and hardware complex for a proposed remotely operated underwater vehicle, which is specially developed for operating in the rivers of northern Siberia.

1. Introduction

The structure of a modern integrated geographic information system (GIS) of a fast-moving river cannot consist exclusively from aerial and acoustic imaging data. An important addition to the counters of the water bodies is data collected within the course of special investigations performed with the help of unmanned underwater vehicles. The need for novel principles in the design of such devices arrives from the fact that existing mass-produced apparatuses are designed for operating in marine coastal environments and in still water reservoirs or those with slow current. Therefore, they cannot be deployed in large fast-moving rivers, such as the Yenisei. Furthermore, the operators of remotely operated underwater vehicles (ROVs) often face a number of challenges associated with visual fatigue. In addition, the operator has to divert his or her attention to maintaining the ROV on course. Therefore, there is always a risk of missing an important underwater feature.

In order to solve this problem, we propose to design a set of programs such as a neural network that will be able to: (1) identify targeted objects; (2) improve visualization quality; (3) classify objects; and (4) reconstruct the event.

2. Proposed design of the ROV

As aforementioned, aerial data does not provide the operator with sufficient information on the hydrography of a waterway. Existing acoustic imaging systems, such as the side scan sonar (SSS), also leave large white spots due to their inability to identify certain objects and determine the nature of the image, which is not optical, and therefore, is often difficult to interpret. This suggests a return to traditional optical imaging systems, which feature moving platforms with video and photographic cameras. There has been a long and successful tradition in using such systems for various kinds of underwater research, including underwater archaeology, zoology, and other areas of Earth and environmental sciences. In this report, we provides an overview of existing technical solutions for the design of miniature ROVs, suitable for use in the proposed study area.

The vast majority of professionally manufactured ROV, available on the market, are designed for marine and lake environments. The water here usually has a sufficient visibility degree (for instance, in the Sea of Azov, visibility ranges from 1 to 1.8 m [1]). The operational depths of ROV working in marine environments range from 100 to 3000 m, with speeds ranging from 1.5 to 2 kn (0.77–1.02 m/s) [2, 3, 4].

Structurally, commercial ROVs can be subdivided into two groups:

1. Quadcopter configurations with radial drivers for four to eight propellers.
2. Frame configuration with rotary reverse propellers and auxiliary stationary vertical displacement fans.

The light source for most models comes from powerful light-emitting diodes (LED).

The use of these ROVs can be complicated in fast-moving rivers (current speeds of up to 7 km/h) with high sediment rates, and low (2–20 m) depths. Thus, (1) currents of up to 1 m/s (3.6 km/h) will reduce the maneuverability of the ROV and waste precious electrical energy; (2) the propellers of the ROV are susceptible to clogging with algae; (3) vertical flows from motionless fixed propellers will raise sediments, further complicating visualization and negatively affecting the orientation of the apparatus.

4. Powerful LED backlights can reduce visibility if sedimentation rates are high.

The proposed solution consists of the following principles:

1. The design of the body of the ROV should exclude protruding components, such as masts, braces, and gratings that increase drag. A more preferable shape is streamlined having a circular or ellipsoidal cross-section

with elongated, torpedo-shaped outlines. The design should be monolithic, smooth, and sustainable to viscous and friction resistance.

2. Vertical fans and propellers should be eliminated. Vertical movements should be executed with special rudders, or by reverse movement of pairs of horizontal propellers.

3. Protection of the propellers from algae is necessary.

4. The light source should be a combination of conventional LEDs and red or infrared emitters that will significantly increase visibility range in turbid water.

3. Neural networks for underwater object recognition

The effectiveness of the ROV operator depends on his or her concentration. When observing a monotonous image, a person's attentiveness decreases after about 15 minutes. As visual fatigue accumulates, the operator can miss an important object that briefly got caught by the camera. Moreover, the operator can be distracted and fail to notice an interesting underwater feature. Most often, this problem is solved using an onboard computer equipped with artificial intelligence (AI) programs operating on the principle of an organized mathematical model of an artificial neural network.

An example of a popular open source cross-platform is the YOLO [5] complex developed on the basis of TensorFlow[6]. In its basic form the neural network is presented along with the training models (training coefficients). However, they are suitable only for solving the simplest problems, such as searching for missing persons, vehicles, vessels, and managing agricultural activities and forestry.

In order to identify and classify more complex objects, it is necessary to use machine learning based on neural network. Examples of such programs are MXNet and Scikit-Learn [7, 8]. Due to the fact that high performance is required from the computer, the use of an external server or a cluster of such servers is advisable. Once the data is collected, the next step is bringing the images to a standard dataset. This is typically a symmetrical matrix with fixed sizes of 1024×1024 pixels and a gray gradient of 256 shades. For this purpose, the EBImage program [9] (distributed under a free license) can be used.

The normalized graphical objects are proposed to be processed through deep learning algorithms and then transferred into the input of a convolutional neural network (CNN). This technology is widely used. For instance, Amazon Information Service automatically generates contextual advertising of goods and services; Google can search through images; Instagram and Pinterest can personalize the personal pages of its users. The technical implementation of CNN is presented in the program ADVENTURES-IN-ML-CODE [6], which is integrated with the PyTorch and TensorFlow frameworks.

The last component of the software package will be implemented directly to the ROV and respond to events and objects appearing in the video stream.

The hardware of the complex should include the following parameters: (1) a 48 core CPU, with a frequency of at least 2.1 GHz; (2) a level 3 cache of at least 22 MB; (3) a memory capacity of at least 4 TB.

4. Acknowledgments

The authors express the gratitude for supporting this project to Reshetnev Siberian State University of Science and Technology and the projects 'The Yenisei – Arctic Geographical Information System'.

References

[1] Goptareva N P (ed) 1991 *Girometeorologiia i gidrokhimiia morei SSSR. T. 5: Azovskoe more* [The Hydrometeorology and Hydrochemistry of the Seas of the USSR (St Petersburg)

[2] Small-sized remote-controlled unmanned underwater vehicles JSC "Tethys Pro" (available at http://www.tetis-pro.ru/catalog/332/)

[3] Marine Geo Service (available at http://marinegeoservice.ru/ROVs.html)

[4] Cable tracking ROV / ECA Group (available at https://www.ecagroup.com/en/solutions/cable-tracking-rov)

[5] YOLO: Real-Time Object Detection (available at https://pjreddie.com/darknet/yolo/)

[6] API Documentation. Tensor Flow (available at https://www.tensorflow.org/api_docs/)

[7] MXNet Architecture (available at http://mxnet.incubator.apache.org/versions/master/architecture/index.html)

[8] Sci Learn | Project Description. PyPl (available at https://pypi.org/project/scikit-learn/)

[9] Bioconductor – EBImage. Image processing and analysis toolbox for R (available at https://bioconductor.org/packages/release/bioc/html/EBImage.html)

IOP Conf. Series: Materials Science and Engineering **822** (2020) 012043 doi:10.1088/1757-899X/822/1/012043

The Cayley graphs of a centralizer of the Burnside group $B_0(2,5)$

Alexander A Kuznetsov[1], **Alexandra S Kuznetsova**[2], **and Vladimir V Kishkan**[1]

[1]Reshetnev Siberian State University of Science and Technology, Krasnoyarsk, 660037, Russian Federation

[2]Krasnoyarsk State Agrarian University, 660049, Russian Federation

E-mail: alex_kuznetsov80mail.ru

Abstract. Let $B_0(2,5) = \langle a_1, a_2 \rangle$ be the largest two-generator Burnside group of exponent five. It has the order 5^{34}. We define an automorphism φ under which every generator is mapped into another generator. Let $C_{B_0(2,5)}(\varphi)$ be the centralizer of φ in $B_0(2,5)$. It is known that $|C_{B_0(2,5)}(\varphi)| = 5^{17}$. We calculated the growth function of this group relative to the minimal generating set and also the symmetric generating set. As the results, the diameters and the average diameters of $C_{B_0(2,5)}(\varphi)$ were computed.

1. Introduction

One of the important tools for defining the structure of a group is the study of its growth with respect to a fixed generating set. Let $G = \langle X \rangle$. We call the ball K_s of radius s of a group G the set of all its elements, which can be presented as a group of words with length s in the alphabet X. For each non-negative integer s, one can define the (spherical) growth function of the group $F(s)$, which is equal to the number of elements of the group G with respect to X, that can be represented as an irreducible group words with the length s. Thus,

$$F(0) = |K_0| = 1,\ F(s) = |K_s| - |K_{s-1}| \ when\ s \in \mathbb{N}.$$

As a rule, the growth function of a finite group is represented in the form of a table which contains non-zero values of $F(s)$.

Also we note that along with computing the growth function of a group, we define some characteristics of the corresponding Cayley graph, for instance, the diameter and the average diameter [1]. Let $F(s_0) > 0$, but $F(s_0+1) = 0$, then s_0 will be the diameter of the Cayley graph of the group G in the generating alphabet X, which we will denote $D_X(G)$. Accordingly, the average diameter $\overline{D}_X(G)$ is equal to $\frac{1}{|G|}\sum_{s=0}^{s_0} s \cdot F(s)$.

Unfortunately, although the computation of the growth function of a large finite group is solvable, it is a rather complicated problem. This is due to the fact that, in general, the task of determining of the minimal word of a graup element, as shown by S. Iven and O. Goldreich [2] is NP–hard. Thus, in the worst case, the number of elementary operations that must be performed to solve this problem is an exponential function of $|X|$.

It should be noted that during the process of computing the growth function of a group, the characteristics of the corresponding Cayley graph are defined in a parallel way – for instance, the diameter and the average diameter. Let $F(s_0) > 0$, but $F(s_0 + 1) = 0$, then s_0 will be the diameter of the Cayley graph of the group G in the alphabet generating X, which we will denote $D_X(G)$. Accordingly, the average diameter $\overline{D}_X(G)$ is equal to $\frac{1}{|G|}\sum_{s=0}^{s_0} s \cdot F(s)$.

As it is known, the Cayley graphs, due to their remarkable properties, have found wide application in the simulating the topologies of multiprocessor computing systems (MCS) – supercomputers [3] and data-centers [4]. It is likely that in the near future, knowledge on very large graphs will be needed for designing distributed systems, the peak capacity of which will approach 1 exaflops and more.

One of the more widely used topologies of MCS is the k–dimensional hypercube. This graph is determined by the group $B(k, 2)$, which is the k–generated Burnside group of period 2. $B(k, 2)$ has a simple structure and is equal to the direct product of k copies of cyclic groups of order 2. In the work [5] the Cayley graphs of the groups $B(k, 3)$, i. e. groups of period 3, were studied and a comparative analysis of these graphs with respect to a hypercube was carried out. This analysis showed that the characteristics of $B(k, 3)$ are more preferable than the characteristics of $B(k, 2)$. It means that while paired comparison graphs $B(k_1, 3)$ and $B(k_2, 2)$ have approximately the same number of vertices, the first ones have a smaller diameters, the average diameters and degrees. We obtained a similar result in [6] during the study of groups of period 4. In this respect the task of the study of the Cayley graphs of finite Burnside groups of other periods is interesting.

Let $B(2, 5) = \langle a_1, a_2\rangle$ be a free two–generating Burnside group of period 5. Until now, it is not known whether this group is finite or infinite. Further, let $B_0(2, 5) = \langle a_1, a_2\rangle$ be the maximum finite group, whose order equals 5^{34} [7]. If $B(2, 5)$ is finite, then $B_0(2, 5) = B(2, 5)$.

To calculate the growth function $B_0(2, 5)$ with respect to the minimal generating set is hardly possible because the number of its elements is very large:

$$5^{34} = 5820766091346740722656 25 \approx 5 \cdot 10^{23}.$$

Earlier, using computer calculations, we have obtained the growth functions of the factor–groups of the group $B_0(2, 5)$ whose order does not exceed 5^{19} [8].

Let us consider the map φ of the following form:

$$\varphi : \begin{cases} a_1 \to a_2, \\ a_2 \to a_1. \end{cases}$$

It is easy to see [10] that φ is an involutive automorphism of the groups $B(2, 5)$ and $B_0(2, 5)$.

Let $C_{B(2,5)}(\varphi)$ and $C_{B_0(2,5)}(\varphi)$ be the centralizers of the automorphism φ in $B(2, 5)$ $B_0(2, 5)$ respectively. According to V. P. Shunkov's theorem [11], if $C_{B(2,5)}(\varphi)$ is a finite group, then the group $B(2, 5)$ is also finite. In other words, if $C_{B(2,5)}(\varphi) = C_{B_0(2,5)}(\varphi)$, then $B(2, 5) = B_0(2, 5)$. Considering this, the study of the growth function of $C_{B_0(2,5)}(\varphi)$ is of great interest. Further, for brevity, we will write C instead of $C_{B_0(2,5)}(\varphi)$.

In [10] the structure of group C has been studied with the following results:

(i) $|C| = 5^{16}$,

(ii) $C = C_0 \times \langle z\rangle$, where $|C_0| = 5^{15}$ and $|\langle z\rangle| = 5$,

(iii) z is the central element of the group $B_0(2, 5)$,

(iv) $C_0 = \langle X_0\rangle$, where $|X_0| = 4$ is the minimal number of the generators of C_0,

(v) power commutator presentation of C_0 is computed.

The objective of this paper is to study the growth function of the group C with respect to the minimal generating set $X = X_0 \cup \{z\}$ and the symmetric – $Y = X \cup X^{-1}$.

Note that in [12] a solution is proposed to a similar problem for the centralizer $C_{B_0(2,5)}(\psi)$ where ψ is the automorphism, which translates generating elements into their inverses.

Since for evaluating of growth function it is required to multiply elements of the group, then for the practical implementation an efficient algorithm for multiplication is needed. There are two ways to calculate the product of elements in the group given by power commutator presentation: the collective process [3, 13] and Hall's polynomials method [14]. Numerous computational experiments showed that the second method allows to multiply the elements in these groups much faster than in a collecting process (at least in exponent) [5, 9, 15, 16].

The documentation of the computer algebra system GAP refers to the possibility of automated calculation of Hall's polynomials in the simplest cases. However, in the general case, this task is not trivial, because is not reduced to routine computation and requires the involvement of programming languages that support complex regular expressions, and also systems of computer mathematics with a wide range of procedures for symbolic computation. In fact, working with a group that has a large order, usually it is required the unique revision of the code that takes into account the feature of the group structure and characteristics of the computer.

In the following section of the article the Hall's polynomials of the groups C_0 are presented. They were calculated using algorithm from [15]. This algorithm in general was implemented in C++, with the exception of symbolic procedures, which were written in the MATLAB language.

In the last section the results of computer calculations of the growth function of the group of C with respect to the generating set X and Y are given.

2. Hall's polynomials of the group C_0

Let $C_0 = \langle X_0 \rangle$, where $X_0 = \{a_1, a_2, a_3, a_4\}$ be the minimal generating set of C_0. The following theorem is proved.

Theorem. *Let $a_1^{x_1} \dots a_{15}^{x_{15}}$ and $a_1^{y_1} \dots a_{15}^{y_{15}}$ be two arbitrary elements in the group C_0 written in the commutator form. Then their product is equal to $a_1^{x_1} \dots a_{15}^{x_{15}} \cdot a_1^{y_1} \dots a_{15}^{y_{15}} = a_1^{z_1} \dots a_{15}^{z_{15}}$, where $z_i \in \mathbb{Z}_5$ are Hall's polynomials, given by the formulas* (1–15).

$$z_1 = x_1 + y_1, \quad (1)$$

$$z_2 = x_2 + y_2, \quad (2)$$

$$z_3 = x_3 + y_3, \quad (3)$$

$$z_4 = x_4 + y_4, \quad (4)$$

$$z_5 = x_5 + y_5 + x_2y_1 + 4x_3y_2, \quad (5)$$

$$z_6 = x_6 + y_6 + x_3y_1 + 4x_3y_2, \quad (6)$$

$$z_7 = x_7 + y_7 + x_4y_1, \quad (7)$$

$$z_8 = x_8 + y_8 + x_4y_2, \quad (8)$$

$$z_9 = x_9 + y_9 + x_4y_3, \quad (9)$$

$$z_{10} = x_{10} + y_{10} + 2x_3y_2 + 2x_4y_1 + x_7y_4 + 3x_4^2y_1 + x_4y_1y_4, \quad (10)$$

$$z_{11} = x_{11} + y_{11} + 4x_3y_2 + 2x_4y_2 + x_8y_4 + 3x_4^2y_2 + x_4y_2y_4, \quad (11)$$

$$z_{12} = x_{12} + y_{12} + 2x_3y_2 + 2x_4y_3 + x_9y_4 + 3x_4^2y_3 + x_4y_3y_4, \quad (12)$$

$$z_{13} = x_{13} + y_{13} + 3x_3y_2 + 2x_4y_1 + 2x_5y_4 + x_7y_2 + 4x_8y_1 + 2x_7y_4 + x_{10}y_4 + \\ + 2x_4^2y_1 + x_4^3y_1 + 3x_7y_4^2 + 3x_4y_1y_4^2 + 3x_4^2y_1y_4 + 2x_2x_4y_1 + 2x_2y_1y_4 + x_4y_1y_2 + 4x_4y_1y_4, \quad (13)$$

$$z_{14} = x_{14} + y_{14} + 3x_3y_2 + 2x_4y_2 + 4x_6y_4 + 2x_7y_3 + 3x_9y_1 + 2x_8y_4 + x_{11}y_4 + \\ + 2x_4^2y_2 + x_4^3y_2 + 3x_8y_4^2 + 3x_4y_2y_4^2 + 3x_4^2y_2y_4 + 4x_3x_4y_1 + 4x_3y_1y_4 + 2x_4y_1y_3 + 4x_4y_2y_4, \quad (14)$$

$$z_{15} = x_{15} + y_{15} + x_3y_2 + 2x_4y_3 + x_8y_3 + 4x_9y_2 + 2x_9y_4 + x_{12}y_4 + 2x_4^2y_3 + \\ + x_4^3y_3 + 3x_9y_4^2 + 3x_4y_3y_4^2 + 3x_4^2y_3y_4 + 2x_3x_4y_2 + 2x_3y_2y_4 + x_4y_2y_3 + 4x_4y_3y_4. \quad (15)$$

3. Computer calculations of the growth functions of the group C

The calculation of the growth function of the group C related to X was carried out according to the algorithm [8]. For efficient multiplication of elements, the Hall's polynomials were used obtained in section 2. The algorithm was implemented in C++. As a tool for parallelization, it was used the library OpenMP. For the calculations, it was used a computer with an 8-core processor and 64 Gb of RAM, running the Linux operating system. The program was compiled by the embedded compiler GCC. Calculating growth functions for the generating set X takes about 1.5 hours, and for Y – 3 hours. Their graphs are shown in Fig. 1–2. For clarity, an approximating Gaussian curve obtained by the method of the least squares is added on each graph.

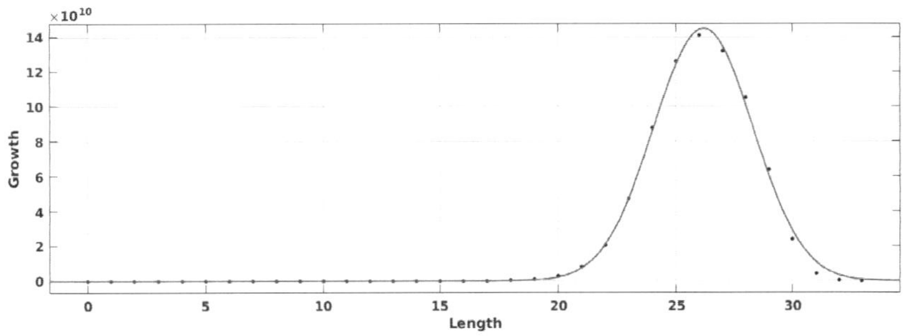

Figure 1. The growth function of C generated by X

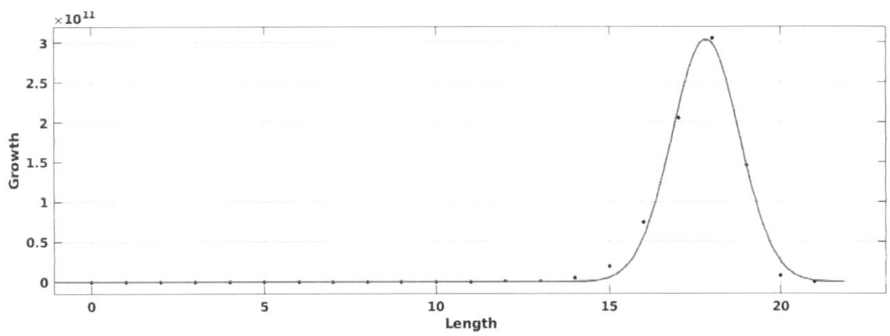

Figure 2. The growth function of C generated by Y

As already mentioned, the growth function of the group that contains information about the characteristics of the corresponding Cayley graph is:

Corollary 1. $D_X(C) = 33$, $\overline{D}_X(C) \approx 26$.

Corollary 2. $D_Y(C) = 21$, $\overline{D}_Y(C) \approx 18$.

References

[1] Kuznetsov A A and Kuznetsova A S 2014 A parallel algorithm for study of the Cayley graphs of permutation groups *Vestnik SibGAU* **1** 34–39 (In Russ)

[2] Even S and Goldreich O 1981 The Minimum Length Generator Sequence is NP–Hard *J. of Algorithms* **2** 311–313

[3] Holt D, Eick B and O'Brien E 2005 *Handbook of computational group theory* (Boca Raton: Chapman & Hall/CRC Press)

[4] Camelo M, Papadimitriou D, Fàbrega L and Vilà P 2014 Efficient Routing in Data Center with Underlying Cayley Graph *Proc. of the 5th Workshop on Complex Networks CompleNet* 189–197

[5] Kuznetsov A A 2015 The Cayley graphs of Burnside groups of exponent 3 *Sib. Elektron. Mat. Izv.* **12** 248–254 (In Russ)

[6] Kuznetsov A A and Kuznetsova A S 2016 Perspective topologies of multiprocessor computing systems based on the Cayley graphs of groups of period 4 *Vestnik SibGAU* **3** 34–39 (In Russ)

[7] Havas G, Wall G and Wamsley J 1974 The two generator restricted Burnside group of exponent five *Bull. Austral. Math. Soc.* **10** 459–470

[8] Kuznetsov A A and Kuznetsova A S 2018 A resource-efficient algorithm for study the growth in finite two-generator groups of exponent 5 *Prikl. Diskr. Mat.* **42** 94–103 (In Russ)

[9] Kuznetsov A A 2016 An algorithm for computation of the growth functions in finite two-generated groups of exponent 5 *Prikl. Diskr. Mat.* **3** 116–125 (In Russ)

[10] Kuznetsov A A and Filippov K A 2010 On an involutive automorphism of the Burnside group $B_0(2,5)$ *Sib. Zh. Ind. Mat.* **13** 68–75 (In Russ)

[11] Shunkov V P 1972 On periodic groups with an almost regular involution *Algebra and Logic* **4** 470–494 (In Russ)

[12] Kuznetsov A A and Safonov K V 2018 On applications of the Cayley graphs of some finite groups of exponent five *J. Sib. Fed. Univ. Math. Phys* **11:1** 70–78

[13] Sims C 1994 *Computation with Finitely Presented Groups* (Cambridge: Cambridge University Press)

[14] Hall P 1988 *Nilpotent groups, Notes of lectures given at the Canadian Mathematical Congress 1957 Summer Seminar, in The collected works of Philip Hall* (Oxford: Clarendon Press)

[15] Kuznetsov A A and Kuznetsova A S 2013 Fast multiplication in finite two-generated groups of exponent five *Prikl. Diskr. Mat.* **1**(19) 110–116 (In Russ)

[16] Kuznetsov A A and Safonov K V 2014 Hall's polynomials of finite two-generator groups of exponent seven *J. Sib. Fed. Univ. Math. Phys* **2** 186–190

AUTHOR INDEX

Institute of Physics
Dirac House, Temple Back
Bristol BS1 6BE UK

ISSN: 1757-8981
ISBN 978-1-7138-3162-4

CURRAN ASSOCIATES INC.
proceedings.com